METHODS IN MOLECULAR BIOLOGY™

Series Editor
John M. Walker
School of Life Sciences
University of Hertfordshire
Hatfield, Hertfordshire, AL10 9AB, UK

For further volumes:
http://www.springer.com/series/7651

Protein Expression in Mammalian Cells

Methods and Protocols

Edited by

James L. Hartley

Protein Expression Laboratory, SAIC-Frederick, Inc., National Cancer Institute, Frederick, MD, USA

Humana Press

Editor
James L. Hartley
Protein Expression Laboratory
SAIC-Frederick, Inc.
National Cancer Institute
Frederick, MD, USA
hartleyjames@mail.nih.gov

ISSN 1064-3745 e-ISSN 1940-6029
ISBN 978-1-4939-6160-3 ISBN 978-1-61779-352-3 (eBook)
DOI 10.1007/978-1-61779-352-3
Springer New York Dordrecht Heidelberg London

Printed on acid-free paper

Humana Press is part of Springer Science+Business Media (www.springer.com)

Preface

For the past decade, biomedical science has been heavily influenced by the development of high-throughput DNA sequencing technologies. Despite the public perception that scientists are the ultimate objectivists and driven by noble and altruistic interest in arcane corners of human knowledge, the reality is that we are as susceptible as anyone to the influence of the "drunk looking for his keys under the lamp post" phenomenon. (If you have lost your keys on a dark city street, you should certainly start your search where the light is the best.) These days, the light shines especially brightly where we can find ways of using next-generation sequencing to answer questions relevant to our larger research interests.

However, in a sense, our present preoccupation with millions of bits of DNA sequence is a sham and a delusion or, to put it more accurately, a postponement of the most difficult and fundamental work that will ultimately be required. We find it so convenient to deal with the uniform and predictable behavior of nucleic acids that we are seduced into thinking of biology as reducible to As, Gs, Cs, Ts, and Us. But, so far as we know today, all those billions and trillions of bases lead nowhere but to the world of PROTEINS: how much, with what activities, where and when expressed, how mutated or modified, and interacting with what partners. Why do we care about exons and splicing; copy numbers and ploidy; deletions and mutations; alternative splicing and promoters; enhancers and micro RNAs and epigenetics? Only because they affect the proteins and protein organelles in cells and how they function.

The previous paragraph would be an overstatement if we could predict the amounts and behaviors of proteins by knowing, say, the complete sequence of the human genome, with all its SNPs, mutations, and variations. But we do know all those base pairs, and the biology of even cells in culture, let alone intact organisms, remains in many fundamental ways deeply mysterious. Thus, even though an encyclopedic knowledge of all the proteins of a cell will not solve all of its mysteries, it is also true that without such knowledge our ability to manipulate cells and intervene in human disease will remain proportionately superficial.

As the chapters in this volume illustrate, mammalian cells are uniquely suited for the expression of mammalian proteins. Because you and I are mammals, protein expression in mammalian cells will grow in importance to the increased understanding of our biology.

Frederick, MD, USA *James L. Hartley*

Contents

Contributors

LUDGER ALTROGGE • *Lonza Cologne GmbH, Cologne, Germany*
ZAHRA ASSUR • *Department of Biochemistry and Molecular Biophysics, Columbia University, New York, NY, USA*
LUCIA BALDI • *Laboratory of Cellular Biotechnology, École Polytechnique Fédérale de Lausanne, Lausanne, Switzerland*
WALTER BERGER • *Department of Medicine I, Institute of Cancer Research, Medical University of Vienna, Vienna, Austria*
MICHAEL J. BETENBAUGH • *Department of Chemical and Biomolecular Engineering, Johns Hopkins University, Baltimore, MD, USA*
ALICIA BIELIK • *New England Biolabs, Ipswich, MA, USA*
MONTSERRAT BUCETA • *Selexis SA, Plan-les-Ouates, Switzerland*
CHRISTINE T. DEMARIA • *Therapeutic Protein Expression, Genzyme Corporation, Framingham, MA, USA*
HAIMANTI DORAI • *Gene Expression, Centocor, Inc., Radnor, PA, USA*
CYNTHIA B. ELIAS • *Bulk Manufacturing/Cells and Viral Media, Sanofi-Pasteur, Toronto, ON, Canada*
DOMINIC ESPOSITO • *Protein Expression Laboratory, SAIC-Frederick, Inc., National Cancer Institute, Frederick, MD, USA*
ALEKSANDRA FLANAGAN • *Division of Structural Biology, Wellcome Trust Centre for Human Genetics, Oxford, UK*
PIERRE-ALAIN GIROD • *Selexis SA, Plan-les-Ouates, Switzerland*
OLIVER GRESCH • *Lonza Cologne GmbH, Cologne, Germany*
MICHAEL GRUSCH • *Department of Medicine I, Institute of Cancer Research, Medical University of Vienna, Vienna, Austria*
ELLEN GUTHRIE • *New England Biolabs, Ipswich, MA, USA*
DAVID L. HACKER • *Laboratory of Cellular Biotechnology and Protein Expression Core Facility, École Polytechnique Fédérale de Lausanne, Lausanne, Switzerland*
REBECCA HAMER • *Department of Statistics, University of Oxford, Oxford, UK*
NIAMH HARRAGHY • *Laboratory of Molecular Biotechnology, University of Lausanne, Lausanne, Switzerland*
JAMES L. HARTLEY • *Protein Expression Laboratory, SAIC-Frederick, Inc., National Cancer Institute, Frederick, MD, USA*
WAYNE A. HENDRICKSON • *Department of Biochemistry and Molecular Biophysics, Columbia University, New York, NY, USA*
STEVEN C.L. HO • *Bioprocessing Technology Institute, Agency for Science, Technology and Research, Singapore*
RALPH F. HOPKINS • *Protein Expression Laboratory, SAIC-Frederick, Inc., Frederick, MD, USA*

FUMIO IMAMOTO • *Laboratory of Molecular Biology, Research Institute for Microbial Diseases, Osaka University, Osaka, Japan*
BARBARA ANN JARDIN • *Biomedical Technology and Cell Therapy Research Laboratory, Department of Biomedical Engineering, Artificial Cells and Organs Research Center, McGill University, Montreal, QC, Canada; Biomedical Technology and Cell Therapy Research Laboratory, Department of Physiology, Artificial Cells and Organs Research Center, McGill University, Montreal, QC, Canada*
EMANUEL KREIDL • *Department of Medicine I, Institute of Cancer Research, Medical University of Vienna, Vienna, Austria*
RENATE KUNERT • *Department of Biotechnology, University of Natural Resources and Life Sciences, Vienna, Austria*
ANDREAS LACKNER • *Department of Medicine I, Institute of Cancer Research, Medical University of Vienna, Vienna, Austria*
PAULA MAGNELLI • *New England Biolabs, Ipswich, MA, USA*
FILIPPO MANCIA • *Department of Physiology and Cellular Biophysics, Columbia University, New York, NY, USA*
MARIATI • *Bioprocessing Technology Institute, Agency for Science, Technology and Research, Singapore*
CARINE MEERSCHMAN • *Protein Expression Core Facility, École Polytechnique Fédérale de Lausanne, Lausanne, Switzerland*
NICOLAS MERMOD • *Laboratory of Molecular Biotechnology, University of Lausanne, Lausanne, Switzerland*
JOANNE E. NETTLESHIP • *Oxford Protein Production Facility UK, Research Complex at Harwell, Rutherford Appleton Laboratory, Oxfordshire, UK*
SAY KONG NG • *Bioprocessing Technology Institute, Agency for Science, Technology and Research (A*STAR), Singapore*
RAYMOND J. OWENS • *Oxford Protein Production Facility UK, Research Complex at Harwell, Rutherford Appleton Laboratory, Oxfordshire, UK*
BARBARA PETER-VÖRÖSMARTY • *Department of Medicine I, Institute of Cancer Research, Medical University of Vienna, Vienna, Austria*
SATYA PRAKASH • *Biomedical Technology and Cell Therapy Research Laboratory, Department of Biomedical Engineering, Artificial Cells and Organs Research Center, McGill University, Montreal, QC, Canada; Biomedical Technology and Cell Therapy Research Laboratory, Department of Physiology, Artificial Cells and Organs Research Center, McGill University, Montreal, QC, Canada*
NAHID RAHMAN-HUQ • *Oxford Protein Production Facility UK, Research Complex at Harwell, Rutherford Appleton Laboratory, Oxfordshire, UK*
ALEXANDRE REGAMEY • *Selexis SA, Plan-les-Ouates, Switzerland*
TINA M. SAUERWALD • *Gene Expression, Centocor, Inc., Radnor, PA, USA*
TAKEFUMI SONE • *Division of Gene Therapy, Research Center for Genomic Medicine, Saitama Medical University, Saitama, Japan*
SABINE SPIEGL-KREINECKER • *Department of Neurosurgery, Wagner Jauregg Hospital, Linz, Austria*
KAROLA VORAUER-UHL • *Department of Biotechnology, University of Natural Resources and Life Sciences, Vienna, Austria*

VANESSA E. WALL • *Protein Expression Laboratory, SAIC-Frederick, Inc., National Cancer Institute, Frederick, MD, USA*
FLORIAN M. WURM • *Laboratory of Cellular Biotechnology, École Polytechnique Fédérale de Lausanne, Lausanne, Switzerland*
YUANSHENG YANG • *Bioprocessing Technology Institute, Agency for Science, Technology and Research, Singapore*
MIRANDA G.S. YAP • *Bioprocessing Technology Institute, Agency for Science, Technology and Research, Singapore*
MATTHEW P. ZUSTIAK • *Department of Chemical and Biomolecular Engineering, Johns Hopkins University, Baltimore, MD, USA*

Chapter 1

Why Proteins in Mammalian Cells?

James L. Hartley

Abstract

Producing recombinant mammalian proteins in native or near-native conformation is fundamental to many aspects of biology. Unfortunately, it is also a task whose outcome is extremely unpredictable. A protein that has been shaped over millions of generations of evolution for expression at a level appropriate to a specific cell type or in a particular developmental stage, may be toxic to a new host cell, or become insoluble (among many possible obstacles) when overexpressed in vitro. The object of this volume, "Protein Expression in Mammalian Cells," is to offer guidance for those who wish (or who have been forced by circumstance) to overexpress a mammalian protein in mammalian cells.

Key words: Protein quality, Stable expression, Transient expression, Protein folding, Secreted proteins

1. Why Proteins in Mammalian Cells?

The aim of this volume of Methods in Molecular Biology is to provide guidance for those wishing to produce recombinant proteins, and who may be considering doing so in mammalian cells. Following the example of monoclonal antibodies, which are produced in thousands of kilograms and are a multibillion dollar segment of the pharmaceutical market (1, 2), it is indeed possible to derive clones of mammalian cells that synthesize large quantities (grams per liter) of important proteins (3). And of course, the need to produce and purify proteins at much smaller scales is fundamental to all phases of modern biology.

But why in mammalian cells? Bacterial hosts, such as *Escherichia coli*, are far more productive per unit cost and per unit labor. The variety of expression tools in *E. coli* is far greater than in mammalian cells. Insect cells using the baculovirus vector technology are in general more quickly and cheaply productive than mammalian cells.

James L. Hartley (ed.), *Protein Expression in Mammalian Cells: Methods and Protocols*, Methods in Molecular Biology, vol. 801, DOI 10.1007/978-1-61779-352-3_1, © Springer Science+Business Media, LLC 2012

These considerations lead directly to the issue of protein quality, which is the simple and accurate answer to this "why" question. Since costs are relatively high using mammalian cells as expression hosts, there must be circumstances in which quality is the overriding concern.

Quality is a matter of the requirements for each protein's ultimate use. For almost all purposes, the protein is desired to be as native as possible, i.e., resemble the protein made in the organism and tissue and folding and activity state found in nature. However, there are many situations where aspects of quality are readily dispensable. For example, glycosylation of mammalian proteins can be important for aspects of biological activity (Magnelli, et al., Chapter 13; (4)), but glycosylation introduces heterogenity that complicates structure determinations and reproducible production. For therapeutic uses, maintaining near-native glycosylation can be of critical importance, while for structure determinations the sites of protein glycosylation are frequently removed by mutating genes before expression so that glycosylation is completely absent.

Aside from the many kinds of posttranslational modifications available in mammalian cells, it is the very old question of protein folding that is the main driver for using them to express recombinant proteins. It is now abundantly clear not only that far more than the linear sequence of amino acids is required to fold proteins correctly, but also that even proteins expressed in nature are frequently misfolded. In fact, it has been pointed out (5) that all proteins are to some extent in equilibrium with multiple (depending on size and domain structure) unfolded states (see Note 1). Thus, there are multiple interacting mechanisms (dubbed the "proteostasis network" (6)) that shepherd a protein from its birth at the ribosome to its death through proteolysis. As our appreciation of these mechanisms grows, it is less surprising that pairing the desired protein with the appropriate expression host is critical for successful production of many proteins. Since human proteins (and their close relatives in model organisms) are of primary importance in biomedical research, it follows that mammalian cells are often the host of choice for producing active, well-folded recombinant proteins.

It is important to observe that negative evidence (i.e., fewer publications) shows that mammalian cells are successful hosts for relatively few cytoplasmic, as opposed to secreted, proteins. A significant contributor to this phenomenon may be the commercial aspects of biopharmaceuticals. Proteins are large compared to "white powder" (i.e., chemically synthesized) drugs (typically with molecular weights less than 500) and proteins do not penetrate cells except through endocytic pathways which often result in their destruction. Thus, the proteins whose expression is optimized for drug use are most often those whose binding partners are either secreted or on the cell surface, and it follows that therapeutic proteins (i.e., proteins that naturally interact with secreted or cell surface proteins) are themselves naturally secreted or are the extracellular domains of cell surface proteins (see Note 2). Another aspect affecting expressing proteins in the cytoplasm of mammalian cells may be that secreted proteins leave the cell and are diluted into the surrounding culture medium, whereas large amounts of protein retained in the cytoplasm have more potential for toxicity. Whatever the cause, the literature of protein expression in mammalian cells is dominated by secreted protein targets (see Note 3). Note, however, that this volume contains a chapter (Hopkins) offering guidance to those who wish to test mammalian cells for cytoplasmic expression.

2. Transient vs. Stable Transfection

Readers experienced in transforming *E. coli* with plasmid vectors will not find equivalent tools available for mammalian cells. As is discussed in more detail below, the fundamental difference between what is possible in the two hosts stems from the lack of a simple, small, high copy number mammalian origin of DNA replication, i.e., there are no mammalian plasmids. In the *E. coli* case, a single plasmid molecule can enter a single cell, increase in copy number using host cell enzymes, and confer drug resistance so that a colony of cells can form overnight. Each cell in the colony is identical; often, we treat each colony as identical to every other colony if a pure plasmid has been used for the transformation.

Transfection of mammalian cells operates according to different principles. Using the most common transfection protocols, (1) thousands of plasmid DNA molecules enter each transfected cell (because using less DNA results in less protein expression); (2) the transfected plasmids do not replicate but instead are linearized and ligated to each other; (3) the plasmids express the protein of interest but are lost by dilution as host cells replicate; and (4) the plasmid molecules may be degraded or modified (silenced) over time so that protein expression is lost. Within the population of transfectants is a small percentage (less than 10%, sometimes less than 1%) of cells in which the entering DNA has integrated randomly into the host cell genome. However, selection with drugs does not yield identical clones, but instead results in clones that vary widely in the productivity of the protein of interest, and even results in clones that are highly resistant to the drug but produce no protein at all. Clones that have desirable characteristics may change over time and lose their value. The time required to achieve a high-quality mammalian clone expressing a protein of interest is typically measured in months.

As a consequence, protein expression in mammalian cells is divided into transient and stable expression. Both modes start with the introduction of plasmid DNA into a cell line, called transfection. For transient expression, the time frame to protein harvest is generally a few days, at which point the yield per volume of (expensive) medium starts to decline. Thus, improving yield in transient expression is governed by improving the percentage of cells that are transfected, decreasing the cost of the transfection reagent, making enough DNA (often hundreds of micrograms per liter), increasing the expression of the protein of interest through vector and gene optimization, and increasing the volume of cells transfected (usually, in suspension cells), from milliliters to liters (see Baldi, Chapter 2).

The commitment to the stable cell line track is a major one, and is ordinarily preceded by small-scale transient testing. With preliminary data in hand, transfection is followed within a day or two by selection conditions (nutritional or drug). Within a week or two of culture, most of the cells die either because they were not transfected or because plasmid DNA was not integrated into their genomes. Resistant cells take over the culture, and at some point individual clones are isolated and assessed for their potential for long-term productivity. Because so many aspects are random, most importantly the chromosomal location of integration and its influence on protein expression, dozens to thousands of clones may be evaluated without any guarantee that a satisfactory one may be found. For commercial production, a number of clones may be followed at moderate scale for many months to ensure high and stable long-term productivity.

3. Introducing DNA into Mammalian Cells

Mammalian cells in culture are constantly sampling their environment not only through specific transport mechanisms, but also by endocytosis, i.e., vacuole formation and internalization from the cell surface. Since cell surfaces are negatively charged (7), complexes of DNA with substances that impart an overall positive charge (DEAE dextran and calcium phosphate (8), cationic polymers (9) and cationic lipids (Kunert, Chapter 14; (10))) adhere to the cell surface and are internalized in membrane bound vesicles. Release from these compartments may occur via fusion of endocytic vesicles to lysosomes, or through osmotic lysis that occurs as a result of neutralization of the cationic DNA complexes (11). Movement of transfecting DNA from cytosol to the nucleus is thought to occur at cell division (12), when dissolution and reformation of the nuclear membrane allows access to the nuclear compartment, because direct injection of naked plasmid DNA achieves much higher levels of gene expression than injection into the cytoplasm (13). However, transfection reagents, such as polyethylenimine and cationic lipids linked to nuclear localization peptides, seem to promote movement of DNA into intact nuclei (14, 15) (see Note 4).

Transfected DNA is not stably maintained in mammalian cells as episomes (nonintegrated DNA molecules that replicate and segregate during host cell division). No simple mammalian origin of replication has been demonstrated (16). Systems such as the Epstein-Barr virus oriP/nuclear antigen that promote "episomal replication" are very inefficient (17). Episomal vectors without viral elements look more promising (18, 19). Transfected DNA that reaches the nucleus is lost by dilution (not replicated, thus fewer and fewer copies of DNA per cell as the cells divide), unless it is incorporated into the genome of the host (below). Selecting for "stable pools" (i.e., putting transfectants under selection using a drug resistance gene on the transfecting plasmid) results in cells that as a pool express lower amounts of the protein of interest as the number of cell divisions increase, even when the drug resistance and protein of interest genes are on the same mRNA (19). However, improved vector elements can help pools of transiently transfected cells maintain expression for useful lengths of time (20).

4. Protein Expression with Viruses

To induce mammalian cells to express a recombinant protein, it is necessary to introduce the gene encoding that protein into the nucleus of the cell (see Notes 5 and 6). Since viruses that infect mammalian cells have evolved ways to efficiently enter cells and their nuclei, viruses such as adenovirus, lentivirus (mostly based on HIV), alphaviruses such as Semliki Forest virus, and vaccinia have been adapted for protein expression (see Note 7). However, mammalian viruses are not widely used for overexpressing recombinant proteins. It is of course more straightforward to immediately transfect naked DNA than to produce viral particles that express the same protein, so using viral particles is often limited to cell types and experiments (such as gene therapy) for which transfection is not efficient or possible. Biosafety concerns regarding mammalian viruses add to up-front complexity because special cell lines or mutant viruses or multiple vectors are often invoked to protect against unintentional delivery of genes to laboratory personnel. In addition, viruses that kill host cells, such as adenovirus and alphaviruses (Semliki Forest Virus and Sindbis virus) limit the time during which protein expression can take place (21).

Interestingly, insect viruses (baculoviruses) are growing in popularity as a way to express proteins in mammalian cells (see Elias, Chapter 4; (22–24)). Insect viruses have an inherent safety advantage over mammalian viruses because they do not replicate in mammalian cells (25) and insect promoters are inactive there (26, 27). Since baculovirus particles are actively taken up by mammalian cells, apparently through the endocytic pathway (23), when a mammalian promoter is included upstream of a gene of interest the gene may be expressed at a useful level. A wide variety of mammalian cell lines have been shown to be transducible with baculovirus, although with considerable differences in efficiency (28).

5. Observations on Transfection and Protein Expression in Mammalian Cells

1. By far the most widely used method of introducing heterologous genes into mammalian cells for protein expression is transfection of naked DNA. For a decade starting in 1977 (29–31), innovative experiments worked out many of the details of how transfection operates. Transfection of mammalian cells achieves useful levels of protein expression with thousands or tens of thousands of DNA molecules per cell (see Note 8). Soon after DNA enters cells, it is randomly linearized (if it is originally circular) and ligated into concatamers (32). Individual transfected cells either lose the transfected DNA over time (by dilution and degradation) or, rarely, incorporate the DNA into random positions in the genome. Thus, individual stable clones of cells from a single transfection experiment differ in the number of tandem copies integrated into their genomes, and the locations of those integration events (33). It is rare for a single clone to have transfecting DNA integrated at more than one chromosomal location (32, 34). The amount and stability of long-term expression from these stable cell lines may be affected not only by characteristics of the transfected DNA itself (promoter, enhancer, CpG content (19), mRNA structure (35), etc.), but also by the chromosomal locus into which it is integrated. Expression from such a "stable" cell line may not be uniform, as over time cells in a supposedly clonal population may develop subpopulations that differ in expression levels (36).
2. Selection of stable cell lines using drugs, such as methotrexate (MTX; see Note 9) or G418 (see Note 10) require that the transfecting DNAs contain the appropriate resistance genes (NeoR for G418, dhfr for MTX). Because transfected DNAs are linearized and ligated into concatamers in vivo (above), it is not necessary for the resistance gene and the gene of interest to be on the same molecule. Thus, it is common for two separate plasmids to be mixed and co-transfected, one to express the target protein and the other to confer drug resistance for use in selection. Excess expression plasmid over drug resistance plasmid (e.g., 10:1) is often used to increase the likelihood of

abundant target protein expression, although subsequent in vivo mechanisms (deletions, silencing, etc.) can confound these efforts.

3. Early reports showed that proteins are expressed at higher levels when supercoiled DNA is used in transient transfections (37, 38). However, there are reports that linear DNAs are equally active as circular molecules (39, 40). Since transfected DNAs are randomly linearized and ligated into concatamers in vivo, if the goal is to isolate high-producing stable cell lines it is sensible to transfect with DNA that has been linearized outside the regions required for protein expression. Typically, this is done with a restriction enzyme digestion in the bacterial origin of replication or drug resistance marker used for selection in *E. coli* (for example, Harraghy, et al., Chapter 7; Ng, Chapter 11; (41, 42)) so that the activity of integrated copies is maximized. However, stable cell lines are readily isolated using circular DNAs (Kunert, Chapter 14; DeMaria, Chapter 3).
4. For stable cell lines (transfected DNA stably integrated into the genome), there is not necessarily a correlation between the number of integrated copies of the transfecting DNA and the amount of protein expressed (33). Even a decline in the amount of protein-specific mRNA over time does not necessarily indicate that less of the protein of interest is being made (43), which is thought to be an indication that processes downstream of transcription (mRNA processing, export to the cytoplasm, translation, movement through the secretory pathway) may be limiting (see Note 11).
5. For pharmaceutical companies, the random nature of insertion of plasmids into the genome to make stable cell lines adds uncertainty to a process with huge financial consequences. Targeted insertion is available (44–46) but reports showing by direct comparison that the productivity of such targeted clones have achieved productivity levels equivalent to that of random clones are lacking.
6. When the protein of interest comprises two polypeptides, the relative expression of each polypeptide required for maximum yield must be established by experimentation. For example, IgG antibodies comprise two (identical) light chains and two (identical) heavy chains, but it has been demonstrated by various groups that excess light chain production results in higher overall yield in mammalian cells (Kunert, Chapter 14; (35)).
7. Evolution has selected for mammalian proteins that are expressed at appropriate amounts in the appropriate tissues at the appropriate times during development or circumstances in the life of the host animal. Thus, it does not follow that the amino acid or nucleotide sequence of the native protein or

gene or mRNA will yield the highest amount of protein desired by the experimenter. Many steps of mRNA transcription, splicing, export from the nucleus, and transport to ribosomes, translation, and degradation (47, 48) might influence protein yields. Small steps toward understanding how changes in gene structure (maintaining the native amino acid sequence but altering the nucleotide sequence) can influence the amounts of protein have been taken. Several companies offer "codon optimization" as part of gene synthesis services, but it should be emphasized that maximal yields do not necessarily follow from simply substituting frequently occurring codons for rare ones. Rather the removal of cryptic splice sites, minimizing secondary structures, balancing GC/AT richness, etc., are part of "gene optimization" (35, 49).

6. A Certain Future for Protein Expression

The continuing revolution in DNA sequencing methods is revealing a fascinating landscape of genetic diversity in human populations and human diseases. However, it is less likely than ever that nucleotide sequences alone will lead to a deep and enabling understanding of the biology of cells. For example, a single amino acid change can completely alter the folding of a small protein (50) or abolish the activity of an enzyme of the TCA cycle and replace it with another (51). So many aspects of evolution, development, and the cellular milieu contribute to the conversion of a DNA sequence to a properly expressed and functional protein that the necessity to use mammalian cells to express mammalian proteins is likely to persist for a very long time.

7. Notes

1. As Lindquist has observed (52), the large amounts of chaperone proteins found in normal cells are probably present for fast response to sudden increases (caused by stochastic or external stresses) in misfolded proteins. Since cancer cells, growing out of control, are permanently stressed, they are often acutely sensitive to agents that target the same chaperone proteins. Several drugs targeting chaperones are in clinical trials for the treatment of human cancers.
2. The expression and purification of integral membrane proteins is a field to itself. Such proteins may have from 1 to 12 or more transmembrane helices, and the lipid environment of these helices in the membrane must be replaced by an artificial lipidic

environment during purification. For a perspective on the challenges of studying membrane proteins, see Kobilka and Schertler (53).

3. While I am not aware of any systematic investigation of the issue, it has been our observation that it is not usually possible to force secretion of a cytoplasmic protein by simply adding an N-terminal signal peptide.
4. There is much still to be learned about optimizing transfection efficiency. For example, it has been shown that free PEI, i.e., not bound to DNA, is necessary for improved transfection of CHO cells (54).
5. An exception to this statement is the situation with alphaviruses, which are RNA viruses of mammals that have no DNA phase in their life cycle. Their RNA genome is replicated and expressed entirely in the cytoplasm of their host. For examples of the use of alphaviruses, notably Semliki Forest Virus, for protein expression, see Lundstrom (55).
6. The reader should also be aware that a large literature exists around introducing heterologous genes into human and model mammalian cells for purposes of gene therapy.
7. Terminology is important when thinking about viruses as delivery vehicles for recombinant genes. A virus may be fully replicative in a particular host cell but not other cells. For example, most adenoviruses used for gene delivery have been engineered for biosafety and will replicate in HEK293 cells, which express adenovirus E1A protein, but will not replicate in other cell types. Also for safety, lentiviruses are typically made in 293 cells by co-transfecting multiple plasmids such that the resulting viral particles can go through only part of their life cycle in a subsequent host. Many viral particles may be defective so that truly replicative particles comprise a fraction, and sometimes a small fraction, of the total particle count. Viral particles that enter a cell and express a heterologous gene without making progeny virus may be said to "transduce" the gene, instead of "infecting" the host. It may be necessary to arrange expression of viral receptor protein on the cells to be used for protein production. For example, adenoviruses transduce CHO cells very inefficiently unless a receptor protein is coexpressed in the cells (56). Alternatively, production of the virus may be arranged so that the particles display ligands for receptors on cells they would not ordinarily infect, which is called "pseudotyping." Lentiviral particles are usually pseudotyped with a protein from vesicular stomatitis virus (VSVg), which has the double advantage of allowing them to transduce many mammalian cell lines (with variable efficiency), and improves their physical stability.

8. One microgram of 5 kb plasmid comprises about 2×10^{11} molecules. Thus, transfecting one million cells with a microgram of DNA is equivalent to more than 100,000 DNA molecules per cell.
9. MTX is an inhibitor of folate metabolism. Its target enzyme is dihydrofolate reductase (dhfr), to which it binds and inhibits stoichiometrically (57). Mammalian cells subjected to gradually increasing concentrations of MTX adapt by increasing the number of copies of their dhfr genes. If cells are transfected with an exogenous dhfr gene, increasing MTX concentrations will result in amplification of either the endogenous or transfected dhfr genes, or both. If the transfected cells lack a native dhfr gene (such cell lines require supplementation of their growth medium with nucleosides, such as hypoxantine and thymine) only the transfected dhfr gene (and any DNA linked to it) will be amplified.
10. G418, neomycin, and kanamycin are related drugs (aminoglycosides). G418, which is also sold under the trade name Geneticin, kills mammalian cells, while neomycin and kanamycin kill bacterial cells but not mammalian cells. All three drugs are inactivated by the neomycin phosphotransferase gene, originally isolated from a bacterial transposon, Tn5. The neomycin phophotransferase gene may be called NeoR or KanR, depending on its use. To be resistant to G418, neomycin, or kanamycin, the gene must be expressed by the appropriate promoter and have the appropriate expression context (translation signals) for the host in which it will be used.
11. In our lab, we make it a practice to analyze cells as well as medium by Western blot for the presence of secreted target protein, since we want to know if we are not successfully secreting all the protein that is made.

References

1. Beck, A., Wurch, T., Bailly, C., and Corvaia, N. (2010) Strategies and challenges for the next generation of therapeutic antibodies. *Nat. Rev. Immunol.* **10**, 345–352.
2. Labrijn, A.F., Aalberse, R.C., and Schuurman, J. (2008) When binding is enough: nonactivating antibody formats. *Curr. Opin. Immunol.* **20**, 479–485.
3. Wurm, F.M. (2004) Production of recombinant protein therapeutics in cultivated mammalian cells. *Nat. Biotechnol.* **22**, 1393–1398.
4. Shields, R.L., Lai, J., Keck, R., O'Connell, L.Y., Hong, K., Meng, Y.G., et al. (2002) Lack of fucose on human IgG1 N-linked oligosaccharide improves binding to human Fcgamma RIII and antibody-dependent cellular toxicity. *J. Biol. Chem.* **277**, 26733–26740.
5. Powers, E.T., Morimoto, R.I., Dillin, A,, Kelly, J.W., and Balch, W.E. (2009) Biological and chemical approaches to diseases of proteostasis deficiency. *Annu. Rev. Biochem.* **78**, 959–991.
6. Sifers, R.N. (2010) Manipulating proteostasis. *Nat. Chem. Biol.* **6**, 400–401.
7. Danon, D., Goldstein, L., Marikovsky, Y., and Skutelsky, E. (1972) Use of cationized ferritin as a label of negative charges on cell surfaces. *J. Ultrastruct. Res.* **38**, 500–510.
8. Graham, F.L., and van der Eb, A.J. (1973) A new technique for the assay of infectivity of human adenovirus 5 DNA. *Virology.* **52**, 456–467.

9. Boussif, O., Lezoualc'h, F., Zanta, M.A., Mergny, M.D., Scherman, D., Demeneix, B., et al. (1995) A versatile vector for gene and oligonucleotide transfer into cells in culture and in vivo: polyethylenimine. *Proc. Natl. Acad. Sci. USA* **92**, 7297–7301.
10. Felgner, P.L., Gadek, T.R., Holm, M., Roman, R., Chan, H.W., Wenz, M., et al. (1987) Lipofection: a highly efficient, lipid-mediated DNA-transfection procedure. *Proc. Natl. Acad. Sci. USA* **84**, 7413–7417.
11. Kichler, A., Leborgne, C., Coeytaux, E., and Danos, O. (2001) Polyethylenimine-mediated gene delivery: a mechanistic study. *J. Gene Med.* **3**, 135–144.
12. Wong, E.A., and Capecchi, M.R. (1985) Effect of cell cycle position on transformation by microinjection. *Somat. Cell. Mol. Genet.* **11**, 43–51.
13. Capecchi, M.R. (1980) High efficiency transformation by direct microinjection of DNA into cultured mammalian cells. *Cell.* **22**, 479–488.
14. Pollard, H., Remy, J.S., Loussouarn, G., Demolombe, S., Behr, J.P., and Escande, D. (1998) Polyethylenimine but not cationic lipids promotes transgene delivery to the nucleus in mammalian cells. *J. Biol. Chem.* **273**, 7507–7511.
15. Hawley-Nelson, P., Lan, J., Shih, P.J, Jessee, J.A., Schifferli, K.P., Gebeyehu, G., et al. (2002) Peptide enhanced transfections, US *Patent.* 6,376,248, 2002.
16. Wang, C.Y., and Sugden, B. (2008) Identifying a property of origins of DNA synthesis required to support plasmids stably in human cells. *Proc. Natl. Acad. Sci. USA* **105**, 9639–9644.
17. Leight, E.R., and Sugden, B. (2001) Establishment of an oriP replicon is dependent upon an infrequent, epigenetic event. *Mol. Cell Biol.* **21**, 4149–4161.
18. Piechaczek, C., Fetzer, C., Baiker, A., Bode, J., and Lipps, H.J. (1999) A vector based on the SV40 origin of replication and chromosomal S/MARs replicates episomally in CHO cells. *Nucleic Acids Res.* **27**, 426–428.
19. Haase, R., Argyros, O., Wong, S.P., Harbottle, R.P., Lipps, H.J., Ogris, M., et al. (2010) pEPito: a significantly improved non-viral episomal expression vector for mammalian cells. *BMC Biotechnol.* **15**, 20.
20. Ye, J., Alvin, K., Latif, H., Hsu, A., Parikh, V., Whitmer, T., et al. (2010) Rapid protein production using CHO stable transfection pools. *Biotechnol. Prog.* **26**, 1431–1437.
21. Casales, E., Aranda, A., Quetglas, J.I., Ruiz-Guillen, M., Rodriguez-Madoz, J.R., Prieto, J., et al. (2010) A novel system for the production of high levels of functional human therapeutic proteins in stable cells with a Semliki Forest virus noncytopathic vector. *Nat. Biotechnol.* **27**, 138–148.
22. Kost, T.A., Condreay, J.P., and Jarvis, D.L. (2005) Baculovirus as versatile vectors for protein expression in insect and mammalian cells. *Nat. Biotechnol.* **23**, 567–575.
23. Hofmann, C., Sandig, V., Jennings, G., Rudolph, M., Schlag, P., and Strauss, M. (1995) Efficient gene transfer into human hepatocytes by baculovirus vectors. *Proc. Natl. Acad. Sci. USA* **92**, 10099–10103.
24. Barsoum, J., Brown, R., McKee, M., and Boyce, F.M. (1997) Efficient transduction of mammalian cells by a recombinant baculovirus having the vesicular stomatitis virus G glycoprotein. *Hum. Gene Ther.* **8**, 2011–2018.
25. Volkman, L.E., and Goldsmith, P.A. (1983) In Vitro Survey of Autographa californica Nuclear Polyhedrosis Virus Interaction with Nontarget Vertebrate Host Cells. *Appl. Environ. Microbiol.* **45**, 1085–1093.
26. Carbonell, L.F., and Miller, L.K. (1987) Baculovirus interaction with nontarget organisms: a virus-borne reporter gene is not expressed in two mammalian cell lines. *Appl. Environ. Microbiol.* **53**, 1412–1417.
27. Boyce, F.M., and Bucher, NL. (1996) Baculovirus-mediated gene transfer into mammalian cells. *Proc. Natl. Acad. Sci. USA* **93**, 2348–2352.
28. Kost, T.A., and Condreay, J.P. (2002) Recombinant baculoviruses as mammalian cell gene-delivery vectors. *Trends Biotechnol.* **20**, 173–180.
29. Bacchetli, S., and Graham, F.L. (1977). Transfer of the gene for thymidine kinase to thymidine kinase-deficient human cells by purified herpes simplex viral DNA. *Proc. Natl. Acad. Sci. USA* **74**, 1590–1594.
30. Maitland, N.J., and McDougall, J.K. (1977). Biochemical transformation of mouse cells by fragments of herpes simples virus DNA. *Cell.* **11**, 233–241.
31. Wigler. M., Silverstein, S., Lee, L.S., Pellicer, A., Cheng, Y., and Axel, R. (1977) Transfer of purified herpes virus thymidine kinase gene to cultured mouse cells. *Cell.* **11**, 223–232.
32. Perucho, M., Hanahan, D., and Wigler, M. (1980) Genetic and physical linkage of exogenous sequences in transformed cells. *Cell.* **22**, 309–317.
33. Derouazi, M., Martinet, D., Besuchet Schmutz, N., Flaction, R., Wicht, M., Bertschinger, M., et al. (2006) Genetic characterization of CHO production host DG44 and derivative recombinant cell lines. *Biochem. Biophys. Res. Commun.* **340**, 1069–1077.

34. Folger, K.R., Wong, E.A., Wahl, G., and Capecchi, M.R. (1982) Patterns of integration of DNA microinjected into cultured mammalian cells: evidence for homologous recombination between injected plasmid DNA molecules. *Mol. Cell Biol.* **2**, 1372–1387.
35. Kalwy, S., Rance, J., and Young, R. (2006) Toward more efficient protein expression: keep the message simple. *Mol. Biotechnol.* **34**, 151–156.
36. Kaufman, W.L., Kocman, I., Agrawal, V., Rahn, H.P., Besser, D., and Gossen, M. (2008) Homogeneity and persistence of transgene expression by omitting antibiotic selection in cell line isolation. *Nucleic Acids Res.* **36**, e111.
37. Weintraub, H., Cheng, P.F., and Conrad, K. (1986) Expression of transfected DNA depends on DNA topology. *Cell.* **46**, 115–122.
38. Chen,C., and Okayama H. (1987) High-efficiency transformation of mammalian cells by plasmid DNA. *Mol. Cell Biol.* **7**, 2745–2752.
39. Liang, X., Teng, A., Braun, D.M., Felgner, J., Wang,Y., Baker, S.I., et al. (2002) Transcriptionally active polymerase chain reaction (TAP): high throughput gene expression using genome sequence data. *J. Biol. Chem.* **277**, 3593–3598.
40. Derouazi, M., Flaction, R., Girard, P., de Jesus, M., Jordan, M., and Wurm, F.M. (2006) Generation of recombinant Chinese hamster ovary cell lines by microinjection. *Biotechnol. Lett.* **28**, 373–382.
41. de la Cruz Edmonds, M.C., Tellers, M., Chan, C., Salmon, P., Robinson, D.K., and Markusen, J. (2006) Development of transfection and high-producer screening protocols for the CHOK1SV cell system. *Mol. Biotechnol.* **34**, 179–190.
42. Stuchbury, G., and Münch, G. (2010) Optimizing the generation of stable neuronal cell lines via pre-transfection restriction enzyme digestion of plasmid DNA. *Cytotechnology.* **62**, 189–194.
43. Barnes, L.M., Bentley, C.M., and Dickson, A.J. (2004) Molecular definition of predictive indicators of stable protein expression in recombinant NS0 myeloma cells. *Biotechnol. Bioeng.* **85**, 115–121.
44. Schebelle, L., Wolf, C., Stribl, C., Javaheri, T., Schnütgen, F., Ettinger, A., et al. (2010) Efficient conditional and promoter-specific in vivo expression of cDNAs of choice by taking advantage of recombinase-mediated cassette exchange using FlEx gene traps. *Nucleic Acids Res.* **38**, e106.
45. Nehlsen, K., Schucht, R., da Gama-Norton, L., Krömer, W., Baer, A., Cayli, A., et al. (2009) Recombinant protein expression by targeting pre-selected chromosomal loci. *BMC Biotechnol.* **9**, 100.
46. Barron, N., Piskareva, O., and Muniyappa, M. (2007) Targeted genetic modification of cell lines for recombinant protein production. *Cytotechnology.* **53**, 65–73.
47. Welch, M., Villalobos, A., Gustafsson, C., and Minshull, J. (2009) You're one in a googol: optimizing genes for protein expression. *J. R. Soc. Interface.* 6 Suppl 4, S467–76.
48. Maquat, L.E., Tarn, W.Y., and Isken, O. (2010) The pioneer round of translation: features and functions. *Cell.* **142**, 368–74.
49. Hung, F., Deng, L., Ravnikar, P., Condon, R., Li, B., Do, L., et al. (2010) mRNA stability and antibody production in CHO cells: improvement through gene optimization. *Biotechnol. J.* **5**, 393–401.
50. Alexander, P.A., He, Y., Chen, Y., Orban, J., and Bryan, P.N. (2009) A minimal sequence code for switching protein structure and function. *Proc. Natl. Acad. Sci. USA* **106**, 21149–21154.
51. Dang, L., White, D.W., Gross, S., Bennett, B.D., Bittinger, M.A., Driggers, E.M., et al. (2010) Cancer-associated IDH1 mutations produce 2-hydroxyglutarate. *Nature.* **462**, 739–744.
52. Whitesell, l., and Lindquist, S. (2005). HSP90 and the chaperoning of cancer. *Nat. Rev. Cancer.* **5**, 761–772.
53. Kobilka, B., and Schertler, G.F. (2008) New G-protein-coupled receptor crystal structures: insights and limitations. *Trends Pharmacol. Sci.* **29**, 79–83.
54. Bertschinger, M., Schertenleib, A., Cevey, J., Hacker, D.L., and Wurm, F.M. (2008) The kinetics of polyethylenimine-mediated transfection in suspension cultures of Chinese hamster ovary cells. *Mol. Biotechnol.* **40**, 136–43.
55. Lundstrom, K. (2003) Semliki Forest virus vectors for large-scale production of recombinant proteins. *Methods Mol. Med.* **76**, 525–43.
56. Gaillet, B., Gilbert, R., Amziani, R., Guilbault, C., Gadoury, C., Caron, A.W., et al. (2007) High-level recombinant protein production in CHO cells using an adenoviral vector and the cumate gene-switch. *Biotechnol. Prog.* **23**, 200–209.
57. Kaufman, R.J., Wasley, L.C., Spiliotes, A.J., Gossels, S.D., Latt, S.A., Larsen, G.R., et al. (1985) Coamplification and coexpression of human tissue-type plasminogen activator and murine dihydrofolate reductase sequences in Chinese hamster ovary cells. *Mol. Cell Biol.* **5**, 1750–1759.

Chapter 2

Large-Scale Transfection of Mammalian Cells

Lucia Baldi, David L. Hacker, Carine Meerschman, and Florian M. Wurm

Abstract

The large-scale transfection of mammalian cells allows moderate (milligram to gram) amounts of recombinant proteins (r-proteins) to be obtained for fundamental or clinical research. In this article, we describe a one-liter transfection using polyethyleneimine (PEI) for DNA delivery into human embryonic kidney (HEK-293) cells cultivated in serum-free suspension to produce a recombinant human monoclonal antibody that yields up to about 1 g/L in a 10-day process. The method is based on a DNA delivery step performed at high cell density (20×10^6 cells/mL) by direct addition of DNA and PEI to the culture. Subsequently, the cells are diluted 20-fold for the 10-day production phase in the presence of valproic acid (VPA), a histone deacetylase inhibitor. The methods for plasmid purification, antibody quantification by enzyme-linked immunosorbent assay (ELISA), and affinity purification with protein A are also described.

Key words: HEK-293, Transfection, ELISA, Protein A, Suspension culture, Recombinant protein, Polyethyleneimine

1. Introduction

Recombinant proteins (r-proteins) have many applications in fundamental and clinical research. Until recently, the main technique for obtaining a sufficient amount of an r-protein from mammalian cells was the generation of a stable cell line through a lengthy procedure that included gene delivery, genetic selection, and cell cloning (1). Large-scale transient gene expression (TGE) is being developed as a faster and more economical alternative for producing r-proteins for various research applications (2, 3). With this approach, milligram to gram quantities of a protein can be produced within days after construction of the expression vector. Proteins localized to the nucleus, cytoplasm, or plasma membrane in addition to secreted proteins have been expressed using TGE (4).

James L. Hartley (ed.), *Protein Expression in Mammalian Cells: Methods and Protocols*, Methods in Molecular Biology, vol. 801, DOI 10.1007/978-1-61779-352-3_2, © Springer Science+Business Media, LLC 2012

Recently, volumetric productivity by TGE has passed the 1 g/L barrier at small scale using HEK-293 cells by increasing the cell density at the time of transfection and by adding VPA, a histone deacetylase inhibitor (5–7).

TGE is a relatively simple procedure that does not require expensive equipment – only a simple cell culture incubator and an orbital shaker are necessary. Here, we describe the procedure as applied to a one-liter culture in an orbitally shaken 5-L glass bottle, but the same method can be performed in other containers, including Erlenmeyer flasks, spinner flasks, Wave and stirred-tank bioreactors (8, 9). The method is fully scalable and can be performed in either smaller or larger volumes than those described here. The process can be divided into three steps (1) cell culture scale-up to the desired biomass; (2) transfection; and (3) production in a batch or fed-batch mode. To produce a r-protein by large-scale TGE, it is necessary to have (1) a simple assay for its quantification to facilitate the optimization of the transfection and (2) a method for its purification. Here, we describe the production of a human monoclonal antibody from two co-transfected plasmids carrying the full-length cDNAs of the IgG light and heavy chain genes (10, 11). A third plasmid carrying the enhanced green fluorescent protein (eGFP) gene is also co-transfected to have a convenient method (eGFP-specific fluorescence) to measure the efficiency of the transfection. The methods for the purification of the antibody by affinity to protein A and its quantification with an enzyme-linked immunosorbent assay (ELISA) are also described. Although the latter methods are specific for recombinant antibodies, the TGE protocol itself can be applied to any r-protein with slight modifications of some parameters.

2. Materials

2.1. Cell Culture

1. HEK-293 cells adapted to cultivation in serum-free suspension (12).
2. Cylindrical and square-shaped glass bottles with nominal volumes of 100 mL to 5 L (Schott Glass, Mainz, Germany).
3. Ex-cell® 293 medium without L-glutamine and phenol red (Sigma-Aldrich, St. Louis, MO).
4. 50× L-glutamine and phenol red solution. A stock solution with 200 mM glutamine and 250 μg/mL phenol red is made by dissolving 29.23 g/L glutamine (Applichem GmbH, Darmstadt, Germany) and 250 mg phenol red (Sigma-Aldrich) in 800 mL water. After the compounds are dissolved, the volume is adjusted to 1 L by further addition of water. The solution is sterilized by filtration through a 0.2-μm Steritop (bottle-top) filter unit (Express Membrane) with a neck size of 45 mm (Millipore AG,

Zug, Switzerland). The solution is transferred into sterile 50-mL centrifuge tubes (TPP, Trasadingen, Switzerland) and kept frozen at −20°C. For each liter of Ex-cell® 293 medium, 20 mL of the stock solution is added.

5. Trypan blue solution (0.4%) (Sigma-Aldrich).

2.2. Plasmids

1. pEGFPN1 (Clontech, Palo Alto, CA) expressing eGFP (13).
2. pXLG^{HEK}-RhLC expressing the anti-Rhesus D IgG light chain cDNA (10, 11).
3. pXLG^{HEK}-RhHC expressing the anti-Rhesus D IgG heavy chain cDNA (10, 11).

2.3. Plasmid DNA Preparation

1. LB agar plates with 50 μg/mL kanamycin (Applichem) or 100 μg/mL ampicillin (Applichem).
2. LB medium (Invitrogen AG, Basel, Switzerland) with 50 μg/mL kanamycin or 100 μg/mL ampicillin.
3. NucleoBond AX 500 anion exchange chromatography column (Macherey-Nagel, Düren, Germany). The kit includes all the buffers.
4. Resuspension buffer S1: 50 mM Tris–HCl, 10 mM EDTA, 100 μg/mL RNase A (Macherey-Nagel), pH 8.0.
5. Lysis buffer S2: 200 mM NaOH, 1% SDS.
6. Neutralization buffer S3: 2.8 M potassium acetate, adjusted to pH 5.1 with acetic acid.
7. Equilibration buffer N2: 100 mM Tris–HCl, 15% ethanol, 900 mM KCl, 0.15% Triton X-100, adjusted to pH 6.3 with H_3PO_4.
8. Wash buffer N3: 100 mM Tris–HCl, 15% ethanol, 1.15 M KCl, adjusted to pH 6.3 with H_3PO_4.
9. Elution buffer N5: 100 mM Tris–HCl, 15% ethanol, 1 M KCl, adjusted to pH 8.5 with H_3PO_4.
10. 95 and 70% ethanol (Applichem).
11. TE (10 mM Tris–HCl, pH 7.4; 1 mM EDTA), sterilized by autoclaving.

2.4. Transfection

1. RPMI 1640 medium containing 25 mM HEPES and 4 mM glutamine (Lonza Ltd, Basel, Switzerland).
2. Pluronic® F-68 (Applichem) dissolved in water at a concentration of 2%. The solution is filter sterilized and stored at room temperature.
3. VPA (2-propyl-pentanoic acid, sodium salt) (Sigma-Aldrich) is prepared as a 500-mM solution in water, filter sterilized, aliquoted into sterile 50-mL tubes, and stored at −20°C. The working solution is maintained at 4°C. *VPA is a known teratogen.*

4. Linear 25 kDa polyethyleneimine (PEI) (Polysciences, Eppenheim, Germany) is dissolved in water at 1 mg/mL at a pH of 7.0. When dissolving, lower the pH with 1 N HCl. When the PEI is in solution, the pH is increased to 7.0 with 1 N NaOH. The solution is filter sterilized, aliquoted into sterile 50-mL tubes, and stored at −20°C. It can be stored frozen for years as long as repeated freeze–thaw cycles are avoided.
5. Ex-cell® 293 medium without L-glutamine and phenol red (see Subheading 2.1).
6. 50× L-glutamine–phenol red solution (see Subheading 2.1).

2.5. IgG Purification

1. Streamline-rProtein A beads (GE Healthcare Europe GmbH, Glattbrugg, Switzerland).
2. Econo-Pac chromatography column (Bio-Rad Laboratories AG, Reinach, Switzerland).
3. Equilibration and washing buffer: 20 mM sodium phosphate (pH 7.5).
4. Elution and regeneration buffer: 100 mM sodium citrate (pH 3).
5. Neutralizing buffer: 1 M Tris–HCl (pH 8.0).
6. Storage buffer: 20% ethanol in 20 mM sodium phosphate (pH 7.5).
7. Centricon Plus-70 (Millipore AG, Zug, Switzerland).
8. Phosphate buffered saline (PBS) at pH 7.1.

2.6. ELISA

1. 96-well ELISA microtiter plates with flat bottom (Becton-Dickinson AG, Basel, Switzerland).
2. Blocking buffer: 0.5% casein hydrolysate (Applichem) and 0.05% Tween 20 (Sigma-Aldrich) in PBS (pH 7.1).
3. Capture antibody: Goat anti-human *kappa* light chain (AbD Serotec, Dusseldorf, Germany).
4. Coating solution: For each 96-well plate, 11 μL of capture antibody is mixed with 11 mL of PBS (pH 8.0).
5. Washing buffer: PBS (pH 8.0) with 2% Tween 20.
6. Detection antibody: Alkaline phosphatase-conjugated goat anti-human *gamma* heavy chain (Biosource).
7. Standard: Human IgG, whole molecule (ChromPure, Jackson ImmunoResearch Europe Ltd., Suffolk, UK). The standard is diluted in blocking buffer to 40 ng/mL and then serially diluted 1:2 with blocking buffer.
8. Substrate buffer: Add 97 mL diethanolamine (Sigma-Aldrich) to 700 mL H_2O and adjust pH to 9.8 with 2 M HCl. Add 0.5 mL 1 M $MgCl_2$ and 2 g NaN_3. Adjust volume to 1 L.
9. Substrate: 4-nitrophenyl phosphate disodium salt (NPP) (Applichem) is dissolved in substrate buffer to 1.5 mg/mL.
10. Stop solution: 3 M NaOH.

3. Methods

3.1. Plasmid Purification

1. *E. coli* strain DH5α is separately transformed with each plasmid by the standard $CaCl_2$ method and spread onto LB agar plates with the appropriate antibiotic (100 μg/mL ampicillin for pXLG^{HEK}-RhLC and pXLG^{HEK}-RhHC; 50 μg/mL of kanamycin for pEGFP-N1) (14) (see Notes 1 and 2).
2. Incubate the plates overnight (16 h) at 37°C.
3. With a toothpick or pipette tip, transfer a single colony from each plate to a sterile round-bottom, polyproplene 14-mL culture tube, (BD Falcon, Cat. #2059) containing 3 mL LB broth with either 100 μg/mL ampicillin (for pXLG^{HEK}-RhLC and pXLG^{HEK}-RhHC) or 50 μg/mL kanamycin (for pEGFP-N1).
4. Incubate at 37°C for 4–6 h with agitation at 220 rpm.
5. Use the 3 mL culture to inoculate a 5-L Erlenmeyer flask containing 1 L of LB broth with 100 μg/mL ampicillin or 50 μg/mL kanamycin depending on the plasmid.
6. Incubate 12–16 h at 37°C with agitation at 220 rpm.
7. Transfer the culture to two 500-mL centrifuge bottles (Costar, Corning, New York).
8. Centrifuge at 5,000 × *g* for 20 min at 4°C and decant the medium into an Erlenmeyer flask. Retain the cell pellets and dispose of the medium after autoclaving or treatment with bleach.
9. Resuspend each cell pellet in 12 mL of buffer S1 from the NucleoBond AX 500 kit (Macherey-Nagel). Completely resuspend the cells by pipetting with a 10-mL pipette.
10. Transfer the resuspended cells into a 50-mL centrifuge tube.
11. Add 12 mL of buffer S2 to the suspension. Close the cap and mix gently by inverting the tube 6–8 times.
12. Incubate the mixture at room temperature (20–25°C) for 2–3 min. Do not vortex, as this will release chromosomal DNA from the cellular debris.
13. Add 12 mL of prechilled (4°C) buffer S3 to the suspension. Close the cap and mix gently by inverting the tube 6–8 times until a homogeneous suspension containing an off-white flocculate is formed. Let the tube stand in ice for 5 min.
14. Centrifuge the suspension at 5,000 × *g* for 30 min at 4°C (Varifuge 3.0R, Heraeus AG, Zürich, Switzerland). Repeat this step if the supernatant contains residual particles after the first centrifugation.
15. Attach the NucleoBond AX 500 column to a support stand and equilibrate the column with 6 mL of buffer N2. Allow the column to empty by gravity flow and discard the flow-through.

16. Load the cleared lysate supernatant from step 14 with a pipette onto the NucleoBond column. Allow the column to empty by gravity flow.
17. Wash the column twice with 32 mL of buffer N3. Collect the flow-through in a beaker and then discard.
18. Elute the plasmid DNA with 15 mL of buffer N5. Collect the eluate in a clean, sterile 50-mL centrifuge tube (TPP, Trasadingen, Switzerland). Any standard 50-mL centrifuge tube can be used if clean and sterile.
19. Add 11 mL of isopropanol at room temperature to precipitate the plasmid DNA. Mix well and centrifuge at 4,000 × *g* for 1 h at 4°C.
20. In a laminar flow hood, carefully pour the supernatant into a waste container.
21. To the pellet, add 15 mL of 70% ethanol. Vortex briefly and centrifuge at 8,000 rpm for 20 min at room temperature.
22. In a laminar flow hood, carefully decant the 70% ethanol. Allow the pellet to air dry in the hood at room temperature.
23. To the pellet add 0.7 mL of sterile TE and incubate at 37°C for 2–3 h on an orbital shaker.
24. Determine the plasmid yield by UV spectrometry. Dilute the DNA sample by adding 5 μL DNA to 495 μL water. Determine the absorbance at 260 and at 280 nm in a spectrophotometer (Biophotometer, Vaudaux-Eppendorf AG, Basel, Switzerland). The concentration of DNA is determined using the conversion factor (1 $A_{260} = 50$ mg/mL DNA). Calculate the A_{260}/A_{280} ratio. Only DNA preparations with ratios ≥1.8 are used for transfection.
25. Analyze the DNA by loading an aliquot of 0.5 μg on a 1% agarose gel in 1× TAE to assess the percentage of supercoiled plasmid under UV light.

3.2. Routine Cell Cultivation

1. HEK-293 cells are subcultivated every 3–4 days (see Note 3) by inoculation in 100 mL Ex-cell® 293 medium (when used for cell culture, the medium contains L-glutamine and phenol red as indicated in Subheading 2.1) (see Note 4) in a 250-mL square-shaped glass bottle at an initial cell density of 0.3×10^6 cells/mL.
2. Determine the cell density and viability by trypan blue staining using a Neubauer hemocytometer chamber and an inverted phase contrast microscope (100× magnification, Telaval 31, Carl Zeiss AG, Feldbach, Switzerland).
3. After cell counting, transfer 3–4 × 10^7 cells into a 50-mL centrifuge tube and centrifuge at 500 × *g* for 5 min in a standard tabletop centrifuge (Labofuge 200, Heraeus AG).

4. The medium is removed by aspiration or decanting. The cell pellet is resuspended in 10 mL of Ex-cell® 293 medium and transferred to a 250-mL square-shaped bottle containing 90 mL of prewarmed Ex-cell® 293 medium.
5. Attach the bottle to a platform mounted on an orbital shaker (model ISF-4-W with a rotational diameter of 5 cm; Kühner AG, Birsfelden, Switzerland) using double-sided adhesive transfer tape (3M Corp., Minneapolis, MN, USA) and agitate at 110 rpm at 37°C in a 5% CO_2 atmosphere without humidity. Keep the cap of the bottle opened about one quarter of a turn.

3.3. Cell Expansion for Transfection

1. One day before transfection, the cells are counted as described in Subheading 3.2 (see Note 5).
2. Transfer 6×10^8 cells into a 250-mL centrifuge bottle (Corning). For this step, it is necessary to have two 100-mL cultures.
3. Centrifuge the cells for 10 min at 1,000 rpm at room temperature (Cryofuge 6000i, Heraeus AG).
4. Remove the medium by aspiration or decanting and gently resuspend the cell pellet in 50 mL of prewarmed Ex-cell® 293 medium.
5. Transfer 25 mL into each of two 1-L square-shaped glass bottles with 275 mL of Ex-cell® 293 medium. The starting cell density of each culture is about 1×10^6 cells/mL.
6. Place the bottles on an orbital shaker as described in Subheading 3.2 (step 6) and incubate at 37°C overnight with agitation at 110 rpm. Keep the bottle caps open one quarter of a turn.

3.4. Transfection

1. The cells in the two 1-L bottles are counted as described in Subheading 3.2.
2. Transfer a total of 1×10^9 cells from the two overnight cultures into two 500-mL centrifuge bottles (Corning) and centrifuge at 1,000 rpm for 10 min at room temperature.
3. Remove the medium by aspiration or decanting and resuspend the cells from the two centrifuge bottles in a total volume of 50 mL by addition of prewarmed RPMI 1640 medium containing 0.1% Pluronic® F-68. The cell density after resuspension is 20×10^6 cells/mL.
4. Transfer the cells to a 250-mL cylindrical glass bottle.
5. Add 1.25 mg of plasmid DNA to the culture and mix gently by swirling the bottle. For the example described here, 612 μg of pXLG^{HEK}-RhLC, 612 μg pXLG^{HEK}-RhHC, and 25 μg pEGFP-N1 (49:49:2 w/w/w ratio) are added to the culture (see Note 6).
6. Add 3.75 mL linear 25-kDa PEI solution (1 mg/mL) to the culture and gently mix by swirling the bottle (see Note 7).

7. The bottle is agitated in an incubator shaker at 110 rpm at 37°C in a 5% CO_2 atmosphere without humidity with the bottle cap slightly opened as described in Subheading 3.2.
8. After 3 h of incubation, transfer the cells to a 5-L cylindrical glass bottle containing 950 mL of prewarmed Ex-cell® 293 medium (see Notes 8 and 9).
9. Add 7 mL of 0.5 M VPA to achieve a final concentration of 3.75 mM.
10. Incubate the culture as in step 7.
11. After 1 day of incubation, the analysis of transfection efficiency is performed on a small aliquot (20 μL) by flow cytometry on a Guava EasyCyte (Millipore). The sample is diluted 1:10 in PBS and the percentage of GFP-expressing cells is determined by counting at least 10,000 cells (see Note 10).

3.5. Analysis of Antibody Production

1. To measure recombinant antibody accumulation over time, 100 μL aliquots are taken daily during the production phase. After centrifugation to remove cells, the antibody concentration in each sample is measured by sandwich ELISA.
2. Coat a 96-well ELISA plate overnight at 4°C with 100 μL of goat anti-human kappa light chain IgG diluted in PBS (coating solution) as described in Subheading 2.6 (item 4).
3. Remove the coating solution with a multichannel pipettor and wash each well three times with 200 μL of washing buffer (see Subheading 2.6, step 5). The final wash is performed just before the samples are loaded into the wells. After the final wash, tap the plate on a paper towel to remove any remaining wash solution.
4. Samples from the culture are diluted 1:10 in blocking buffer, and then 200 μL of each is loaded in triplicate onto the plate. Two serial two-fold dilutions in blocking buffer (100 μL sample + 100 μL blocking buffer) are done directly on the plate (see Note 11).
5. Load the antibody standard as serial 1:2 dilutions (see Subheading 2.6, step 7) in triplicate on the plate.
6. Incubate the plate for 1 h at 37°C and then remove the samples using a multichannel pipetter.
7. Wash each well three times with 200-μL washing buffer as in step 3.
8. Add AP-conjugated goat anti-human gamma chain IgG diluted 1,000-fold in PBS to each well. For each plate, dilute 11 μL of antibody in 11 mL of PBS and add 100 μL to each well.
9. Incubate the plate for 1 h at 37°C.
10. Remove solutions and wash each well three times in 200-μL washing buffer as in step 3.

11. Add 100 μL substrate solution to each well and cover the plate with aluminum foil.
12. Incubate the plate for 15 min at room temperature with gentle agitation.
13. Stop the reaction by addition of 100 μL of 3 M NaOH.
14. Measure the absorbance at 405 nm using a microplate reader (SPECTRAmax™340; Molecular Devices, Palo Alto, CA, USA).
15. Determine the antibody concentration in each sample after generation of the standard curve from the absorbance of the standard antibody samples.

3.6. Harvest and Purification of the Recombinant Antibody

1. At the end of the production phase of 7–12 days (see Note 12), cells are harvested by transferring the culture into two 500-mL centrifuge bottles (Corning).
2. Centrifuge at 2,000 × *g* for 15 min at 4°C.
3. Recover the supernatant by decanting into a 1-L bottle.
4. Remove any additional cell debris by filtration with a 500-mL filter unit with a 0.45-μm membrane (Corning) and transfer the filtrate into two 500-mL centrifuge bottles (Corning).
5. Equilibrate the Streamline-rProtein A beads. For 1 L of supernatant containing about 1 g of antibody, pipette 20 mL beads into a 250-mL centrifuge bottle. Wash the beads three times by addition of 200 mL of 20 mM sodium phosphate (pH 7.5) to the bottle. Mix gently by inverting several times. Let the beads settle to the bottom of the bottle and carefully remove most of the buffer by decanting, paying attention not to lose the beads.
6. With a 10-mL pipette, transfer the washed beads along with the remaining buffer to the two 500-mL centrifuge bottles with the cell culture medium and incubate overnight with agitation on an STR4 rotator (Stuart Scientific, UK) at 4°C.
7. Let the beads settle by gravity to the bottom of the conical centrifuge bottles and then transfer them to a chromatography column (Econo-Pac, Bio-Rad Laboratories, Reinach, Switzerland) with a 10-mL pipette. Let the beads deposit at the bottom of the column. Remove the tip of the column to allow liquid flow.
8. After the medium has drained from the column, wash the beads with 10 column volumes (CVs) of 20 mM sodium phosphate (pH 7.5) and collect the flow-through in a clean beaker.
9. Elute the antibody from the beads with 5 CVs of 100 mM sodium citrate (pH 3.0) and collect in a 250-mL bottle.
10. Add one volume of 1 M Tris–HCl (pH 8.0) to the eluate. Store at 4°C until the solution is ready to be concentrated.

11. Regenerate the column with 5–10 CVs of 100 mM sodium citrate buffer (pH 3.0).
12. Reequilibrate the column with 5 CVs of 20 mM sodium phosphate (pH 7.5).
13. Remove the beads from the column by adding 20 mM sodium phosphate buffer (pH 7.5) with 20% ethanol and pouring them in a 50-mL centrifuge tube.
14. Store at 4°C (see Note 13).
15. Concentrate the partially purified antibody and change the buffer to PBS using a Centricon Plus-70 (Millipore) according to the manufacturer's protocol.
16. Determine the antibody concentration by measuring absorbance at 280 nm. Convert absorbance units to concentration with the conversion factor (1 U A_{280} = 1 mg/mL antibody).

4. Notes

1. Since a significant amount of plasmid DNA is necessary for TGE at large scale, it is important to maximize plasmid yields by choosing a vector with a high copy number origin of replication. This will also make the recovery of plasmid DNA easier since the ratio of plasmid DNA to contaminants, such as genomic DNA, RNA, and protein, will be greater (15).
2. For TGE in HEK-293 cells, the gene of interest is usually cloned under the control of a constitutive viral or cellular promoter. Our highest expression levels have been achieved with the human cytomegalovirus (HCMV) major immediate early promoter/enhancer.
3. To assure reproducibility, we do not recommend keeping cells in culture longer than 3 months (20–25 passages).
4. Ex-cell® 293 medium contains plant-derived peptone hydrolysates. Although valuable as a medium additive, peptones may be a source of lot-to-lot variation of the medium. Peptone hydrolysates may also be the source of unknown factors that have a negative impact on transfection (16). Because of the possible variation in medium lots, HEK-293 cells should be adapted to a new lot for at least two passages prior to transfection.
5. The physiological status of the cells at the time of transfection influences the transfectability of HEK-293 cells (8). To demonstrate this effect, cells cultivated for a number of hours after passage were transfected in a final volume of 10 mL in Culti Flask® 50 tubes (Sartorius Stedim AG, Göttingen, Germany), and the transfection efficiency was evaluated 24 h posttransfection with a

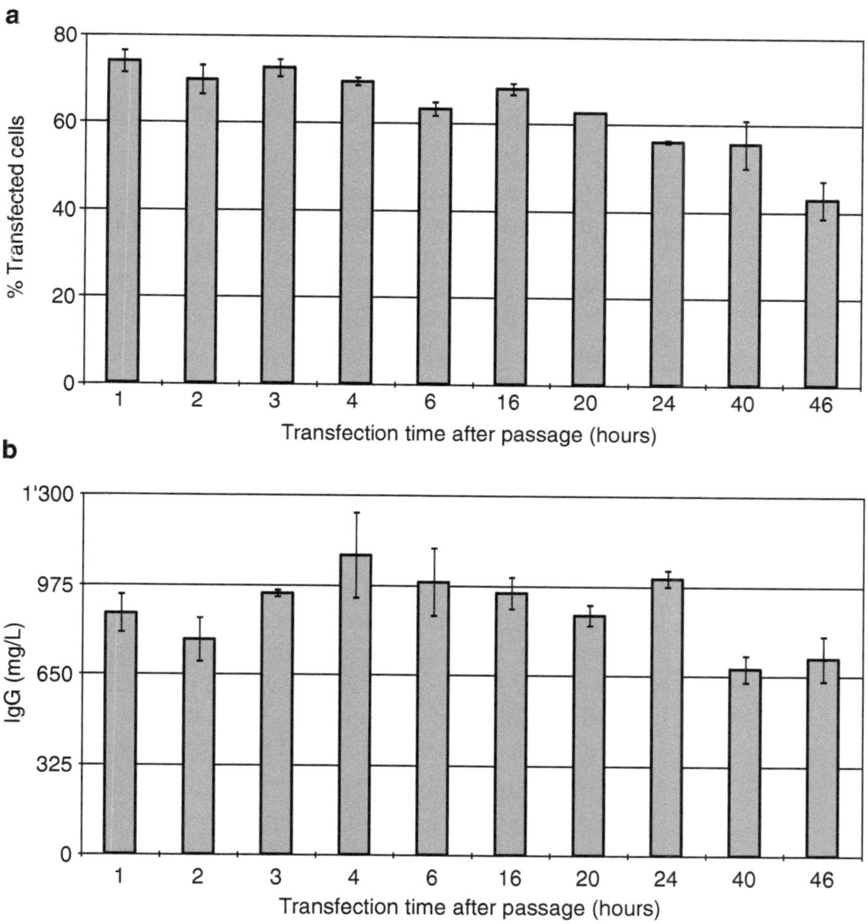

Fig. 1. Effect of time after passage on TGE. Cells were inoculated in Ex-cell® 293 medium and incubated at 37°C for the times indicated. The cells were then transfected in duplicate at a final volume of 10 mL in CultiFlask® 50 tubes. (**a**) The transfection efficiency (% of GFP-positive cells) was measured at day 1 posttransfection. (**b**) The recombinant antibody concentration was measured by ELISA at 10 days posttransfection.

benchtop flow cytometer. The transfection efficiency (Fig. 1a) and the recombinant protein yield (Fig. 1b) decreased with the age of the culture at the time of transfection. Growing cells may shed metabolites and/or cellular macromolecules that affect DNA uptake (17, 18).

6. Here, two plasmids coding for the heavy and light IgG chains were co-transfected. For such cases, the plasmid ratio needs to be empirically determined (19). We have tested various heavy/light vector combinations and found that a 1:1 ratio generally gives the best yields but this must be tested for each new combination of light and heavy chain vectors (4, 20). In addition,

expression of the IgG heavy and light chain cDNAs from a single plasmid has been shown to yield the same amount of recombinant antibody as with co-transfection with separate plasmids (Kiseljak and Wurm, unpublished data).

7. The volumes of transfection and the amounts of plasmid DNA and PEI can be easily determined for the scale-up and scale-down of the transfection method described here. In each case, the cell density at the time of transfection is 20×10^6 cells/mL. The amounts of DNA and PEI are 1.25 and 3.75 μg, respectively, for each mL of transfection volume after dilution. For a 10 mL transfection, for example, resuspend 10×10^6 cells in 0.5 mL of RPMI 1640 and add 12.5 μg plasmid DNA and 37.5 μL PEI (1 mg/mL stock solution). The transfection is then diluted by addition of 9.5-mL Ex-cell® 293 medium.
8. The gene delivery phase of transfection is a time-dependent process, but the time needed to achieve the highest DNA uptake is quite short (<2 h) (21). We tested the influence of the culture dilution by transfecting parallel cultures of HEK-293 cells at the same time and then diluting them at different times after DNA addition. The maximal transfection efficiency was reached after 1 h of transfection (Fig.2a), and recombinant protein yield did not change substantially by extending the time between transfection and dilution beyond 1 h (Fig.2b).
9. Affinity purification of a recombinant protein from medium containing large amount of peptones may be difficult, particularly for his-tagged proteins (unpublished observations). In this situation, it is recommended to dilute the transfected cells in either Freestyle® (Invitrogen) or Pro293s (Lonza, Verviers, Belgium) media. It is also possible to obtain batches of Ex-cell® 293 medium without peptones through direct contact with the manufacturer.
10. The efficiency of transfection is typically 60–70% for the method described here.
11. The dilution must be determined empirically for each recombinant protein.
12. The time of harvest needs to be determined empirically for each recombinant protein. For intracellular and membrane proteins, the maximum level of protein is usually reached by day 3 posttransfection.
13. The Streamline-rProtein A beads can be used multiple times for purification of the same antibody.

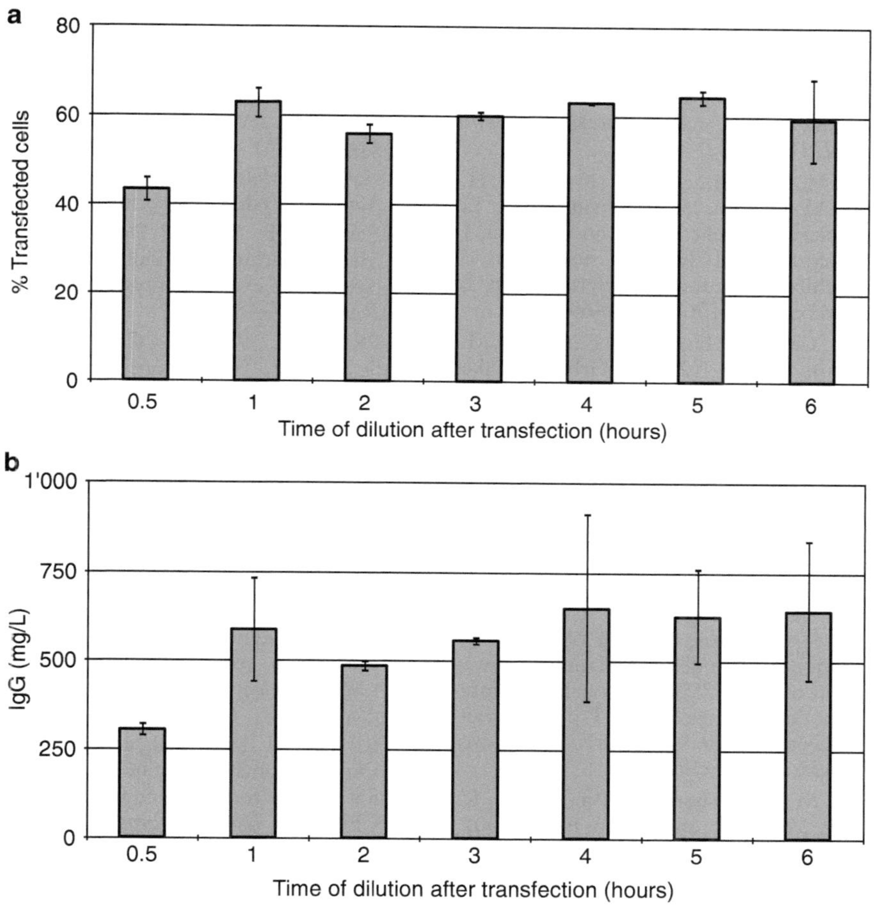

Fig 2. Effect of dilution time on transient transfection. Cells were prepared and transfected according to the protocol described here. For each transfection, 10^7 cells were resuspended in 0.5 mL RPMI 1640 in CultiFlask® 50 tubes. The cells were then diluted with 9.5 mL of Ex-cell® 293 medium at various times after transfection. (**a**) The transfection efficiency (% of GFP-positive cells) was measured at day 1 posttransfection. (**b**) The recombinant antibody concentration was measured by ELISA at 10 days posttransfection.

References

1. Wurm, F. M. (2004) Production of recombinant protein therapeutics in cultivated mammalian cells, *Nat Biotechnol 22*, 1393–1398.
2. Pham, P. L., Kamen, A., and Durocher, Y. (2006) Large-scale transfection of mammalian cells for the fast production of recombinant protein, *Mol Biotechnol 34*, 225–237.
3. Baldi, L., Hacker, D. L., Adam, M., and Wurm, F. M. (2007) Recombinant protein production by large-scale transient gene expression in mammalian cells: state of the art and future perspectives, *Biotechnol Lett 29*, 677–684.
4. Baldi, L., Muller, N., Picasso, S., Jacquet, R., Girard, P., Thanh, H. P., Derow, E., and Wurm, F. M. (2005) Transient gene expression in suspension HEK-293 cells: application to large-scale protein production, *Biotechnol Prog 21*, 148–153.
5. Backliwal, G., Hildinger, M., Hasija, V., and Wurm, F. M. (2008) High-density transfection with HEK-293 cells allows doubling of transient titers and removes need for a priori DNA complex formation with PEI, *Biotechnol Bioeng 99*, 721–727.

6. Backliwal, G., Hildinger, M., Kuettel, I., Delegrange, F., Hacker, D. L., and Wurm, F. M. (2008) Valproic acid: a viable alternative to sodium butyrate for enhancing protein expression in mammalian cell cultures, *Biotechnol Bioeng 101*, 182–189.
7. Gottlicher, M., Minucci, S., Zhu, P., Kramer, O. H., Schimpf, A., Giavara, S., Sleeman, J. P., Lo Coco, F., Nervi, C., Pelicci, P. G., and Heinzel, T. (2001) Valproic acid defines a novel class of HDAC inhibitors inducing differentiation of transformed cells, *EMBO J 20*, 6969–6978.
8. Muller, N., Girard, P., Hacker, D. L., Jordan, M., and Wurm, F. M. (2005) Orbital shaker technology for the cultivation of mammalian cells in suspension, *Biotechnol Bioeng 89*, 400–406.
9. Stettler, M. (2007) Bioreactor processes based on disposable materials for the production of recombinant proteins from mammalian cells, EPFL Thesis no. 3947, Lausanne, Switzerland, http://library.epfl.ch/theses/?nr=3947.
10. Backliwal, G., Hildinger, M., Chenuet, S., Wulhfard, S., De Jesus, M., and Wurm, F. M. (2008) Rational vector design and multi-pathway modulation of HEK 293E cells yield recombinant antibody titers exceeding 1 g/l by transient transfection under serum-free conditions, *Nucleic Acids Res 36*, e96.
11. Pick, H. M., Meissner, P., Preuss, A. K., Tromba, P., Vogel, H., and Wurm, F. M. (2002) Balancing GFP reporter plasmid quantity in large-scale transient transfections for recombinant anti-human Rhesus-D IgG1 synthesis, *Biotechnol Bioeng 79*, 595–601.
12. Meissner, P., Pick, H., Kulangara, A., Chatellard, P., Friedrich, K., and Wurm, F. M. (2001) Transient gene expression: recombinant protein production with suspension-adapted HEK293-EBNA cells, *Biotechnol Bioeng 75*, 197–203.
13. Zhang, G., Gurtu, V., and Kain, S. R. (1996) An enhanced green fluorescent protein allows sensitive detection of gene transfer in mammalian cells, *Biochem Biophys Res Commun 227*, 707–711.
14. Sambrook, J., Fritsch, E. F., and Maniatis, T. (1989) *Molecular Cloning: A Laboratory Manual*, 2 ed., Cold Spring Harbor Laboratory.
15. Eastman, E. M., and Durland, R. H. (1998) Manufacturing and quality control of plasmid-based gene expression systems, *Adv Drug Deliv Rev 30*, 33–48.
16. Pham, P. L., Perret, S., Cass, B., Carpentier, E., St-Laurent, G., Bisson, L., Kamen, A., and Durocher, Y. (2005) Transient gene expression in HEK293 cells: peptone addition posttransfection improves recombinant protein synthesis, *Biotechnol Bioeng 90*, 332–344.
17. Tuvesson, O., Uhe, C., Rozkov, A., and Lullau, E. (2008) Development of a generic transient transfection process at 100 L scale, *Cytotechnology 56*, 123–136.
18. Schlaeger, E. J., and Christensen, K. (1999) Transient gene expression in mammalian cells grown in serum-free suspension culture, *Cytotechnology 30*, 71–83.
19. Schlatter, S., Stansfield, S. H., Dinnis, D. M., Racher, A. J., Birch, J. R., and James, D. C. (2005) On the optimal ratio of heavy to light chain genes for efficient recombinant antibody production by CHO cells, *Biotechnol Prog 21*, 122–133.
20. Bentley, K. J., Gewert, R., and Harris, W. J. (1998) Differential efficiency of expression of humanized antibodies in transient transfected mammalian cells, *Hybridoma 17*, 559–567.
21. Bertschinger, M., Schertenleib, A., Cevey, J., Hacker, D. L., and Wurm, F. M. (2008) The kinetics of polyethyleneimine-mediated transfection in suspension cultures of Chinese hamster ovary cells, *Mol Biotechnol 40*, 136–143.

Chapter 3

Selection of High Expressing Mammalian Cells by Surface Display of Reporters

Christine T. DeMaria

Abstract

A flow cytometry method using a nonfluorescent reporter protein was developed for rapid, early-stage identification of cells producing high levels of a recombinant protein of interest. A cell surface reporter protein is coexpressed with the protein of interest, and the reporter protein is detected using a fluorescently labeled antibody. The genes encoding the reporter protein and the protein of interest are linked by an IRES so that they are transcribed in the same mRNA but are translated independently. Since they each arise from a common mRNA, the reporter protein's expression level accurately predicts, on a per cell basis, the relative expression level of the protein of interest. This method provides an effective process for selecting cells that express high levels of recombinant proteins, with the benefits of rapid and accurate 96-well plate clone screening (that is both quantitative and qualitative) and elimination of unstable clones during subsequent scale up and culture. Furthermore, because this method does not rely on the availability of a detection reagent specific for the protein of interest that is expressed, it can be easily implemented into any cell line development process.

Key words: Flow cytometry, Clone screen, Fluorescent reporter, Internal ribosome entry site

1. Introduction

Mammalian cell lines expressing high levels of recombinant proteins are needed for a variety of research and biotechnology applications. The process of choosing a cell population expressing high levels of a protein of interest typically involves the scale up of clones in 96-well plates to larger cell culture vessels from which assays for protein expression can be reliably performed. Often, the high expressing cells are a small percentage of the total cell population to be seeded into 96-well plates, making it necessary to expand hundreds of wells for evaluation at larger scale. If the protein of

James L. Hartley (ed.), *Protein Expression in Mammalian Cells: Methods and Protocols*, Methods in Molecular Biology, vol. 801, DOI 10.1007/978-1-61779-352-3_3, © Springer Science+Business Media, LLC 2012

interest is difficult to express in the host cell type, even more wells may need to be expanded and evaluated to identify those with adequate expression levels for downstream needs. Cell line scale up and maintenance, as well as the subsequent analysis of protein expression, becomes very time-consuming and labor intensive as the number of clones to be assessed increases into hundreds or more. In addition, a significant amount of time and space is often spent expanding and assessing clones that, in the end, do not have the desired protein expression levels. For these reasons, the availability of a highly efficient and accurate screening method early in the process would substantially reduce the effort needed to obtain high expressing cells.

Implementing an effective screen at the 96-well plate stage of development is very desirable as it focuses all subsequent work on the clones with high expression levels. Analysis of the cell culture medium from each well can be used at this stage to identify clones secreting high levels of the recombinant protein. However, use of this analysis requires normalization to cell number and medium volume, both of which vary across wells in a 96-well plate. If normalization is not included, the assay may not accurately identify the high expressing clones. Also, additional effort is often required to develop and/or optimize a new high-throughput assay for each protein of interest.

Flow cytometry, on the other hand, has the advantage of assessing per cell expression of a protein, removing the need for normalization to cell number and media volume. Therefore, we utilized flow cytometry in conjunction with a nonfluorescent reporter protein as a method to rapidly screen clones producing high levels of a protein of interest. The cell surface protein CD20, not normally expressed on our host cells, is coexpressed with the protein of interest and detected using a fluorescent anti-CD20 antibody. The genes encoding CD20 and the protein of interest are linked by an internal ribosome entry site (IRES) so that they are transcribed in the same mRNA but are translated independently (1). The lower efficiency of IRES-mediated translation relative to 5′cap-mediated translation ensures that cellular resources are utilized mainly for synthesis of the protein of interest rather than the reporter protein (2). However, since they arise from the same mRNA, the CD20 reporter expression level accurately predicts, for each clone, the relative expression level of the protein of interest (3). In the screening method, cells coexpressing the reporter protein with the protein of interest are seeded in 96-well plates. As wells reach the optimal cell confluence, cells are removed from each well and are stained with a fluorescent antibody specific for the reporter protein. Flow cytometry is used to analyze each sample, and the cells with highest fluorescence are selected as the high expressing clones.

An important consideration in selecting high expressing clones for long-term use is that the recombinant protein expression is sustained at the desired level throughout clone scale up and continued passage. For example, upon removing selection pressure (such as a drug in the medium), recombinant protein expression may decline with increased passage number (4); this is often referred to as unstable expression. Having a rapid screening tool to assess protein expression in real time is therefore useful for making final decisions on the optimal clones. We have shown that the reporter protein-based flow cytometry screen effectively identifies clones that have unstable protein expression. Another advantage of this screening method is the ability to visually inspect the qualitative characteristics of each clone via the flow cytometry profiles. In doing this, cell populations that show a large degree of heterogeneity can be avoided. Overall, this screening method provides an effective process for selecting clones expressing high levels of recombinant proteins, with the benefits of rapid and accurate 96-well plate clone screening and elimination of unstable clones at an early stage in the development process.

2. Materials

2.1. Plasmid DNA Cloning

1. Molecular cloning reagents (including restriction enzymes, DNA ligase).

2.2. Plasmid DNA Preparation and Transfection

1. QIAGEN Plasmid Maxi kit or QIAfilter Plasmid Maxi kit (QIAGEN Inc, Valencia, CA).
2. GenePulser Xcell™ (BioRad Laboratories, Hercules, CA) for electroporation to transfect cells or the appropriate transfection reagent(s) for the cell type being used.
3. Growth medium for the cell type being used. For the DHFR-deficient, suspension CHO cells used in our lab, the growth medium is CD DG44 (Invitrogen, Carlsbad, CA) supplemented with 4 mM L-glutamine (Invitrogen) and 0.18% Pluronic F-68 (Invitrogen).
4. Selection medium for the cell type being used. For the suspension CHO cells used in our lab, the selection medium is CD CHO (Invitrogen) supplemented 4 mM L-glutamine (Invitrogen). A second selection step, if necessary, is performed using this same medium supplemented with 20 nM methotrexate (Calbiochem/EMD Biosciences, San Diego, CA).

2.3. Analyzing Reporter Expression in Transfected Pools

1. 5 ml polystyrene round-bottom tubes (BD-Falcon, Franklin Lakes, NJ) for harvest and preparation of cells.
2. Ice bucket and/or tube rack with ice bath.
3. Flow cytometry wash (and incubation) buffer: Phosphate-buffered saline (PBS) pH 7.2 (Invitrogen) with serum added

to 2% final concentration (to block nonspecific binding of the antibody), and then 0.22 μm filter sterilized. The final buffer preparation should be kept on ice or at 4°C.

4. Fluorescently labeled antibody that is specific for the reporter protein being used, stored per the manufacturer's instructions. When CD20 is used as the reporter protein, we use either FITC- or PE-conjugated mouse anti-human CD20 (BD-Pharmingen, Franklin Lakes, NJ). These antibodies are stored at 4°C in the dark.
5. Aluminum foil.

2.4. 96-Well Plate Cloning

1. Sterile 96-well flat-bottom plates (BD-Falcon).
2. Cloning medium utilized for the cell type being used.
3. Posttransfection (or, if applicable, postselection) growth medium utilized for the cell type being used.

2.5. Flow Cytometry Screening of 96-Well Plates

1. Nonsterile, polypropylene 96-well V-bottom plates (BD-Falcon) for harvest and preparation of cells for assay; 96-well plate adapters for centrifuge.
2. Multichannel pipettor.
3. Flow cytometry wash and incubation buffer: PBS pH 7.2 with serum added to 2% final concentration, and then 0.22 μm filter sterilized. The final buffer preparation should be kept on ice or at 4°C.
4. Fluorescently labeled antibody specific for the reporter protein expressed on the cells, stored per the manufacturer's instructions. When CD20 is used as the reporter protein, we use PE-conjugated mouse anti-human CD20 (BD-Pharmingen), stored at 4°C in the dark.
5. Ice bucket or refrigerator at 4°C.
6. Aluminum foil.

3. Methods

The reporter protein-based flow cytometry screen relies on the coexpression of a reporter protein with the protein of interest. This is enabled by transfecting the host cell line with a mammalian DNA expression vector that contains the following nucleotide sequence, from 5′ to 3′: open reading frame (ORF) encoding the protein of interest, weakened IRES, and ORF encoding a reporter protein (Fig. 1). The purpose of the IRES is to enable ribosome binding and protein translation of the reporter mRNA (1, 2). It is important

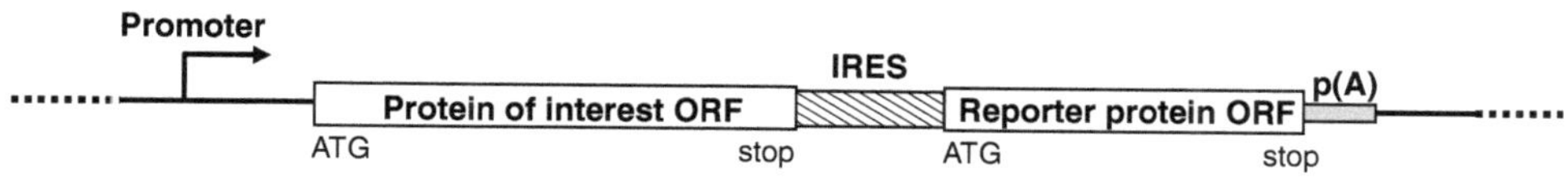

Fig. 1. Diagram of the gene expression cassette. In the mammalian expression vector used, the DNAs encoding the therapeutic protein and the reporter protein are linked by an internal ribosome entry site (IRES) so that they are transcribed in the same mRNA but are translated independently. *ORF* open reading frame, *p(A)* polyadenylation sequence, *ATG* initiation codon, *stop* stop codon.

that the IRES sequence used has substantially lower efficiency of protein translation than the cap-mediated translation occurring at the 5′ end of the mRNA, from which the protein of interest is translated. Therefore, the IRES sequence used should be one that has naturally low translation efficiency or one that has been mutated to have weakened translation efficiency. In our system, a weakened version of the EMCV IRES was used (5, 6).

Requirements for the reporter protein are as follows: there must be a fluorescently labeled detection reagent available (see Note 1), the reporter must be detectable on the surface of your host cell type by flow cytometry (see Note 2), and the reporter must be suitable for coexpression with your protein of interest (see Note 3). We have used a CD20 reporter sequence in this system. All of our work has been done with suspension CHO cells, but this method can be utilized with a cell line of your choice as long as it meets the above criteria.

3.1. Construction of DNA Expression Vectors

1. Construct a vector backbone that contains a multiple cloning site followed by an IRES-reporter sequence. Start with the vector backbone typically used to express proteins in your host cell type. The vector backbone should already include, as part of the gene expression cassette, a transcriptional promoter and a polyadenylation sequence. Immediately 3′ of the promoter sequence and 5′ of the polyadenylation sequence, there should also be a multiple cloning site sequence that contains several unique restriction enzyme sites. If such a multiple cloning site does not exist, it is recommended to create one and insert it into your vector backbone. At the 3′ end of the multiple cloning sites, insert the IRES-reporter sequence that you have chosen for use in the final expression vector. In this IRES-reporter DNA fragment, the ATG within the 3′ end of the IRES that can serve as the translation initiation site must be in frame with the ATG initiation codon of the reporter open reading frame. The IRES-reporter DNA can be synthesized as a single fragment, or the individual sequences can be synthesized and/or isolated from another plasmid DNA then separately ligated into the vector backbone, as long as the in frame requirement noted above is maintained.

2. Insert the ORF encoding your protein of interest into the IRES-reporter expression vector backbone created in step 1, utilizing the restriction enzyme sites in the multiple cloning sites. If your protein of interest is expressed as more than one polypeptide (such as two subunits), see Note 4. In the final DNA vector, the gene expression cassette should be set up as diagrammed in Fig. 1.

3.2. Preparation of Plasmid DNA and Generation of Cells Expressing the Protein of Interest

1. The final DNA expression vector created in step 2 in Subheading 3.1 should be propagated in a bacterial cell host of your choice using your preferred methods. From a bacterial clone expressing your plasmid, generate a sterile, large-scale plasmid DNA preparation for transfection using a Plasmid Maxi kit, with the final ethanol removal, drying of the DNA pellet, and DNA resuspension all done in a sterile hood. Resuspend the DNA in your preferred solvent (such as sterile H_2O or Tris buffer) at an appropriate concentration for your transfection method (typically 1–2 μg/μl).
2. Transfect the plasmid DNA prepared above into your host cells using the protocol that is optimal for your host cell line. Our work has used a suspension CHO DXB11 cell line that is transfected by electroporation. Allow cells to recover from transfection in the appropriate growth medium for 48–72 h.
3. If a selection step(s) to obtain pools of stably transfected cells is required, begin the selection process after the transfection recovery period has ended. For our suspension CHO cells, posttransfection selection is done using nucleotide-deficient CD CHO medium supplemented with 4 mM L-glutamine to generate stable pools. If necessary, this is followed by selection in the same medium supplemented with 20 nM methotrexate.
4. Assess the expression levels of your protein of interest in the stable pools to determine the best pool(s) to isolate clones from. Utilize the procedure available for assessing expression of your protein of interest (such as ELISA or Western blot). The pools must also be evaluated for expression of the reporter protein by flow cytometry (Subheading 3.3). If there is no assay available for your protein of interest at this stage, go directly to Subheading 3.3 for analysis of reporter protein expression levels.

3.3. Determining Reporter Protein Expression in Pools

Prior to implementing the clone screen, the pool(s) to be cloned from must be assessed for reporter protein expression by flow cytometry. This is important for the subsequent comparison of clone results to that of the original pool. It is also useful to know the average level of reporter expression in a pool relative to the average expression level of the protein of interest.

1. For cells growing in mid-log phase, harvest 1×10^6 viable cells by centrifugation in a 5 ml round-bottomed tube. Negative control cells of the same cell type (typically, the untransfected host cell line) should also be prepared and analyzed. From this point on, keep the tubes on ice or in a rack with an ice bath so that the samples are kept cold.
2. Wash each cell pellet with 2 ml cold wash buffer (PBS/2% serum), centrifuge at $140 \times g$, and pour off the supernatant. Repeat this wash step and after pouring off the supernatant, resuspend the cells in the residual wash buffer by gently flicking the bottom of the tube.
3. Add the appropriate volume (per the manufacturer's recommendations) of the fluorescently labeled antibody specific for your reporter protein. For PE- or FITC-conjugated anti-CD20 antibody (BD-Pharmingen catalog #s 555623 or 555622, respectively), the volume to use is 20 μl antibody per 1×10^6 cells. Incubate for 30 min on ice, covering the tubes with foil.
4. Perform two washes as described in step 2. Resuspend the cell pellet in 1 ml of cold wash buffer and acquire the samples on an analytical flow cytometer, such as a FACSCalibur (BD Biosciences, San Jose, CA). If desired, samples can be analyzed in a 96-well plate format by transferring the samples prepared in the tubes to a 96-well plate, such as that used in Subheading 3.5. The flow cytometer instrument settings to use will be dependent on your cell type and the fluorescent label being used. For all sample acquisitions, 10,000 live events should be collected. The average reporter protein expression of the pool is determined as the geometric mean fluorescence intensity of the total live population, measured in relative fluorescence units (RFU).

3.4. Cloning in 96-Well Plates

1. Using a pool(s) identified in Subheadings 3.2 and 3.3, seed cells into 96-well plates at a seeding density of 0.3–0.5 cell/well and with a volume of 100 μl/well. Use the appropriate single cell cloning medium for your cell type. Continue to culture the pool(s) that is used for cloning, as it will also be used as a control(s) in the clone screen (in Subheading 3.5).
2. Incubate plates in a 37°C, 5% CO_2 incubator for 7 days. At 7 days postplating, visually assess wells for cell growth (marking such wells, if desired), and then add 100 μl growth media to all wells.
3. Incubate plates in a 37°C, 5% CO_2 incubator for 3–4 more days. At this point, visually assess wells, mark all wells that are positive for cell growth, and feed such wells as follows. If cells are nonadherent, make sure that they are settled at the bottom of the well and remove approximately 150 μl of medium from

the well, leaving a residual volume in the well; then, add 100 μl fresh growth medium into the well. If cells are adherent, aspirate the medium from each well containing cells and add 100 μl fresh growth medium into the well.

4. Continue to visually inspect wells every 1–3 days, depending on how rapidly the cells are growing. It is very important to identify wells for screening before the cells become too confluent (see Note 5). While waiting for screening to begin, continue to feed the wells every 3–4 days as described above (add 100 μl one time, remove media and add 100 μl the next time, and so on).

3.5. Flow Cytometry Screening of Clones in 96-Well Plates

1. While cells are growing in the 96-well plates, frequently monitor the wells to assess degree of confluence. Upon reaching 60–70% confluence, cells should be screened. If the cells are allowed to reach a higher degree of confluence, the assay result could be affected (see Note 5).
2. For each well that is ready to be screened, half of the cells from the well are transferred to a single well of a 96-well polypropylene V-bottom plate as follows. If cells are nonadherent, pipet up and down in the well to evenly mix the cells and medium and transfer half of the total volume (cells + medium) to the V-bottom well; then, replace the volume removed from the original well by adding fresh growth medium to that well. If cells are adherent, perform a cell detachment procedure appropriate for the cell type, transferring half of the cell volume to the V-bottom plate; then, add a sufficient volume of growth medium to the remaining cells in the original well. Place the original 96-well plates back into the incubator.
3. Add control cells (that have been maintained in culture) to the V-bottom plate, with each control placed into duplicate wells at approximately 40,000 cells/well. The controls to use are the untransfected parental cell line (negative control), the original pool that was cloned from (positive control) and if available, a cell line or pool with high expression level of the reporter protein (positive control). From this point on, the samples in the V-bottom plate should be kept cold. If any of the following steps (4–10) are not performed right away, place the plate on ice or at 4°C until the procedure can be continued.
4. For any wells in the V-bottom plate with volumes < 100 μl, add cold wash buffer to bring the total well volume to ≥ 100 μl and ≤ 200 μl. Centrifuge the plate at 140 × *g* for 3 min to pellet the cells. Remove plate from the centrifuge and, in one quick motion, invert the plate to remove most of the supernatant from each well (see Note 6).
5. Using a multichannel pipet, add 100 μl of cold wash buffer per well and pipet up and down two to three times. Centrifuge the

plate and invert to remove the liquid as described in step 4. Tap plate on a paper towel with enough force to remove most of the liquid without dislodging cells (see Note 6).

6. Briefly centrifuge the plate (30 s) to concentrate wash buffer at the bottom of each well. After this centrifugation, the wells should have approximately 10 μl wash buffer remaining. If any wells have significantly larger volumes remaining, remove the extra volume by pipetting.
7. Based on the total number of wells to assay, prepare a working stock of fluorescently labeled antibody that will be used to bind the reporter protein on the cell surface. Per well to be assayed, prepare 100 μl total volume with the amount of antibody used according to either the manufacturer's recommendations and/or previous antibody titration experiments with your starting pool or other positive control cells (see Note 7). For the PE-conjugated anti-human CD20 antibody (BD catalog #555623), 10 μl of antibody plus 90 μl of wash buffer is used per well. Therefore, if there are 50 wells to assay, the working stock preparation would be for 55 wells (50×1.1), which equals 550 μl of antibody plus 4.95 ml cold wash buffer.
8. Add 100 μl of the antibody working stock per well of the V-bottom plate, and mix with the cells by pipetting up and down two to three times. Cover the plate with aluminum foil and incubate on ice or at 4°C for 30 min.
9. Add 100 μl cold wash buffer per well and pipet up and down two to three times. Centrifuge and remove supernatant as described above. Repeat this wash procedure one time, and after tapping the plate to remove the liquid, resuspend the cells with 150 μl cold wash buffer per well.
10. Run the plate on a flow cytometer with 96-well plate loading capability, such as a FACSArray™ Bioanalyzer (BD Biosciences), acquiring 5,000 events from each well. The instrument settings to use will be dependent on your cell type and the fluorescent label being used.

3.6. Data Analysis and Clone Scale Up

1. Analyze the results from the flow cytometry clone screen using flow cytometry analysis software, such as CellQuestPro (BD Biosciences) or FlowJo (Tree Star Inc., Ashland, OR). On a forward scatter (FSC) vs. side scatter (SSC) dot plot, draw a gate around the live cell population. For the total live population gated in the dot plot, generate a histogram profile of the relative fluorescence of the population.
2. From the fluorescence histogram of each clone, you can determine: (1) a quantitative value of the reporter protein expression, represented as the geometric mean fluorescence intensity of the entire live population and measured in RFU (see Fig. 2a); and (2) a qualitative assessment of the population, such as its degree of heterogeneity (see Fig. 2b).

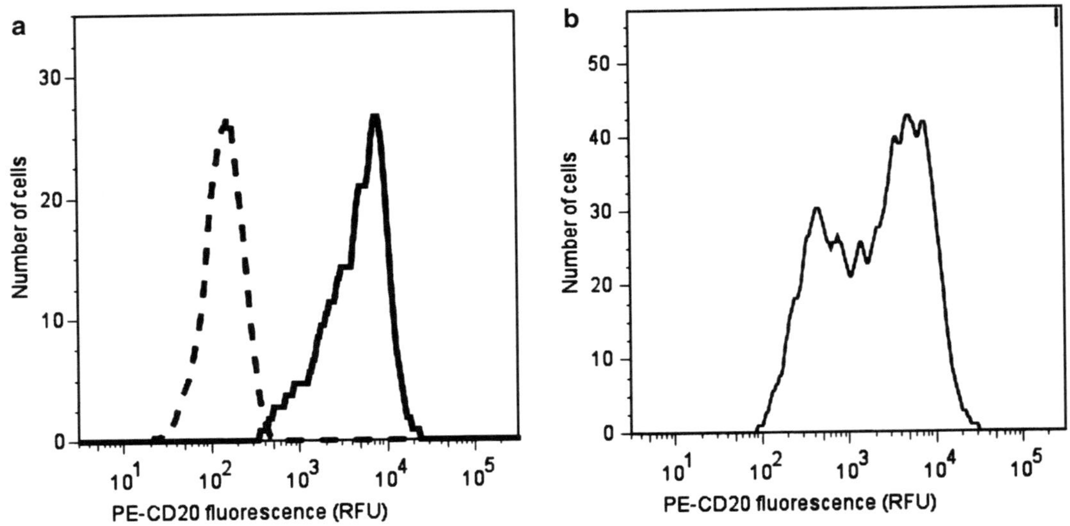

Fig. 2. Representative 96-well plate flow cytometry screen results. (**a**) The flow cytometry profile of the untransfected host cell line, which does not express the CD20 reporter protein, is shown by the *dashed line*. The flow cytometry profile of a clone expressing the CD20 reporter protein is shown by the *solid line*. The geometric mean fluorescence intensities of the two populations (read on the *x*-axis) are 130 and 3,990 relative fluorescence units (RFUs), respectively. (**b**) The flow cytometry profile for one well of a 96-well plate illustrates how qualitative evaluation can also be used for clone selection. The profile reveals a high degree of heterogeneity that could be the result of a nonclonal population or an unstable clone.

3. Based on the results of your screen(s), select clones that you would like to expand. This could be a defined number, such as the top ten clones from each screen, or it could be a percentage (such as top 10%) of the clones screened each day. A best practice is to expand cells 1–2 days after screening, i.e., before they become overgrown.
4. As per the procedure specific to your host cell line, expand the remaining cells in the well of the original 96-well plate to a larger vessel (such as a six-well plate). Continue with subsequent expansion of the desired clones.
5. Either during the expansion process or after scale up is complete and clones are maintained in culture, cells may be analyzed by flow cytometry to confirm stability of protein expression. Antibody staining and analysis may be done in tubes (per Subheading 3.3) or in 96-well plates (per instructions for control cells in Subheading 3.5). If reporter protein expression levels decrease significantly after continued passage, observed as a decrease in the geometric mean fluorescence of the live cell population, the protein expression is likely unstable and the clone may not be suitable for long-term use to express the protein of interest (see ref. 6 for examples).

4. Notes

1. This screening method uses flow cytometry to assess cells based on the degree of fluorescence on the cell surface. The antibody (detection reagent) that is utilized for binding to the cell surface reporter protein must, therefore, be fluorescently labeled. (Also, some antibody suppliers will note, for each fluorescently labeled antibody, whether or not it is recommended for use in flow cytometry.) Different fluorescent conjugates will have different fluorescent intensities, and one may be more preferable than another for your screening purposes. For example, anti-CD20 antibodies conjugated with FITC, PE, or APC each resulted in a different geometric mean fluorescence intensity for the same cell population expressing CD20. For this reason, the same fluorescent antibody should be used for an entire experiment, from the pool stage through the clone stage, to ensure accurate comparisons. Also, prior to the experiment, confirm that the fluorescent conjugate you choose will be detectable on the flow cytometer you will be using.
2. Before utilizing this method for clone screening, it is important to make sure that the reporter protein can be expressed and detected on the cell surface of your cell type. Transient transfection of a DNA vector expressing only the reporter protein (i.e., without an IRES or protein of interest) is generally suitable for this. Following the transfection, the reporter protein expression can be assessed by flow cytometry using the fluorescent antibody described in Note 1.
3. In some instances, stable pools utilizing the IRES-CD20 reporter system have had lower expression levels of the protein of interest than stable pools utilizing the respective nonreporter expression system (i.e., same vector backbone without the IRES-CD20). This suggests that for some proteins, the reporter system may specifically affect expression, an effect that could be related to the IRES element and/or a sequence-dependent interaction for a specific combination of open reading frames. As an initial test of the suitability of a reporter for expression with a protein(s) of interest, transient transfections comparing expression from the reporter-containing vector and the reporterless vector can be performed.
4. This method has been used to screen clones expressing a protein of interest as two polypeptides (6), and in the development and optimization of the method the reporter was coexpressed with one of the polypeptides. It would be possible to link a separate reporter to each polypeptide to be expressed and use two different antibodies for the flow cytometry screen; however, this was not developed as part of the method and

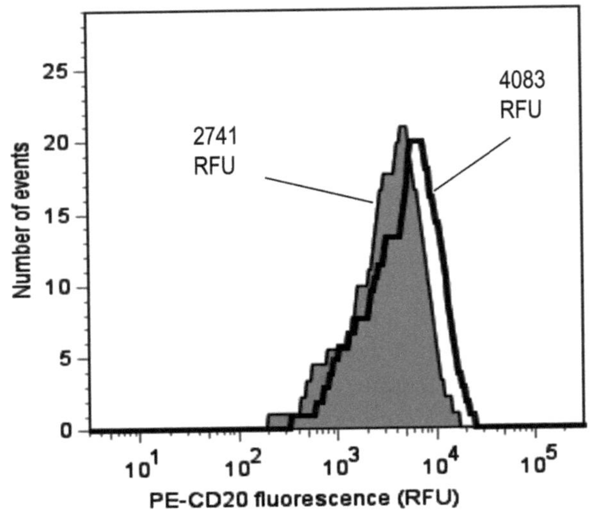

Fig. 3. Choosing an optimal cell density for the 96-well plate screen. For a previously established clone, cells were plated in a 96-well plate at increasing seeding densities per well. Four days later, the wells were assessed and the cells were assayed by flow cytometry. The profiles for two wells that had not yet reached 100% confluence are shown. The profile shaded in *gray* represents a well containing ~65,000 cells (~80–90% confluent) and the *solid black line* represents a well containing ~40,000 cells (~60–70% confluent). The well with the lower cell density had a higher geometric mean fluorescence intensity than the well with the higher cell density. The geometric mean fluorescence intensity values, shown on the figure, are measured in relative fluorescent units (RFU).

may not be preferred due to the additional utilization of cell resources to make another reporter protein.

5. We have sometimes observed that, as cell density increases over time in culture, the live cell population gradually shifts to lower fluorescence intensity. (This is accompanied by a gradual shift of the live population on the FSC vs. SSC plot to slightly decreased FSC and slightly increased SSC, despite the fact that a viable cell count by trypan blue exclusion remained constant at ≥95% viable.) While the difference in fluorescence intensity may seem slight upon visual inspection of the histogram profiles, the geometric mean values at the two time points can be substantially different (for example, see Fig. 3). The most significant decreases in fluorescence intensity were observed when cells had reached 90% confluence or greater. Therefore, we suggest harvesting cells for the screen at approximately mid-log growth, the point at which (1) comparisons of fluorescence intensity values between wells will be more accurate and (2) there are enough cells in the well to allow removal of half for the screen. For the suspension CHO cells we use, this optimal time point was chosen as 60–70% confluent in a well of a 96-well plate. The optimal confluence or cell density may differ for other cell types, but the main points are that cells should be screened before they reach maximum density and that each well should be screened upon reaching the same optimal point

of growth. It is likely that the wells will not all reach this optimal point at the same time, which necessitates clone screens being done on more than 1 day (with the appropriate controls included each time).

6. At the 96-well plate wash steps, first time users are sometimes concerned about potential cell loss upon inversion or tapping to remove the liquid; however, we have never encountered cell loss from the V-bottom wells with the cell type we use. The postwash plate inversion should be done in a quick motion, with a sharp flick of the wrist so that it only needs to be done once for each wash removal. For the tapping step (onto a paper towel) to remove residual liquid prior to the antibody incubation, use a gentle motion to release the liquid onto the paper towel. After each plate inversion or tapping step, the cell pellet should remain intact in the V-bottom well, and this can be observed by holding the plate up to the light. The control samples, which generally have more cells, should be easily visible as pellets in the wells and the test samples will usually be visible as smaller pellets.
7. The amount of antibody to use per well can be based on a previous titration experiment using a pool or clone expressing the reporter protein. The titration experiment should use the maximum number of cells expected per well in the screen (i.e., half of the number in a well at optimal confluence for screening) and should test increasing antibody concentrations to determine when antibody excess is reached. However, if the cells used for titration have much lower reporter expression than is expected for the clones to be screened, a higher concentration of antibody may need to be used to be in antibody excess during the screen.

References

1. Gurtu, V.; Yan, G.; Zhang, G., IRES bicistronic expression vectors for efficient creation of stable mammalian cell lines. *Biochem. Biophys. Res. Comm.* **1996**, *229*, 295–298.
2. Mizuguchi, H.; Xu, Z.; Ishii-Watabe, A.; Uchida, E.; Hayakawa, T., IRES-dependent second gene expression is significantly lower than cap-dependent first gene expression in a bicistronic vector. *Molecular Therapy* **2000**, *1*, 376–382.
3. Liu, X.; Constantinescu, S.N.; Sun, Y.; Bogan, J.S.; Hirsch, D.; Weinberg, R.A.; Lodish, H.F., Generation of mammalian cells stably expressing multiple genes at predetermined levels. *Anal. Biochem.* **2000**, *280*, 20–28.
4. Barnes, L.M.; Bentley, C.M.; Dickson, A.J., Stability of protein production from recombinant mammalian cells. *Biotechnol. Bioeng.* **2003**, *81*, 631–639.
5. Davies, M.V. and Kaufman, R.J.; The sequence context of the initiation codon in the encephalomyocarditis virus leader modulates efficiency of internal translation initiation. *J. Virol.* **1992**, *66*, 1924–1932.
6. DeMaria, C.T.; Cairns, V.; Schwarz, C.; Zhang, J.; Guerin, M.; Zuena, E.; Estes, S.; Karey, K.P., Accelerated Clone Selection for Recombinant CHO Cells Using a FACS-Based High-Throughput Screen. *Biotechnol. Prog.* **2007**, *23*, 465–472.

Chapter 4

Expression of a Secreted Protein in Mammalian Cells Using Baculovirus Particles

Barbara Ann Jardin, Cynthia B. Elias, and Satya Prakash

Abstract

There are many methods presently available to produce recombinant proteins in mammalian systems. The BacMam system is a simple straightforward method which overlaps two well-established technologies, namely the BEVS insect cell system and the transduction of mammalian cells in vitro. This chapter describes a method for the study of gene expression in mammalian cells in a series of simple steps. Protocols outlined include the design and construction of the recombinant baculovirus, cell culture techniques required to maintain both insect and mammalian cells, generation of baculovirus stocks, and methods to obtain maximal and reproducible gene expression in mammalian cells. Currently available statistical techniques using factorial design of experiment to optimize conditions for recombinant protein in vitro are outlined. Then details with respect to process scale-up in disposable bioreactors are included.

Key words: CMV promoter, BacMam, Baculovirus, Full factorial design, Design of experiment, Hollow fiber, Tangential flow filtration

1. Introduction

The last couple of decades have seen great advances in the use of baculovirus technology, including the demonstration that modified baculoviruses have the ability of to enter and transduce a variety of mammalian cells lines, technology known as BacMam (1). The potential for its use as a gene delivery system has been investigated and directly compared to the well-established adenoviral gene delivery systems (2, 3). Some advantages of this transduction method include a good biosafety profile (because baculoviruses cannot replicate in mammalian cells); the capacity of baculovirus vectors for very large inserts (no known limit on the insert size); the ease with which high-titer baculovirus stocks can be produced; and the relatively high efficiency with which many cell lines can be transduced.

James L. Hartley (ed.), *Protein Expression in Mammalian Cells: Methods and Protocols*, Methods in Molecular Biology, vol. 801,
DOI 10.1007/978-1-61779-352-3_4, © Springer Science+Business Media, LLC 2012

Many BacMam designs have been generated and widely studied (1). Briefly, through established molecular cloning techniques, baculoviruses have been engineered to contain promoters active in mammalian cell systems. These promoters are designed to be upstream of foreign transgenes to drive efficient expression of active recombinant proteins. Vectors have been designed for transduction of a large variety of cell lines, such as hepatocytes (4–9), fibroblasts, mesenchymal cells (10, 11), endothelial cells (12), pancreatic cells (13), and retinal cells (14).

A description of the complete process begins with the construction of the recombinant baculovirus and methods of cell culture are briefly examined while quantification of the titer of generated BacMam stocks is described by using an automated cell counting device and determination of viable cell density. Scale-up of these stocks is outlined along with a baculovirus concentration and diafiltration methodology. Transduction is described along with discussion on approaches to optimizing and maximizing recombinant protein expression. The parameters investigated were the multiplicity of infection (MOI) used, the cell density used at the time of transduction, and the concentration of serum and the protein expression enhancer trichostatin A (TSA) (15). Statistical design of experiment (DOE) methods (16) were employed to evaluate the effect of these parameters aimed at the optimization of the scaled-up process. A brief presentation of hollow fiber bioreactors (HFBR) and Wave bioreactor methods are presented as possible disposable bench scale alternatives for this Bac-Mam mediated gene expression in mammalian cells.

2. Materials

2.1. Insect Cell and Mammalian Culture

1. Two incubators are required for the cell cultures, one maintained at 27°C for the insect cell culture and second humidified and adjusted to 37°C which can be supplemented with 5% CO_2 (for mammalian cells).
2. Rotating shaker/incubator maintained at 27°C and 110 rpm (Multitron, New Brunswick).
3. Stationary flasks, T-75 flask (Corning) and 96-well plates.
4. Polycarbonate shake flasks (Corning, 250 mL, Cat. No. 430183) or Kimax glass (1 L, K265051000) with screw caps are used for Sf9 cells.
5. Calibrated micropipettors (2, 20, and 1,000 μL).
6. Certified biological safety cabinet with certified HEPA filter.
7. Potential mammalian cells include human embryonic kidney (HEK 293A) cells which can be purchased from Invitrogen (Carlsbad, CA) (see Note 1).

8. Dulbelco's modified Eagle's medium (DMEM) cell culture medium.
9. Heat-inactivated fetal bovine serum (FBS). Heat inactivate at 56°C for 30 min.
10. The insect cell line Sf9 (derived from *Spodoptera frugiperda*) can be purchased from Invitrogen Life Technologies (Carlsbad, CA) and produces more baculovirus per unit volume than other commonly used insect cell types.
11. Insect cell medium SF900II (Invitrogen; Cat. No. 10902-088) and transfection basal medium (BM), such as IPL41 (SAFC; Cat. No. 56923C), with no supplements (see Note 2).
12. Inverted light microscope.
13. Hemocytometer and coverslips (Hauser Scientific, Horshaw, PA; Cat. No. 4000).
14. Circulating water bath able to be maintained at 27, 37, and 56°C.
15. Cell counter/sizer (for example: Z2 from Coulter Beckman, Cedex from Innovatis, Bielfield Germany or Vicell from Beckman Coulter).
16. Sterile, screw-capped, 1.5-mL tubes (VWR Canlab; Cat. No. 89004-290).
17. Glass Pasteur pipettes, sterilized by autoclaving.

2.2. Vector Preparation

1. pM1–SEAP expression vector (Roche Applied Sciences; Cat. No. 03 045 633 001).
2. pVL1393 transfer vector from BD Biosciences (Cat. No. 21486P).
3. *Spe*I, *Bgl*II, and *Xho*I (New England Labs, Beverly, MA).
4. T4 DNA ligase (Thermo Fisher Scientific, Whitby, ON).
5. Plasmid amplification in DH5α-competent *Escherichia coli* Cat. No. 18265-017 (Invitrogen, Carlsbad, CA).
6. Plasmid purification QIAprep spin Miniprep kit (Cat. No. 27104, Qiagen Sciences, MD).
7. Plasmid cotransfection with linearized baculovirus DNA (Cat. No. 554739, BD BaculoGold).
8. Cellfectin (Cat. No. 10362-100, Invitrogen Life Technologies, Carlsbad, CA).
9. PCR, agarose, power supply for run gel, forward/reverse primers.

2.3. Design of Experiment

1. Design-Expert software from Stat Ease ver. 7.1.

2.4. Baculovirus Concentration and Diafiltration

1. Variable-speed peristaltic pump with sterile silicone tubing and reservoirs.
2. Ultrafiltration membrane 50-kDa low protein-binding and ultrafiltration apparatus, such as the Minimate tangential flow filtration (TFF) System with reservoir Assembly from Pall Filtron Corporation, Canada (for smaller volumes) or Millipore Corporation, Billerica, MA (for larger volumes).
3. Diafiltration apparatus (Millipore Corporation, Billerica, MA). The equipment is the same as the one used for ultrafiltration, with an extra reservoir to add mammalian medium as insect medium is removed (Fig. 1b).
4. Low protein-binding, 0.22-μm membrane for final sterile filtration (Corning, Inc., Corning, NY).

2.5. Protein Production in Hollow Fiber Bioreactor

1. Hollow fiber cartridge from Fibercell (Fibercell systems, Frederick, MD) with molecular weight cutoff of 5 kDa (Cat. No. 4300-C2008).

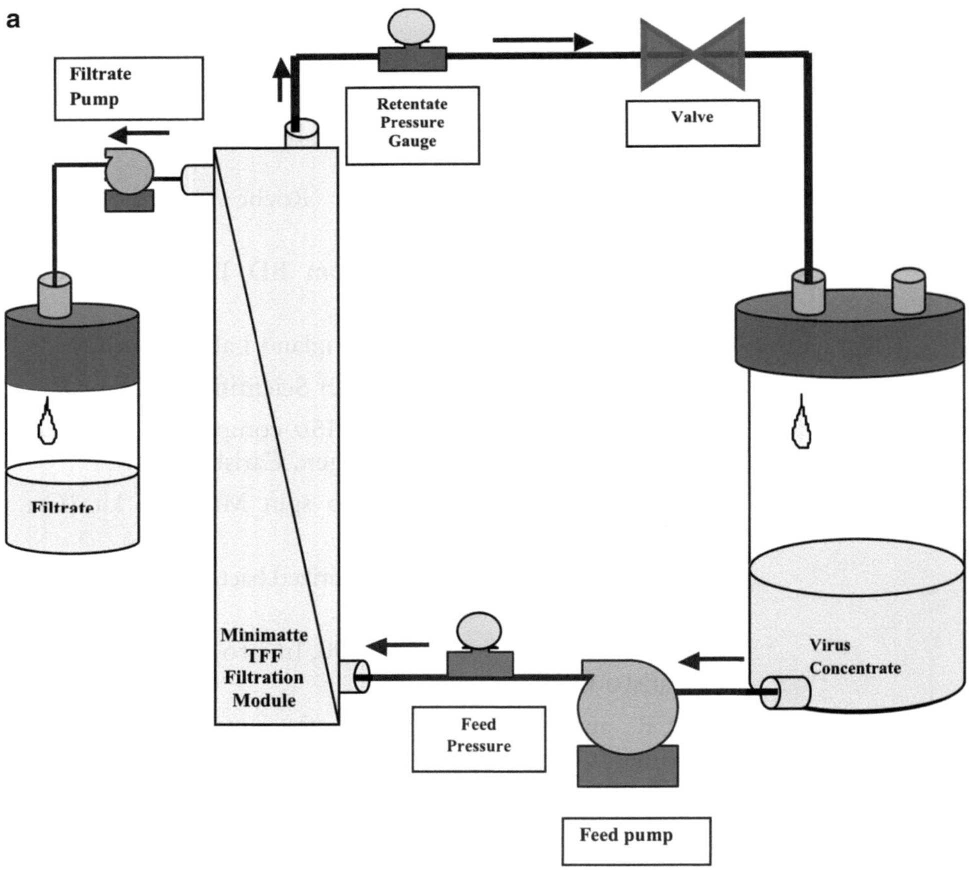

Fig. 1. (**a**) Ultrafiltration Minimate setup for baculovirus concentration. (**b**) Diafiltration system which is the exact system as for the concentration, but with an added buffer reservoir for exchanging insect cell medium with mammalian cell medium (i.e., DMEM).

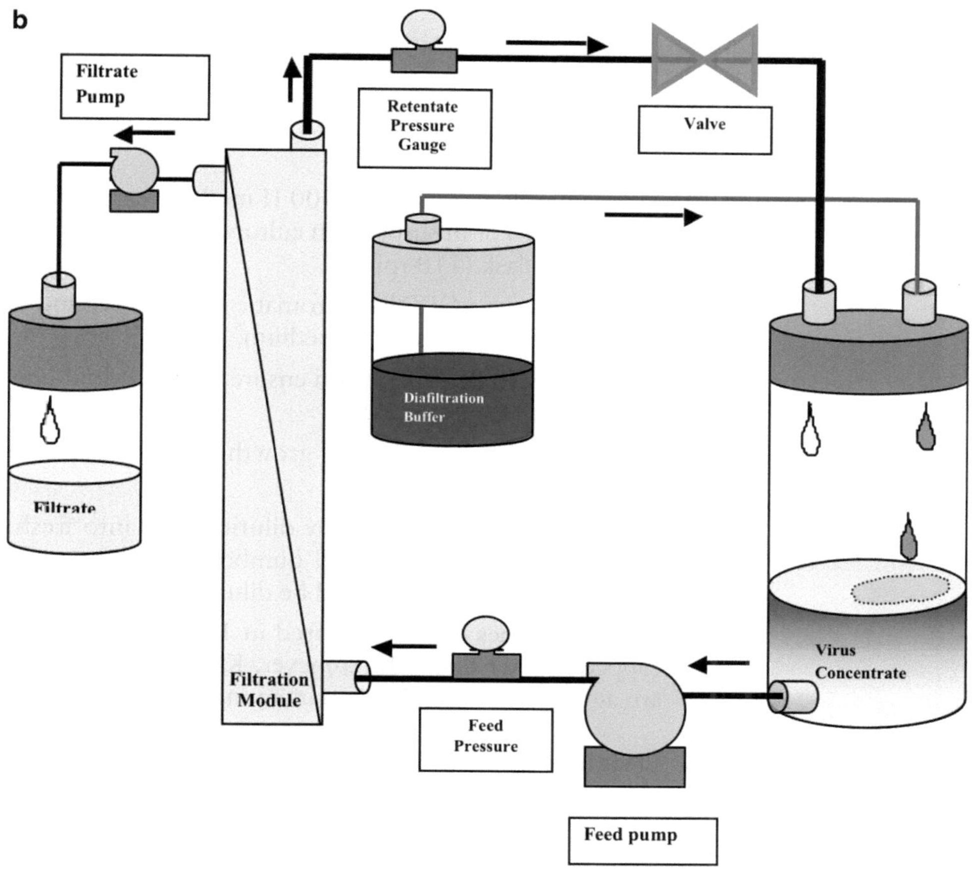

Fig. 1. (continued)

2. DMEM without FBS and DMEM with 10% FBS, phosphate-buffered saline (PBS).
3. Sterile 60 cc Syringes (BD Biosciences) with Luer-lock connections.
4. Pumping system from Fibercell (Cat. No. P3202; Fibercell systems, Frederick, MD).
5. Accessories, including bottle caps and fittings.

2.6. Metabolite and Protein Analysis

1. Glucose, lactate, glutamine, and glutamate bioanalyzer (e.g., YSI 7100 MBS, Yellow Springs, OH).
2. LDH analyzed with Hitachi 911 blood analyzer.
3. SEAP analysis 96-well microtiter plates.
4. Microplate reader able to carry out kinetic absorbance analysis at 405 nm.

2.7. Chemicals

Trypan blue solution, diethanolamine, *p*-nitrophenyl phosphate (pNPP), PBS available from Sigma–Aldrich.

3. Methods

3.1. Cell Culture

3.1.1. Insect Cells

1. Thaw Sf9 cells (Cat. No. 11496015), Invitrogen Life Technologies, Carlsbad, CA.
2. Maintain cells at 27°C in SF900 II medium either in stationary T75 (10 mL) or in suspension culture flasks 50–60 mL in 250-mL shaker flask (110 rpm).
3. Count cells with CEDEX automatic cell counter and dilute to 0.5×10^6 cells/mL in fresh medium.
4. Count daily to determine and ensure that cell-doubling time is below 24 h (see Note 3).
5. Maintain cells in exponential growth phase ($2.5–5 \times 10^6$ cells/mL).
6. Subculture twice per week by diluting cells into fresh media. A guideline is that when cell number reaches between 3 and 5×10^6 cells/mL, then should be diluted to 0.5×10^6 cells/mL.
7. Larger volumes can be prepared in 1–2,000-mL glass, screw-capped, shaker flasks (Erlenmeyer, Kimax, see Note 4), which are agitated at 110–120 rpm in an incubator shaker.

3.1.2. Mammalian Cells (HEK293A Cells)

1. Heat inactivate FBS by incubating in a water bath at 56°C for 30 min and store 40-mL aliquots in 50-mL tubes at less than −20°C. This is done in order to eradicate the activity of heat-labile complement proteins which are cell growth inhibitors.
2. Prepare DMEM with 10% v/v heat-inactivated FBS.
3. Rapidly thaw HEK 293A cells at 37°C water bath for 1–2 min. Add 10-mL fresh DMEM + 10% FBS into a 15-mL centrifuge tube. Centrifuge to remove DMSO, remove supernatant, and resuspend cells in 7–10 mL fresh medium. Add directly into T-75-cm^2 stationary flasks.
4. Grow HEK 293A cells in DMEM supplemented with 10% FBS. Initial cell density for this cell line should be $1–2 \times 10^5$ cells/mL and may reach 2×10^6 cells/mL using this serum-supplemented medium. Subculturing dilutions can range between 1:5 and 1:15.
5. 7–10 mL of HEK 293A cells are maintained in T-75 stationary flasks incubated at 37°C in a humidified controlled environment with 5% CO_2.
6. Subculture two times per week by diluting 1–2 mL cells in 9–12 mL fresh medium. Cells can be detached by trypsinization using a trypsin–EDTA solution (SAFC; Cat. No. 59430). (The trypsin–EDTA solution can also be purchased from other vendors, such as Invitrogen Life Technologies.) Briefly, medium

is removed and the cell monolayer is rinsed once with PBS to remove residual serum from the medium. 2–3 mL trypsin–EDTA solution is added and incubated at 37°C for enough time (generally, takes about 5 min) for the cells to detach from the surface of the T-flask. During this period, close attention should be paid to the flask and should be observed under the microscope intermittently. When cells are seen to be detaching and lifting up from the surface, add medium containing serum. The serum is inhibitory to trypsin action and stops further action. Centrifuge cells at 500 × *g* for 5 min. Remove supernatant, resuspend cells in fresh medium, and add to fresh T-flask.

3.2. Counting Cells with a Microscope

The insect cells infected with baculovirus increase in diameter and lose viability over time, which allows baculovirus production to be monitored. The Trypan blue staining procedure may be used to assess cell number and viability. Cells are added to the dye followed by counting the cells using a hemocytometer with a microscope. The Trypan blue method is not a direct measure of viability, it is considered to be a measure of membrane integrity, and this method is the most commonly used technique to approximate the cell viability, especially in the case where infected cells are being evaluated. The value of this protocol assumes that loss of viability occurs prior to the failure of membrane integrity which is characteristic of nonviable cells. Alternatively, automated devices, such as the Cedex Innovatis (Bielefeld, Germany) or the Vicell (Beckman Coulter) which both use the Trypan blue staining method, may be used. The Coulter Particle Counter counts the total number of cells and the average cell diameter, but does not assess cell viability. A Nucleocounter (YC-100, Chemometric, Denmark) which counts nuclei may also be used to obtain an estimate of live and dead cells by performing differential counts of nuclei in the presence and absence of cell lysis agents.

The manual Trypan blue method is briefly described below.

1. Aseptically aliquot cells from a culture flask with sterile pipette under aseptic conditions and place in test tube. Immediately return the flask to the incubator.
2. Dilute the sample with Trypan blue in order to have between 30 and 100 cells per square, a range which gives most accurate cell counts with a hemocytometer (1:4 to 1:8 times).
3. Mix the solution gently with a Pasteur pipette (see Note 5). The cell count should be performed within 3–4 min after adding the Trypan blue.
4. Pull out a small volume of the cell dye suspension with a pipette, gently place the tip of the pipette on the slot of a clean hemocytometer, and allow the suspension to fill the chamber between the slide and the coverslip.

5. Record the quantity of stained (nonviable) and unstained (viable) cells in the four large squares on each corner of the chamber. Note that each of these squares has a volume of 0.1 mm^3 ($1 \times 1 \times 0.1$ mm [depth]).
6. The total and viable cell density can be calculated using Eqs. 1 and 2, respectively:

$$\frac{\text{TOT cells}}{\text{mL}} = \frac{(N + N') \times 10^4 \times D}{n} \tag{1}$$

$$\frac{\text{Viable cells}}{\text{mL}} = \frac{N' \times 10^4 \times I}{n} \tag{2}$$

where N = total number of stained (i.e., dead) cells counted, N' = total number of unstained (i.e., viable) cells counted, n = number of squares in which cells were counted (usually 4), and D = dilution factor.
7. The percent viability can be calculated using Eq. 3:

$$\%\text{Viability} = \frac{N'}{N + N'} \times 100\% \tag{3}$$

3.3. BacMam Vector Construction

There are a variety of methods to construct a BacMam vector that expresses a recombinant protein in mammalian cells. An overview is described below with the example protein-secreted alkaline phosphatase (SEAP).

1. The vector used (pM1–SEAP) harbors the CMV promoter and SEAP gene. This expression vector was digested with *Spe*I, *Bgl*II, and *Xho*I to jointly excise the promoter and SEAP DNA fragments, resulting in a fragment of 3,300 bp, which was gel purified.
2. The pVL1393 transfer vector was digested with *Xba*I and *Bgl*II. The parent vector and the insert were purified and cloned by compatible end produced by *Spe*I toward *Xba*I restriction site and the presence of common restriction site, *Bgl*II.
3. Insertion of the desired pCMV-SEAP DNA was achieved by compatible-end cloning into the pVL1393 transfer vector. Ligation was carried out with T4DNA ligase.
4. The recombinant pCMV–SEAP/pVL1393 plasmid was then transformed into the high-efficiency DH5α-competent *E. coli* by heat-shock method.
5. Colonies were selected and bacteria were grown with standard molecular biology methods.
6. The pCMV–SEAP/pVL1393 plasmids were then extracted and purified with QIAprep spin Miniprep kit (Qiagen Sciences, MD).

7. Preparations were quantified by 260/280 nm absorbance technique to determine DNA content and purity of Miniprep samples. Typically, 1 μg is needed for transfection of 10^6 Sf9 cells. The DNA concentration should be greater than 300 μg/mL.

3.4. Viral Stock Propagation

3.4.1. Passage 0 Viral Stock

All manipulations are to be carried out under sterile conditions.

1. Prepare a dilution of 0.45–0.5 × 10^6 Sf9 cells/mL in SF900II medium.
2. In a six-well plate, add 0.9–1 × 10^6 cells/well (2 mL).
3. Allow cells to attach (see Note 6).
4. Prepare solution A: While allowing cells to attach, begin cotransfection by combining 1 μg pCMV-SEAP/pVL1393 plasmid DNA or your expression construct and 0.25 μg linearized baculovirus DNA (BD BaculoGold) to 100 μL IPL41 basal medium in a microfuge tube.
5. Prepare solution B: Add 6 μL Cellfectin to 100 μL IPL41 BM. If you are doing *N* transfections, make more of solution B and then dispense accordingly.
6. Gently vortex each tube for 5 s. Centrifuge for 2 s in small, portable microfuge to ensure that all liquid is at the bottom of the tube.
7. Combine solutions B to A, and immediately gently vortex for 5 s each tube. Centrifuge for 2 s in small, portable microfuge to ensure that all liquid is at the bottom of the tube and allow to incubate at room temperature for 45 min. During this time, complexes of DNA and transfection reagent form.
8. After a 45-min incubation, remove SF900II medium from six-well plates by tilting and aspirating the medium with a sterile Pasteur pipette connected to a tube and vacuum pump. Then, add 1–2 mL IPL41 medium to each well (see Note 7).
9. Prepare solution C: Add 800 mL IPL41 basal medium to the A + B mixture and vortex on low for 2 s. Spin in low-spin microfuge for a few seconds to make sure that all liquid is at the bottom of the tube.
10. Remove rinsing medium from the cells in each well by aspiration as before with Pasteur pipette.
11. Add 1 mL solution C gently and dropwise to cells in six-well plate.
12. Incubate the cells with the transfection mixture(s) for 5 h at 27°C without agitation.
13. Remove transfection solution C from cells, immediately add 2 mL SF900II medium to cells, and incubate for 96–120 h at 27°C without agitation. During this time, several cycles of

virus replication occur resulting in cell death, virus release, and infection of new cells.

14. Collect the medium from each transfection at 96–120 h post transfection to obtain the Passage 0 (P0) virus stock.
15. Centrifuge to remove cell debris and filter through 0.22-μm filter immediately and dispense a 0.5-mL aliquot. Usually, one aliquot is used immediately to make the Passage 1 (P1) viral stock (below). Remaining 0.5-mL aliquot should be stored at –80°C in screw-cap vials (see Note 8).

3.4.2. Passage 1 Viral Stock Preparation

1. Dilute exponentially growing Sf9 cells to 1×10^6 cells/mL in SF900II.
2. Add 10 mL of cell suspension to a T75 stationary flask.
3. Add 200–500 μL Passage 0 stock and mix by rocking gently back and forth a few times.
4. Check flask 72 hours post infection (hpi) for signs of infection which include arrested cell growth, decreased viability, increased cell diameter, and increased occurrence of floating cells.
5. Harvest 10 mL supernatant by centrifugation (1,500 × *g*, 10 min, 4 C).
6. Filter immediately through 0.22-μm, low protein-binding filter.
7. Store at 4°C for further use (see Note 8).

3.4.3. Passage 2 Virus Stock Preparation

1. Dilute exponentially growing Sf9 cells to 2×10^6 cells/mL (see Note 9).
2. Add 270 mL to a sterile, glass, 1,000-mL shaker flask (see Note 4).
3. Add 3–6 mL P1 stock and incubate in 27°C shaker at 110 rpm. Start a first stock with 3 mL addition and follow the next steps. If there is very little indication of infection, then increase to 6 mL.
4. Count cells at 24 hpi to see if cells are increasing in number (see Note 10). Viability should be decreasing by 20–30% at 72 hpi.
5. Harvest when cell viability is between 70 and 90% viability (usually between 72 and 96 hpi) (see Note 11).
6. Harvest 270 mL supernatant by centrifugation (1,500 × *g*, 10 min, 4°C).
7. Filter immediately through 0.22-μm, low protein-binding filter and concentrate as per Subheading 3.6 if required.
8. Store at 4°C and determine titer as per Subheading 3.5 (see Note 12).

3.5. BacMam Viral Particle Quantification

1. Dilute Sf9 cells to 0.5×10^6 cells/mL in a volume of 500 mL in a sterile 2,000-mL, glass, shake flask. Incubate at 27°C with shaking at 110 rpm. Use the cells after 72 h, when cell density is ~4×10^6 cells/mL.
2. Prepare flasks as per table below. Add components into 125 ml Corning shake flasks, prepare flasks in duplicate.

Flask	Cells (μL)	Medium (μL)	Virus (μL)	Dilution	Diameter @ 24 h
A	10	20	0	0	$d_{\text{uninfected}}$
B	10	20	6	2×10^{-4}	d_{24hpi}
C	10	20	12	4×10^{-4}	d_{24hpi}
D	5	10	15000	0.5	d_{infected}

3. Incubate in the 27°C shaker and determine the average viable cell density and the average cell size (diameter) between 20 and 24 hpi (see Note 13).
4. Calculate the virus titer by first calculating the number of infected cells:

$$\begin{aligned}\text{Total number of infected cells} &= \frac{(\text{Total number of cells}) \times (d_{\text{24hpi}} - d_{\text{uninfected}})}{(d_{\text{infected}} - d_{\text{uninfected}})}\\ &= \frac{(3.4 \times 10^6 \text{ cells/mL}) \times (15.4 - 13.92)}{(17.12 - 13.92)}\\ &= 1.6 \times 10^6 \text{ cells}\end{aligned},$$

where

$d_{\text{uninfected}}$ = average diameter of Sf9 cells in uninfected control after 24 hpi;

d_{infected} = average diameter of Sf9 cells in infected control after 24 hpi; and

d_{24hpi} = average diameter of Sf9 cells in sample after 24 hpi.

5. Then, calculate the number of infectious virions:

$$\begin{aligned}&\text{Total number of infectious virions}\\ &= \frac{\text{Total number of infected cells}}{\text{dilution}}\\ &= \frac{1.6 \times 10^6 \text{ cells/mL}}{4 \times 10^{-4} \text{ virus}}\\ &= 4 \times 10^9 \text{ infectious virus particle (IP)/mL}\end{aligned}.$$

Calculation carried out for each dilution and average calculated taken (see Note 14).

3.6. Concentrating BacMam Particles

The titers of viral stocks produced using routine methods are usually in the range of 10^8 and 10^9 infectious viral particles (IVP/mL). It is necessary to have higher virus titers when using high MOIs as is the case for mammalian cell transduction with BacMam. Concentration of baculoviruses is not generally required if using to infect insect cells, however, this is done to avoid the addition of large volumes of spent insect cell medium to the mammalian cell system. The composition of the insect cell medium is quite different than that for the mammalian cells and the osmolarity is between 50 and 100 U higher. There is also a consequent dilution of medium because of large volumes of viral stock required for effective transduction. Therefore, virus stock concentration is often required and can be accomplished with an ultrafiltration step using a 50,000-molecular-weight-cutoff membrane, and then a diafiltration step to remove small, solute materials in the viral stock. A schematic of these steps is shown in Fig. 1a, b (see Note 15; for an explanation of how diafiltration relates to ultrafiltration, see Note 16). Add 270 mL viral stock to the reservoir of a Minimate ultrafiltration apparatus (Pall Filtron Corporation, Canada).

1. Ultrafilter through the Omega membrane 50-kDa low protein-binding membrane as per manufacturer's guidelines until the volume reaches 10% of initial volume (around 25 mL in this example).
2. Diafilter the concentrate against four volumes of DMEM (i.e., bring the 25 mL to 100 mL) culture medium as follows. The concentrated baculovirus is now in the retentate reservoir. Further removal of the insect cell medium is carried out by medium exchange with mammalian cell culture medium (DMEM). This is accomplished by constant volume diafiltration. Constant volume diafiltration refers to the process wherein the retentate volume is held constant by maintaining the rate of feed addition, in this case DMEM, equal to that of the rate of removal of filtrate, in this case SF900II. A vacuum is created in the closed retentate reservoir, and a sterile tubing is connected to a fresh supply of DMEM which is introduced at the same flow rate as the removal of the SF900II. This allows for a constant concentration of virus to be maintained, thus minimizing aggregation of the baculovirus. The volume of DMEM added to the retentate is 100 mL in this case and brings the final volume back to 40–50% of the original volume which was 270 mL in this case.
3. Sterilize by filtration through a 0.22-μm filter before use. Determine titer as in Subheading 3.5.

3.7. Transduction of Mammalian Cells with Baculovirus Particles

3.7.1. Initial Studies

1. For HEK293A cells, add 100 μL of freshly diluted cells at a concentration of $0.3–0.5 \times 10^6$ cells/mL to each well of a 96-well plate (see Note 17).
2. Wait for the cells to adhere (see Notes 17 and 18).
3. Add varying amounts of Passage 2 (P2) viral stock in order to achieve MOI of 10–1,000 IP/cell (see Note 19).
4. Carry out experiment in triplicate in order to achieve statistical validity of the results. See the SEAP example below.
5. Refer to Design of Experiment section.

3.8. Recombinant Protein Quantification: Analysis of SEAP

This assay is based on the activity measurement method developed in 1988 by Berger et al. (17).

1. Prepare buffered substrate pNPP.
2. Prepare 1 M solution of diethanolamine.
3. Add $MgCl_2$ to 1 mM.
4. Adjust pH to 9.8 with 0.1 M HCl.
5. Add pNPP to 20 mM.
6. Remove 50 mL of sample (i.e., supernatant from transduced mammalian cells) and add to one well of a 96-well plate. This sample may need to be diluted in H_2O if the concentration is too high. We had to dilute between 10 and 50 times to obtain best results.
7. Add 50 mL buffered pNPP to the well and monitor kinetically the absorbance at 405 nm on a plate reader.
8. Calculate the V_{max} of reaction which can be calculated automatically with the plate reader in "kinetic" mode.

3.9. Design of Experiment

Process optimization can be effectively achieved with DOE methods. The power of the factorial design lies in the fact that many parameters can be addressed simultaneously. The advantage of this methodology is that fewer experiments can be carried out to optimize process while concomitantly identifying interaction effects. Objectives include identification of critical factors to increase productivity of the recombinant protein production by BacMam transduction. The full factorial experimental design and statistical analysis were aided by the statistical software Design-Expert.

Initially, when approaching the factorial design, determination of the measurable outcome of the experiment is vital. In the present study, the selected response was the secreted protein production. The selection of parameters to be studied is the next step. Generally, the factors are chosen from previous experiments and critical review of published factors which influence productivity.

Demonstrated here is the study of two-level factorial designs with three factors. This design permits assessment of key effects and interaction effects, thus pinpointing significant factors. Briefly, two-level design means that a numerically high level and a low level for each parameter are chosen dependent on previous experience. Midpoints are generated by the average to the high and low values. The results from the midpoints are the pivotal values because they are used to evaluate the effects of factors. Four midpoints were chosen.

Once these experiments are finished and results have been analyzed, further experimentation can be carried out with the response surface methods to determine best settings for the given factors which afford maximal production.

3.9.1. Select the Factors and Response

The following parameters are varied to optimize production of the target protein.

1. MOI is a term borrowed from insect cell terminology and relates to the number of infectious viral particles (IP = infectious particles) added per cell.
2. Initial cell density of cells.
3. TSA concentration.
4. Response factor is SEAP activity.
5. Choose high and low values of each variable to be used from previous results or extracted from published results.

3.9.2. Selecting Initial Values

1. Initial cell density was chosen according to the most reasonable concentrations observed under routine culture conditions. It is challenging to select the most effective quantity because during the course of recombinant production, there is a time-dependent growth period, associated with consumption of critical nutrients depleted in the medium. We avoided this limitation by opting for lower cell densities. For instance, a cell concentration of 0.1×10^6 cells/mL is used for subculturing protocol and was chosen as the low value. The selected high cell concentration was 0.3×10^6 cells/mL, which is in mid-exponential growth phase.
2. Documentation of the significant effect of the infectious particle viral loading to ensure optimal recombinant protein production has been studied in detail (1, 18). Values cover a wide range, and often the limit is attained with respect to volumes of spent insect cell medium being added to the mammalian cell systems during transduction. Generally, an MOI of 1,000 IP/cell is the maximal amount which can be added. This is based on the best case scenario, whereby maximal viral titers of virus ~5×10^9 IP/mL can be consistently attained (see Note 12).
3. TSA is a chemical which interferes with normal cell division and growth by arresting cell cycling. This chemical has been shown to assist the introduction of the foreign DNA and gene

Table 1
Concentrations used for three different components in the factorial experiment

	Midpoint	High	Low
Virus (MOI)	300	500	100
HEK293A (MTC/mL)	0.2×10^6	0.3×10^6	0.1×10^6
TSA concentration (mg/mL)	8	12	4

replication in the cell. The dose selected was chosen according to published data (15, 18).

4. See Table 1 for a typical experimental setup (replaced with external table).
5. Following the first round of results from the DOE, a second round should be carried out in order to confirm results and further develop the process. Initially, the significant factors are identified; nevertheless, further maximization of productivity can be achieved by carrying out more factorial experiments.

3.9.3. Setting Up DOE Runs

The runs are set and randomized by the statistical program. The numbers of runs for the sample experiment illustrated here in was composed of 23 experimental conditions, (23 = 8 in 2 blocks) plus midpoints. The response (SEAP activity) was determined between 5 and 10 days.

3.9.4. Interpreting DOE Results

Statistical analysis of variance (ANOVA) responses were calculated and analyzed by software which allows the determination of the significance of the results. In this case, the viral load and the TCA added have the most dramatic positive effect on the recombinant SEAP production. A sample run setup and results are shown below in Fig. 2. The final equation of the factors influencing recombinant SEAP production derived from these data is the following:

$$\text{SEAP [U/mL]} = 168 + 63 \times [\text{Virus [IP]}] + 17 \times [\text{TCA[mg/mL]}]$$

3.10. Scale-Up of Protein Production Bioreactor Setup and Transduction

Batch production of recombinant protein can be carried out at different scales in a bioreactor. The bioreactor selection depends on a variety of issues, such as the type of cells being used, cost issues, space availability, and others. In the case with the HEK293 cells being transduced with the BacMam virus, two methods were investigated. The first was a batch process with a disposable Wave bioreactor (GE Healthcare Life Sciences, Piscataway, NJ). Bioreactor preparation was similar to the protocol outlined in the manufacturer's guideline for HEK 293 cells. The process is described briefly below.

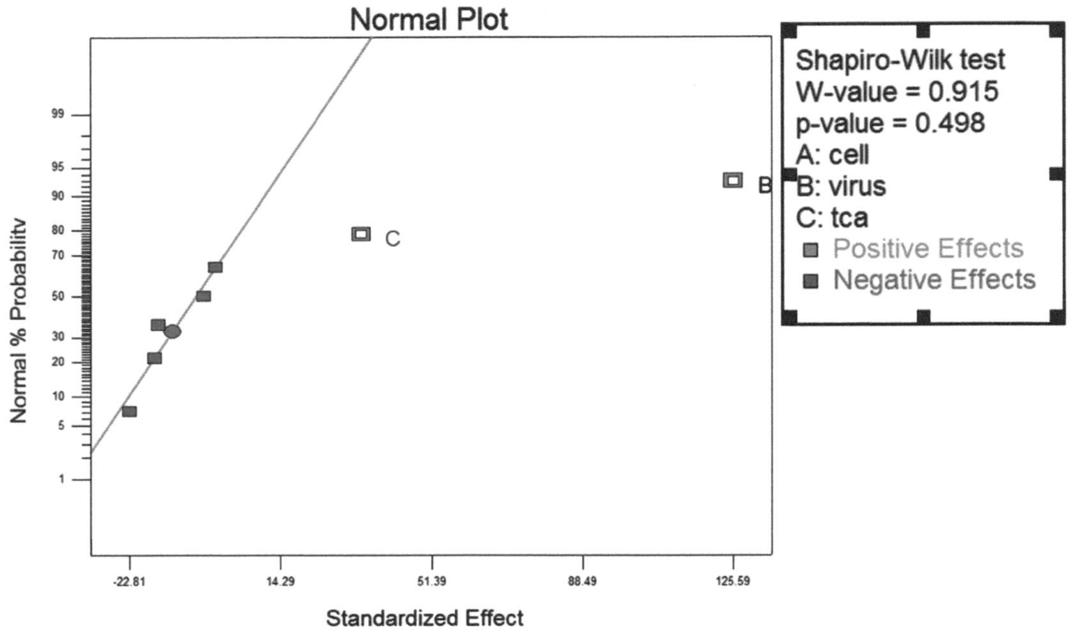

Fig. 2. Normal plot generated following ANOVA analysis. Created by Stat-Ease software Design-Expert®.

3.10.1. Wave Bioreactor

The Wave bags are sold presterilized and equipped with different-sized Luer-lock connections.

1. Autoclave the inoculation bottle with Luer-lock complementary to that on the bag. Let it cool before use.
2. All connections to the bag should be done in a biological containment hood (BCH) if possible prior to placing on the Wave rocking platform.
3. Prewarm DMEM supplemented with 10% FBS to 37°C.
4. Connect a reservoir for the medium to 10 L Cellbag, add 400 mL media, and clamp the inlet and exhaust filters.
5. Addition of the medium can be done by connecting the tubing to a peristaltic pump or alternately just by gravity.
6. Clamp the Cellbag on Wave apparatus. Start rocking the Cellbag at 15 rocks per minute (rpm) and an angle of 7°; set temperature to 37°C.
7. Allow the temperature and pH to equilibrate.
8. Inoculate the Cellbag by first counting cells and adding enough cells in an additional 100 mL of media in order to achieve a target initial cell concentration between 0.2 and 0.3×10^6 cells/mL.
9. Rock system at 12 rpm and an angle of 5–7° (see Note 20).

10. Add 5% CO_2 in air to the headspace and keep the bag well-inflated.
11. Ensure good contact between the temperature probe and the culture to avoid overheating, which occurs frequently at low culture volumes.
12. Check the culture by counting the cells and quantifying lactate, glucose, glutamine, glutamate, and lactate dehydrogenase (LDH) values (see Note 21).
13. After 2–3 days of cultivation (depending on doubling rate), the cell count should reach about $1–2 \times 10^6$ cells/mL.
14. Add 1,500 mL of fresh media to bring the total volume to 2 L.
15. Add BacMam at MOI of 1,000 and add TSA.
16. Sample daily and monitor glucose, lactate, cell density, and recombinant protein production.
17. Harvest culture when maximal protein production has been achieved. This may occur between 5 and 7 days.

3.10.2. Hollow Fiber Bioreactor

An HFBR consists of a cylinder through which pass hundreds of small fibers made of semipermeable membrane resulting in the compartmentalization of the reactor into the intracapillary space (ICS) which is the luminal space of the capillaries and the extracapillary space (ECS) which is the space between the fibers and the cylindrical housing. Cells are grown in the ECS, and growth medium is circulated through the lumen or ICS of the fibers to supply nutrients and remove metabolites. The molecular weight cutoff of the membrane is chosen to retain the secreted protein of interest within the ECS. The advantage of this culture system is that it is run in a perfusion mode which can allow for sustained extended periods of recombinant protein production. See Jardin et al. (19) for side-by-side comparative studies between batch, fed batch, and perfusion bioprocesses.

This system was acquired from Fibercell (Fibercell Systems, Fredrick, MD) and included a hollow fiber membrane packaged and presterilized, along with a pump. Bottles and connectors can be assembled in the laboratory. The molecular weight cutoff of the membrane was 5 kDa and the surface area was 2,100 cm^2. Setup was carried out as per manufacturer's specifications and is described briefly below. All manipulations are carried out using stringent sterile techniques. A simplified schematic of the HFBR is shown in Fig. 3.

1. Wet membrane with PBS for a period of 2–3 days prior to use. Change PBS 1–3 times during this period.
2. Add DMEM supplemented with 10% FBS to the ECS with 2×60-mL syringe on either side of the ECS inlet as in Fig. 3.

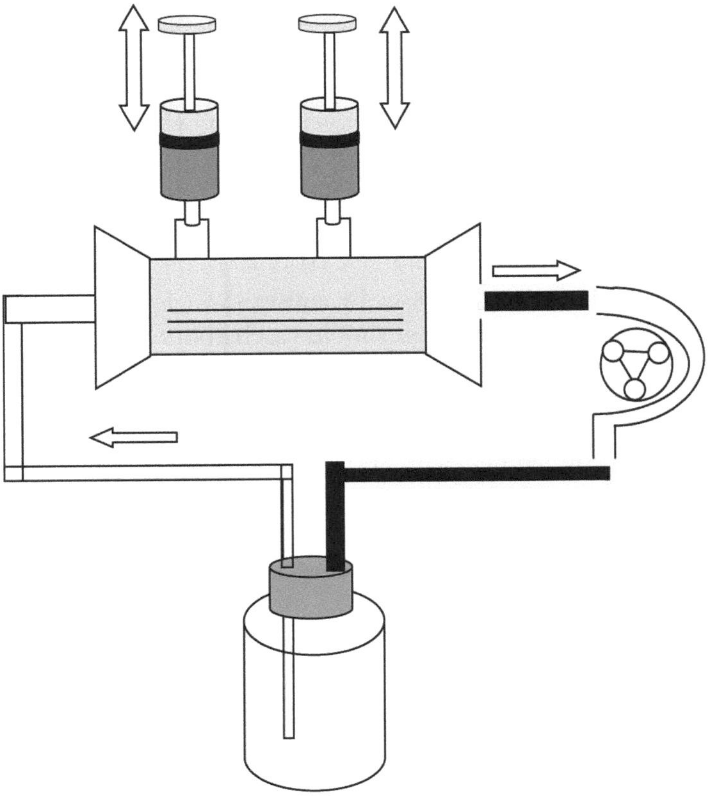

Fig. 3. HFBR setup with syringes attached.

3. Fill lumen (ICS) space with DMEM devoid of serum.
4. Feed bottle with DMEM was attached to the pump and sent through the lumen (ICS side).
5. Cells resuspended in completely fresh medium (DMEM + 10% FBS).
6. Add cells to ECS. Approximately 15–20 mL can be added to the C2008 unit used here.
7. Feed medium is replaced when glucose levels reach 50% of the original value; thus, ICS was sampled to determine glucose concentration.
8. Sample from ECS to determine concentration of nutrients (glucose, glutamine), metabolites (lactate), and LDH activity.
9. Transduction is carried out at 2 days post inoculation by the addition of BacMam to ECS; recombinant protein (SEAP) was evaluated by sampling from ECS.

4. Notes

1. This report describes the use of the HEK293A cell line to determine the feasibility of the process. However other mammalian cell lines can be transduced and thus the choice of cell line will depend on the objective of each individual project.
2. Many insect cells serum-free media types are presently commercially accessible and are equivalently suitable for Sf9 cells. These might include the Excel series of medium from SAFC (St. Louis, MO), ESF-21 (Expression Systems, Woodland, CA), Thermo Scientific HyClone SFX-Insect Cell Culture Media™ (Thermo Fisher, Scientific), Insectagro™ (Krackelar Scientific), Insect- XPRESS™ (Lonza, SW), and BD BaculoGold Max-XP (BD Biosciences, Pharmingen). The list is representative and not exhaustive.
3. Insect cell (Sf-9) do not normally generate metabolic by products (lactate or ammonia) (20, 21), which are the main cause of pH change in animal cell cultures. If there is a significant change in the pH of Sf-9 cell cultures there usually is a problem either with contamination or oxygen starvation.
4. All flasks used for serum free cell growth should be cleaned thoroughly by cleaning them immediately following use. Appropriate cleaning agents are CIP 100 or 200 (VWRCanlab) diluted in hot water. All flask surfaces are brushed with a bottle brush and rinsed with hot water. Three rinses with distilled water are carried out to ensure that all traces of detergent are removed prior to steam sterilization.
5. Pasteur pipettes are used to mix dilutions because the bore size of micropipettes is small and may cause high shear resulting in inaccurately low viabilities.
6. Insect cell attachment may occur as soon as 15 min but it is best to wait for 45–60 min. Observe under microscope, to look for cell attachment. Cells adhering to the flask will be seen as flat and spreading slightly and will not move when the flask is moved gently.
7. Transfection of the insect cells is best carried out with medium without any serum or serum replacement additives. Some which are available are Grace's un-supplemented medium and ILP41 basal medium. Both contain some amino acids, vitamins, minerals sugars and are used during the 5 h incubation of the Sf9 cells with the transfection mixture.
8. Viral stock stability may be maximized by creation of a master virus bank obtained from the early passages P0 and P1 at −80°C and if possible stored in liquid nitrogen (cryo-protectant

beneficial effects remain to be studied). The working stock can then be generated from these frozen aliquots. Immediately following harvest, adding some FBS at 2% has been shown to improve stability of virus stocks stored at 4°C.

9. Cells are two times higher density during the amplification of the viral stock at passage two compared to passage one. Sf9 cells have the ability to reach 10×10^6 cells/mL with the rich SF900II medium and will grow well when seeded at this cell density. P2 titers of more than 1×10^8 per mL are often obtained.
10. The progress of the infection can be followed by measuring the cell number and average cell size (22–24). At low MOI, there is an initial increase in cell concentration following infection. The cell diameter should begin increasing by 24–48 hpi. Harvest should only take place following the concomitant decrease in cell viability and the increase in cell diameter. After infection, available nutrients are depleted by virus replication, protein production, and cell growth and maintenance. It is important to use cells from the exponential growth phase, to use a baculovirus inoculum that does not exceed 10% of the working volume, and that the cells have a density of $2–2.5 \times 10^6$ cell/mL.
11. If the viability is not decreasing, restart the 270 ml culture with the addition of a higher amount of P1 viral stock.
12. Only use fresh virus for studies. For best results it should be less than 1 month old and not older than 6 months. This is done in order to avoid instability of high titer Bacmam particles which is a key bottleneck of this process. Also, do not refilter cloudy virus because the titer will drop significantly. Cloudy virus indicates possible contamination or virus aggregation. The filtration of the freshly made virus stock ensures sterility and removes cell debris. The filtration step must be done immediately because of the possibility of a dramatic decrease in titer following virus stocks settling over 1 week period. The instability of high titer virus stocks is the primary limitation on the titer of BacMam particles mainly due to virus aggregation and greater losses during the diafiltration steps.
13. The ViCell from Beckman Coulter, Inc. and Cedex are commercially available instruments that determine cell viability, cell number, and cell diameter. For example, the CEDEX from Innovatis (Bielefeld, Germany) is an automatic cell counting device. Both adequately determine cell number and diameter. A Coulter counter (Beckman Coulter, Inc., Fullerton, CA) may be used to determine the total cell concentration and the average cell diameter. It is rapid, yet,is not able to quantify cellular viability.

14. For further details see Janakiraman et al. (25).
15. The 5-ft^2 Pellicon tangential flow filtration unit (Millipore Corporation, Billerica, MA) can be used for larger volumes (greater than 300 mL) and smaller volumes can be concentrated using a 0.75-ft2 Minisette Filtron unit (Pall Corporation, East Hills, NY). These protocols also require a peristaltic pump and a reservoir vessel.
16. Ultrafiltration uses pressure to push water and small molecules (the filtrate) through a membrane, while retaining molecules above a specified weight in the retentate. Diafiltration is the same process except that a new buffer or medium, of a preferred composition in this case DMEM, is added to the retentate during the filtration with either a pump or vacuum. Thus ultrafiltration is used for concentrating, and diafiltration is used for buffer or medium exchange. In a typical arrangement the solution to be concentrated or diafiltered is pumped over the surface of the filter (tangential flow) and the retentate is returned to a reservoir. In ultrafiltration (Fig. 1a) the removal of filtrate results in increasing the concentration of the material of interest in the retentate (here, viral particles). If the solution / rententate reservoir is sealed and provided with a port, the decrease in volume during filtration creates a vacuum that draws in air (during ultrafiltration) or new liquid (during diafiltration, Fig. 1b) Further details may be obtained from the Manufacturer's (Millipore Corp) instructions.
17. For the mammalian cell transduction setup the exact cell densities and pre-transduction incubation periods will vary depending on cell line used. One must keep in mind the possible level of nutrient depletion during this pre-transduction period and must take precautions to ensure that eventual recombinant protein production is not stalled due to the lack of a limiting substrate. Adequate results were obtained with HEK293A cells when pre-transduction time ranged from 2 until over 24 h.
18. Mammalian cells should be left at least 8 hours for effective adherence to culture plates to ensure beginning of growth. Please note that some cells such as suspension cells may not adhere. If cells do not adhere transfection may be carried out as usual but the medium containing the baculovirus may be removed by a centrifugation step whereby the cells are pelleted at 700–1000g are resuspended in fresh medium.
19. The MOI is defined as the ratio of pfu per cell at the time of infection or number of baculovirus per cell in the case of transfection of the baculovirus into the mammalian cells. Briefly, the likelihood of a cell being infected can be described by a Poisson distribution given by e^{-MOI}; thus, an MOI of one will result in 36% of the cells remaining uninfected. Cells are also known to

be infected best when in exponential growth phase, presumably because dissolution of nuclear membranes is required for transduction. Thus the distribution of the cell population in different phases of growth may additionally play a role in achieving maximum transduction. Thus, it is essential to determine in shake flask experiments the MOI required necessary to maximize recombinant protein production.

20. These parameters may be modified as required for given cell line.
21. To ensure there is no O2 limitation and sufficient levels of glucose, a dissolved oxygen probe is available with this system and may be used. However, it is preferable to verify proper oxygen levels of the culture by monitoring lactate accumulation. The mixing in this bioreactor is generally sufficient and no oxygen limitations were observed for the HEK293A cells under the conditions described above. Depletion of glutamine would require additional supplementation and glutamate levels increase may be indicate extreme glutamine degradation. Increasing level of LDH would indicate increased stress in the culture and may indicate that decrease in shaking speed or tilt angle may be required.

References

1. Boyce FM, Bucher NLR. Baculovirus-mediated gene transfer into mammalian cells. Proceedings of the National Academy of Sciences of the United States of America 1996;93:2348–52.
2. Lehtolainen P, Tyynela K, Kannasto J, et al. Baculoviruses exhibit restricted cell type specificity in rat brain: a comparison of baculovirus- and adenovirus-mediated intracerebral gene transfer in vivo. Gene Therapy 2002;9: 1693–9.
3. Grassi G, Kohn H, Dapas B, et al. Comparison between recombinant baculo- and adenoviral-vectors as transfer system in cardiovascular cells. Archives of Virology 2006;151:255–71.
4. Martyn JC, Dong X, Holmes-Brown S, et al. Transient and stable expression of the HCV envelope glycoproteins in cell lines and primary hepatocytes transduced with a recombinant baculovirus. Archives of Virology 2007; 152:329–43.
5. Salminen M, Airenne KJ, Rinnankoski R, et al. Improvement in nuclear entry and transgene expression of baculoviruses by disintegration of microtubules in human hepatocytes. Journal of Virology 2005;79:2720–8.
6. Bilello JP, Cable EE, Myers RL, Isom HC. Role of paracellular junction complexes in baculovirus-mediated gene transfer to nondividing rat hepatopytes. Gene Therapy 2003;10:733–49.
7. Barsoum J, Brown R, Mckee M, Boyce FM. Efficient transduction of mammalian cells by a recombinant baculovirus having the vesicular stomatitis virus G glycoprotein. Human Gene Therapy 1997;8:2011–8.
8. Sandig V, Hofmann C, Steinert S, et al. Gene transfer into hepatocytes and human liver tissue by baculovirus vectors. Human Gene Therapy 1996;7:1937–45.
9. Nicholson LJ, Philippe M, Paine AJ, et al. RNA interference mediated in human primary cells via recombinant baculoviral vectors. Molecular Therapy 2005;11:638–44.
10. Lee HP, Ho YC, Hwang SM, et al. Variation of baculovirus-harbored transgene transcription among mesenchymal stem cell-derived progenitors leads to varied expression. Biotechnology and Bioengineering 2007;97:649–55.
11. Ho YC, Chung YC, Hwang SM, et al. Transgene expression and differentiation of baculovirus-transduced human mesenchymal stem cells. Journal of Gene Medicine 2005;7:860–8.
12. Airenne KJ, Hiltunen MO, Turunen MP, et al. Baculovirus-mediated periadventitial gene transfer to rabbit carotid artery. Gene Therapy 2000;7:1499–504.

13. Ma L, Tamarina N, Wang Y, et al. Baculovirus-mediated gene transfer into pancreatic islet cells. Diabetes 2000;49:1986–91.
14. Kinnunen K, Kalesnykas G, Mähönen AJ, et al. Baculovirus is an efficient vector for the transduction of the eye: comparison of baculovirus- and adenovirus-mediated intravitreal vascular endothelial growth factor D gene transfer in the rabbit eye. J Gene Med 2009.
15. Spenger A, Ernst W, Condreay JP, et al. Influence of promoter choice and trichostatin A treatment on expression of baculovirus delivered genes in mammalian cells. Protein Expression and Purification 2004;38:17–23.
16. Gheshlaghi R, Scharer JM, Moo-Young M, Douglas PL. Medium optimization for hen egg white lysozyme production by recombinant Aspergillus niger using statistical methods. Biotechnology and Bioengineering 2005;90: 754–60.
17. Berger J, Hauber J, Hauber R, et al. Secreted Placental Alkaline-Phosphatase - A Powerful New Quantitative Indicator of Gene-Expression in Eukaryotic Cells. Gene 1988;66:1–10.
18. Condreay JP, Witherspoon SM, Clay WC, Kost TA. Transient and stable gene expression in mammalian cells transduced with a recombinant baculovirus vector. Proceedings of the National Academy of Sciences of the United States of America 1999;96:127–32.
19. Jardin BA, Montes J, Lanthler S, et al. High cell density fed batch and perfusion processes for stable non-viral expression of secreted alkaline phosphatase (SEAP) using insect cells: Comparison to a batch Sf-9-BEV system. Biotechnology and Bioengineering 2007;97 :332–45.
20. Bedard C, Tom R, Kamen A. Growth, Nutrient Consumption, and End-Product Accumu-lation in Sf-9 and Bti-Eaa Insect-Cell Cultures - Insights Into Growth Limitation and Metabolism. Biotechnology Progress 1993;9: 615–24.
21. Elias CB, Carpentier E, Durocher Y, et al. Improving glucose and gluamiue metabolism of human HEK 293 and Trichoplusia ni insect cells engineered to express a cytosolic pyruvate carboxylase enzyme. Biotechnology Progress 2003;19:90–7.
22. Zeiser A, Bedard C, Voyer R, et al. On-line monitoring of the progress of infection in Sf-9 insect cell cultures using relative permittivity measurements. Biotechnology and Bioengineering 1999;63:122–6.
23. Zeiser A, Elias CB, Voyer R, et al. On-line monitoring of physiological parameters of insect cell cultures during the growth and infection process. Biotechnology Progress 2000; 16:803–8.
24. Jardin BA, Zhao Y, Selvaraj M, et al. Expression of SEAP (secreted alkaline phosphatase) by baculovirus mediated transduction of HEK 293 cells in a hollow fiber bioreactor system. Journal of Biotechnology 2008;135:272–80.
25. Janakiraman V, Forrest WF, Chow B, Seshagiri S. A rapid method for estimation of baculovirus titer based on viable cell size. Journal of Virological Methods 2006;132:48–58.

Chapter 5

Transfection of Difficult-to-Transfect Primary Mammalian Cells

Oliver Gresch and Ludger Altrogge

Abstract

Primary cells are a valuable tool for researchers and are often preferred over transformed or immortalized cell lines since they are biologically more relevant and resemble the in vivo situation much closer. Unfortunately, efficient gene transfer in primary cells is still limited. Whereas viral strategies are time consuming and involve safety risks, nonviral methods are often inefficient for most primary cells. Nucleofection has been proven to overcome these limitations. Here, we describe the Nucleofection protocol for efficient transfection of human umbilical vein endothelial cells. Using a combination of a cell type-specific solution and electrical conditions, transfection efficiencies up to 90% can be achieved while survival rate is more than 70%.

Key words: Transfection, Primary cells, Nucleofection, HUVEC, Luciferase

1. Introduction

In the past, mainly cell lines have been used for transfection studies since they are easily available and usually more amenable to transfection. However, cell lines are artificial model systems that do not necessarily reflect the biochemical status of a given primary cell. Transfection of primary mammalian cells is an essential tool for scientific and therapeutical applications, such as functional genomics, drug development, and gene-based medicine (1–3). Transient transfection of primary cells can be achieved by several methods, including viral methods (e.g., retroviruses, adenoviruses) and nonviral techniques. Viral vectors usually result in high gene transfer efficiencies, but suffer from limitations, such as the time-consuming production of vectors, high safety requirements, and potential immunogenic reactions in clinical applications (4). However, none

James L. Hartley (ed.), *Protein Expression in Mammalian Cells: Methods and Protocols*, Methods in Molecular Biology, vol. 801,
DOI 10.1007/978-1-61779-352-3_5,

of the classical nonviral transfection methods, such as lipofection and electroporation, gains high transfection rates combined with low posttransfection mortality and a good preservation of cell-specific functionality. The Amaxa Nucleofector Technology offers the first nonviral, easy-to-use transfection method unifying these three aspects (5–10). It is based on the combination of dedicated cell type-specific Nucleofector Solutions together with optimized and unique electrical parameters for each cell type. This combination ensures transfer of the DNA not only in the cytoplasm, but as well in the nucleus of the cells. Thus, resting cells, such as primary T cells or neurons, can be transfected efficiently. Here, we describe an optimized protocol for transfection of primary human umbilical vein endothelial cells (HUVECs). HUVECs are a commonly used endothelial cell type for various research areas, such as cardiovascular diseases and cancer (11, 12). For example, HUVECs are popular primary cells to study highly regulated processes, such as tumor angiogenesis (13) (see Note 1).

2. Materials

2.1. Cell Culture Materials

1. Cryopreserved HUVEC (Lonza, cat. no. CC-2519; see Note 2). Within the vial, there are 5×10^5 cells which can be expanded several times before Nucleofection.
2. For cultivation of cells: EGM-2 BulletKit (Lonza) containing basal medium and supplements.
3. Six-well culture dish (Corning, cat. no. 3516).
4. 50-ml Tubes (Corning, cat. no. 430829).
5. 5-ml Reaction tubes (Sarstedt, cat. no. 72.690).
6. ReagentPack Subculture Reagents (Lonza, cat no. CC-5034), including trypsin/EDTA, trypsin-neutralizing solution, and HEPES-buffered saline solution (HBSS).
7. Phosphate-buffered saline (PBS) (Lonza, cat. no. 17-516F).
8. BSA (Sigma–Aldrich, cat. no. A7906).
9. Pipetman P1000 Micropipette (Gilson, cat. no. F123602).
10. D1000 Tipack tips for Micropipette (Gilson, cat. no. F161672).

2.2. Transfection Materials

1. Nucleofector II device (Lonza, cat. no. AAD-1001).
2. Nucleofector Kit for HUVECs (Lonza, cat. no. VPB-1002) containing Nucleofector solution, supplement, pmaxGFP expression plasmid, cuvettes, and pipettes. Nucleofector solution, supplement, and pmaxGFP should be stored at 4°C. For long-term storage, pmaxGFP is ideally stored at –20°C. Once the Nucleofector supplement is added to the Nucleofector solution, it is stable for 3 months at 4°C (see Note 3).

3. DNA plasmids: You use about 0.5–5 μg of plasmid DNA per reaction. DNAs should be in deionized water or TE buffer (10 mM Tris–HCl, 1 mM EDTA, pH 8.0) with a concentration of 1–5 mg/ml.
4. pmaxGFP (Lonza, provided with the Nucleofector Kit).
5. pRL-CMV-Renilla (Promega, cat. no. E2261).
6. Your experimental plasmid (see Note 4).

2.3. Analysis

1. Fluorescence microscope.
2. Flow cytometer.
3. Luciferase assay (14).
4. Dual-Glo luciferase assay system (Promega) containing Dual-Glo luciferase buffer, substrate (lyophilized), Stop & Glo buffer, and Stop & Glo substrate. Store Dual-Glo substrates at −20°C and Dual-Glo buffers below 25°C (see Note 5). The lyophilized Dual-Glo luciferase substrate contains dithiothreitol (DTT) and is, therefore, classified as hazardous. The reconstituted reagent is not known to present a hazard, as the concentration of DTT is less than 1%. However, we recommend the use of gloves, lab coats, and eye protection when working with these and all chemical reagents. In addition, the Dual-Glo Stop & Glo substrate contains ethanol, a highly volatile solvent. Pipet carefully and close the cap tightly after use.
5. 96-Well plates with white flat bottom (Corning, cat. no. 3917).

3. Methods

HUVECs are cultured as adherent cells in 25-cm^2 flasks. On the day of transfection, cells are trypsinized, counted, and resuspended with DNA in specific Nucleofector solution. After Nucleofection with the appropriate electrical program, cells are cultured in 6-well dishes until analysis.

3.1. Preparation of HUVECs

1. To set up cultures, calculate the number of vessels needed based on the recommended seeding density and the surface area of the vessels being used (see below). Add the appropriate amount of medium to the vessels (1 ml/5 cm^2) and allow the vessels to equilibrate in a 37°C, 5% CO_2, humidified incubator for at least 30 min. Do not seed cells into a well plate directly out of cryopreservation.
2. Wipe cryovials with ethanol or isopropanol before opening to avoid contamination of cells. In a sterile field, briefly twist the cap a quarter turn to relieve pressure and then retighten. Quickly thaw the cryovial in a 37°C water bath being careful not to submerge the entire vial. Watch your

cryovial closely; when the last sliver of ice melts, remove the vial from the water bath. Do not submerge it completely. Thawing the cells for longer than 2 min results in less than optimal results.

3. Immediately resuspend the cells in the cryovial and using micropipette, dispense the calculated number of cells into the prewarmed culture vessels set up earlier. Gently rock the culture vessel to evenly distribute the cells and return to the incubator (see Note 6).

3.2. Cell Culture

1. The optimal seeding conditions are 5–6 × 10^4 cells per 25-cm^2 flask.
2. Medium should be replaced 2–3 times per week. Use 2–3 ml medium per 25-cm^2 flask. The cells should be passaged after reaching 80–90% confluency. HUVECs are adherent cells, so trypsinization is required.
3. Cells should be preferably passaged 2 days before Nucleofection (see Note 7). The optimal confluency prior to Nucleofection is 90%.

3.3. Trypsinization

1. Remove medium from the cultured cells and wash cells once with HBSS; use at least the same volume of HBSS as culture medium.
2. For harvesting, incubate the cells ~1–3 min at 37°C with recommended volume of indicated trypsinization reagent (see Subheading 2, above). If necessary, prolong the incubation time for two more minutes at 37°C.
3. Neutralize trypsinization reaction with trypsin-neutralizing solution once the majority of the cells (>90%) have been detached.

3.4. Transfection

Cultivate the required number of cells (5 × 10^5 cells per sample) as follows.

1. Prepare 0.5–5 μg plasmid DNA per sample (see Note 4) in deionized water or TE buffer (10 mM Tris–HCl, 1 mM EDTA, pH 8.0) in a concentration of 1–5 mg/ml.
2. Add the entire Nucleofector supplement to the Nucleofector solution and prewarm to room temperature (see Note 3). This should be done by standing at room temperature.
3. Prepare 6-well plates by filling appropriate number of wells with 1.5 ml of supplemented basal medium and preincubate/equilibrate plates in a humidified 37°C/5% CO_2 incubator for at least 20 min.
4. Harvest the cells by trypsinization (see Subheading 3.3).
5. Count an aliquot of the trypsinized cells and determine cell density.
6. Centrifuge the required number of cells (5 × 10^5 cells per sample) at 90 × *g* for 10 min at room temperature. Carefully remove and discard supernatants using a pipet.

7. Resuspend the cell pellet carefully in 100 μl room temperature Nucleofector solution per sample by pipetting up and down twice.
8. Combine 100 μl of cell suspension with 0.5–5 μg DNA (see Note 8). Final volume should not exceed 110 μl.
9. Transfer cell/DNA suspension into certified Nucleofection cuvette; sample must cover the bottom of the cuvette without air bubbles.
10. Select Nucleofector program A-034 for HUVECs from Lonza or U-001 as an alternative for self-isolated HUVECs (see Note 9).
11. Insert the cuvette with cell/DNA suspension into the Nucleofector cuvette holder and apply the selected program.
12. Take the cuvette out of the holder once the program is finished.
13. Add ~500 μl of the preequilibrated culture medium from above to the cuvette and gently transfer the sample immediately into the six-well plate (final volume 1.5 ml media per well/sample). Use the supplied pipettes and avoid repeated aspiration of the sample.
14. Incubate the cells in a humidified, 37°C/5%, CO_2 incubator until analysis. Gene expression or downregulation, respectively, is often detectable after only 4–8 h. Normally, cells are analyzed after a 24- to 48-h incubation period (see Note 10).

3.5. Analysis by Fluorescence Microscopy (See Fig. 1)

1. Remove culture medium from each well.
2. Wash transfected cells with culture medium or PBS.
3. Cover transfected cells with fresh culture medium.
4. Depending on the reporter plasmid, expression can be examined by fluorescence microscopy directly (e.g., maxGFP).

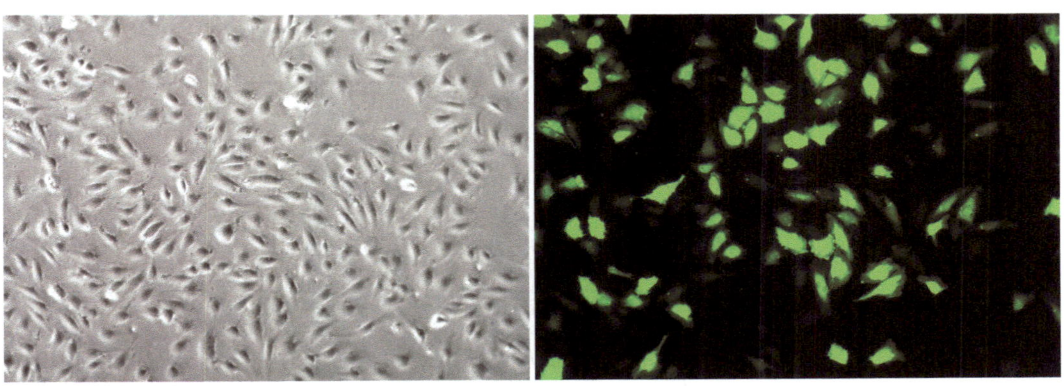

Fig. 1. HUVECs were transfected with 2 μg pmaxGFP using the HUVEC Nucleofector Kit and program A-034. Twenty-four hours post Nucleofection, cells were analyzed by light microscopy (*left*) or fluorescence microscopy (*right*).

3.6. Analysis by Flow Cytometry for Measuring GFP and Viability (See Fig. 2)

1. Detach cells via trypsinization (see Subheading 3.3).
2. Rinse cells from culture plate and transfer into a 1.5-ml reaction tube.
3. Centrifuge at 90 × *g* for 10 min at 4°C.
4. Remove supernatant and resuspend pellet in 400 μl PBS containing 0.5% BSA.
5. Add 1 μl propidium iodide (10 μg/ml) to stain dead cells and mix well.
6. Analyze samples by flow cytometry.

3.7. Analysis by Flow Cytometry for Antibody Staining

1. Detach cells via trypsinization (see Subheading 3.3).
2. Rinse cells from culture plate and transfer into a 1.5-ml reaction tube.
3. Centrifuge at 90 × *g* for 10 min at 4°C.
4. Remove supernatant and resuspend pellet in 100 μl of a fluorescent dye-conjugated antibody against the respective surface marker appropriately diluted in PBS containing 0.5% BSA.
5. Incubate for 10 min on ice in the dark.
6. Wash each sample with 1 ml PBS containing 0.5% BSA.
7. Centrifuge at 90 × *g* for 10 min at 4°C.

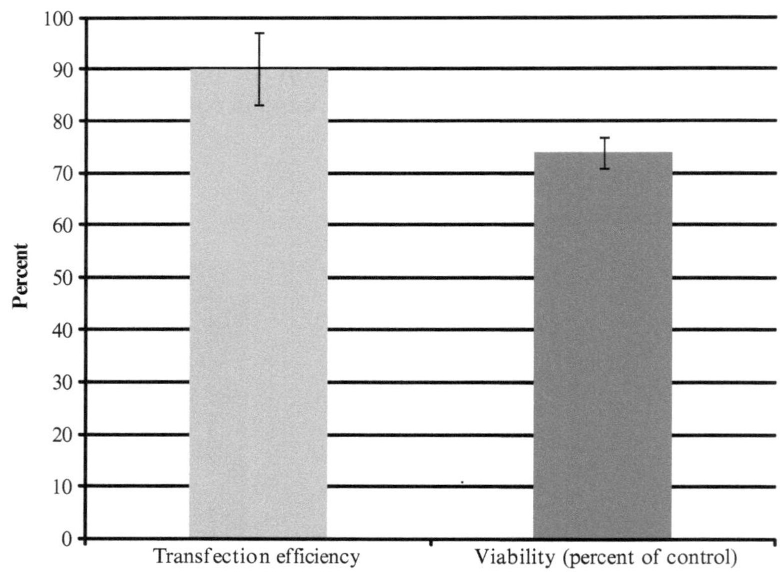

Fig. 2. Analysis of transfected HUVECs by flow cytometry. HUVECs were transfected with 2 μg pmaxGFP using the HUVEC Nucleofector Kit and program A-034 and analyzed at 24 h post transfection.

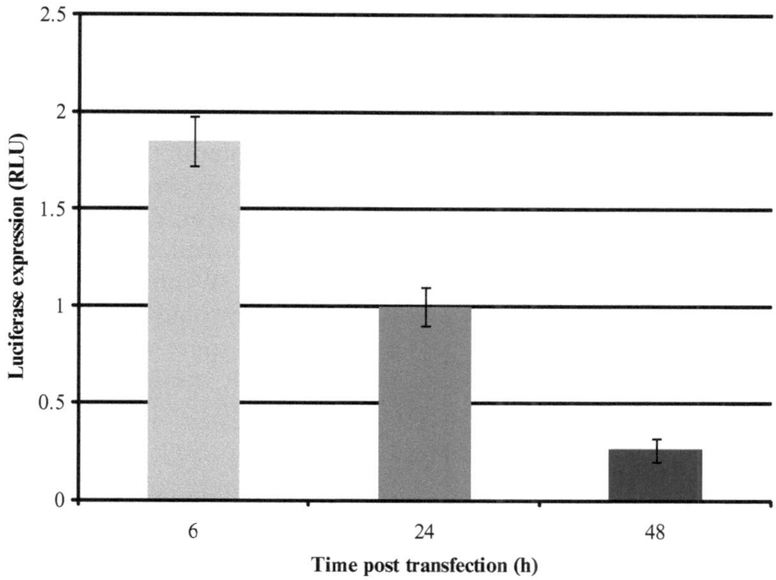

Fig. 3. Analysis of transfected HUVECs by luciferase assay. HUVECs were transfected by Nucleofection with the Renilla luciferase expression plasmid pRL-CMV (Promega). Luciferase expression was analyzed 6, 24, and 48 h post Nucleofection by Dual-Glo Assay (Promega). The 24-h value was set to 1.

8. Remove supernatant and resuspend pellet in 400 μl PBS containing 0.5% BSA per sample.
9. Analyze samples by flow cytometry.

3.8. Luciferase Assay

1. Refer to Fig. 3 and Note 10.
2. Transfer the contents of one bottle of Dual-Glo luciferase buffer to one bottle of Dual-Glo luciferase substrate to create the Dual-Glo luciferase reagent. Mix by inversion until the substrate is thoroughly dissolved (see Note 11).
3. Calculate the amount of Dual-Glo Stop & Glo reagent needed to perform the desired experiments. Dilute the Dual-Glo Stop & Glo substrate 1:100 into an appropriate volume of Dual-Glo Stop & Glo buffer in a new container (see Note 12).
4. Remove culture plates containing mammalian cells from the incubator. Make certain that the plates are compatible with the type of luminometer being used. We recommend 96-well plates with white flat bottom (Corning, cat. no. 3917).
5. Measuring Renilla luciferase activity: Add a volume of Dual-Glo Stop & Glo reagent equal to the original culture medium volume to each well and mix (see Note 13).
6. Wait at least 10 min, and then measure luminescence. Optimal results will be generated if the luminescence is measured within 2 h of addition of Dual-Glo Stop & Glo reagent.

4. Notes

1. A common experimental scheme using HUVECs is to test the activity of a promoter in the presence or absence of an agonist. The Dual-Glo system (Promega) is described below because it allows the activity of an experimental promoter (driving firefly luciferase) to be normalized against a control promoter (driving Renilla luciferase). Whatever experimental design is adopted, one should bear in mind that transfection rates from reaction to reaction can be quite varied. Controls for this variation are essential for reliable results.
2. HUVECs are tested negative for HIV-1, mycoplasma, Hepatitis-B, Hepatitis-C, bacteria, yeast, and fungi. Cell viability, morphology, and proliferative capacity are measured after recovery from cryopreservation.
3. The ratio of Nucleofector solution to supplement is 4.5:1. For a single reaction, use 18 μl of supplement plus 82 μl of solution to make 100 μl total reaction volume.
4. The quality and the concentration of DNA used for Nucleofection in general play a central role for the efficiency of gene transfer. We strongly recommend using endotoxin-free prepared DNA. Endotoxin-free Kits are available from several suppliers (such as QIAGEN EndoFree Plasmid Kit, cat. no. 12362). The presence of endotoxins can increase cell mortality and this is especially true of sensitive cells, such as primary neurons.
5. Buffer storage at room temperature is recommended to prevent the need for temperature equilibration when the reagents are reconstituted. Use the reconstituted Dual-Glo luciferase reagent on the day it is prepared or store at −70°C after preparation for up to 1 month. Prepare the Dual-Glo Stop & Glo reagent on the day it is to be used.
6. Centrifugation should not be performed to remove cells from cryoprotectant cocktail. This action is more damaging than the effects of residual DMSO in the culture.
7. Do not use cells after passage number 10 as this may result in substantially lower gene transfer efficiency and viability.
8. It is recommended to use 2 μg pmaxGFP or appropriate amount of siRNA (30–300 nM or 3–30 pmol/sample) or other substrates.
9. It is recommended to test two Nucleofector programs when using self-isolated HUVECs: A-034 and U-001 in parallel samples, as U-001 has shown sometimes higher transfection efficiency and/or viability. For HUVECs from Lonza, it is always recommended to use program A-034 only.

10. Kinetics of gene expression are significantly different between luciferase and GFP. It is highly recommended to perform luciferase analysis 6 to 16 h post Nucleofection since expression levels usually decrease after this time frame. The optimal time point for GFP expression is 10 to 48 h post Nucleofection.
11. Light intensity is a measure of the rate of catalysis by the luciferases and is, therefore, temperature sensitive. The temperature optimum for the activity of both luciferases is approximately room temperature (20–25°C), so it is important that the reagents be equilibrated to room temperature before beginning measurements. To avoid the need to temperature equilibrate reagents before use, store the Dual-Glo Luciferase buffer and the Dual-Glo Stop & Glo buffer at room temperature. If reagents are colder than room temperature, place them in a room-temperature water bath to equilibrate before use.
12. Assay reagents are stable at room temperature for several hours. Freezing the reagent can reduce the loss of activity of the Dual-Glo Luciferase reagent. Do not thaw the reconstituted reagent at temperatures above 25°C. Mix well after thawing. The most convenient and effective method for thawing is to place the reagent in a room-temperature water bath. Prepare only the amount of Dual-Glo Stop & Glo reagent required. For best results, prepare the Dual-Glo Stop & Glo reagent immediately before use.
13. Dual-Glo Stop & Glo reagent should be added to plate wells within 4 h of addition of Dual-Glo Luciferase reagent.

References

1. Haleem-Smith, H., Derfoul, A., Okafor, C., et al. (2005) Optimization of high-efficiency transfection of adult human mesenchymal stem cells in vitro. Mol. Biotechnology 30, 9–20.
2. Kimmelman, J. (2005) Recent developments in gene transfer: risk and ethics. BMJ 330, 79–82.
3. Williams, D., and Baum, C. (2003) Gene Therapy – New Challenges Ahead. Science 302, 400.
4. Thomas, C. E., Ehrhardt, A., and Kay, M. A. (2003) Progress and problems with the use of viral vectors for gene therapy. Nat. Rev. Genet. 4(5), 346–58.
5. Lakshmipathy, U., Buckley, S., and Verfaillie, C. (2007) Gene transfer via nucleofection into adult and embryonic stem cells. Methods Mol. Biol. 407, 115–26.
6. Zeitelhofer, M., Vessey, J.P., Xie, Y., et al. (2007) High-efficiency transfection of mammalian neurons via nucleofection. Nat. Protoc. 2(7),1692–704.
7. Landi, A., Babiuk, L.A., and van Drunen Littel-van den Hurk S. (2007) High transfection efficiency, gene expression, and viability of monocyte-derived human dendritic cells after nonviral gene transfer. J Leukoc Biol. 82(4), 849–60.
8. Gerber, S.A., and Pober, J.S. (2008) IFN-alpha induces transcription of hypoxia-inducible factor-1alpha to inhibit proliferation of human endothelial cells. J. Immunol. 181(2), 1052–62.
9. Kapoor, M., Zhou, Q., Otero, F., et al. (2008) Evidence for annexin II/S100A10 complex and plasmin in mobilization of cytokine activity of human TrpRS. J. Biol. Chem. 283(4), 2070–2077.
10. Gresch, O., Engel, F.B., Nesic, D., et al. (2004) New non-viral method for gene transfer into primary cells. Methods 33, 151–163.
11. Browne, C.D., Hindmarsh, E.J., and Smith, J.W. (2006) Inhibition of endothelial cell

proliferation and angiogenesis by orlistat, a fatty acid synthase inhibitor. FASEB J. 20(12), 2027–35.

12. Benny, O., Fainaru, O., Adini, A., et al. (2008) An orally delivered small-molecule formulation with antiangiogenic and anti-cancer activity. Nat Biotechnol. 26(7), 799–807.

13. Chen, Y., Wei, T., Yan, L. (2008) Developing and applying a gene functional association network for anti-angiogenic kinase inhibitor activity assessment in an angiogenesis co-culture model. BMC Genomics. 9, 264–279.

14. Fan, F., and Wood, K.V. (2007) Bioluminescent assays for high-throughput screening. Assay Drug Dev. Technol. 5, 127–136.

Chapter 6

Stable Protein Expression in Mammalian Cells Using Baculoviruses

Andreas Lackner*, Emanuel Kreidl*, Barbara Peter-Vörösmarty, Sabine Spiegl-Kreinecker, Walter Berger, and Michael Grusch

Abstract

The baculovirus *Autographa californica* multiple nuclear polyhedrosis virus (AcMNPV) has been widely used in biotechnology for protein expression in insect cells. Baculoviruses use arthropods as their natural hosts and are unable to replicate in mammalian cells. However, AcMNPV is able to enter many mammalian cell types and can be used for transgene expression if engineered to contain suitable expression cassettes. In this chapter, we describe the construction and application of a recombinant baculovirus containing a bicistronic expression cassette that can be used for stable protein expression in mammalian cells. As an example, the generation of glioblastoma and hepatocellular carcinoma cell lines stably expressing green fluorescent protein after puromycin selection is shown.

Key words: Baculovirus, BacMam, Stable protein expression, IRES, Green fluorescent protein

1. Introduction

Baculoviruses are a class of insect viruses with a large, double-stranded DNA genome. They have been used as biopesticides since the 1940s (1), and since the early 1980s the baculovirus *Autographa californica* multiple nuclear polyhedrosis virus (AcMNPV) has been engineered into one of the most widely used eukaryotic protein expression systems (2). Compared to expression in bacteria, protein folding and processing in insect cells more closely resemble the human situation, which is often crucial for protein function. Differences in glycosylation, however, exist between insect cells and

*Andreas Lackner and Emanuel Kreidl contributed to this chapter equally.

James L. Hartley (ed.), *Protein Expression in Mammalian Cells: Methods and Protocols*, Methods in Molecular Biology, vol. 801, DOI 10.1007/978-1-61779-352-3_6, © Springer Science+Business Media, LLC 2012

mammalian cells and may cause problems with correct functioning for some secreted proteins (3). In 1995 and subsequent years, it was discovered that AcMNPV, while being unable to replicate in mammalian cells or express proteins from the polyhedrin promoter commonly used for transgene expression in insect cell systems, is nevertheless able to enter hepatocytes and many other cultured mammalian cell types and can be used for efficient transgene expression when adapted to contain expression cassettes with suitable promoters, such as those from cytomegalovirus (CMV) or Rous sarcoma virus (RSV) (4–6). These baculoviruses intended for use in mammalian cells are often referred to as BacMam (3). Production and propagation of recombinant baculoviruses are relatively easy and cost-effective. Due to their inherent replication deficiency in mammalian cells, there are fewer biosafety concerns than with adeno- or lentiviral expression systems (7–9). AcMNPV is a lytic virus in insect cells and does not integrate into the host cell genome as part of its life cycle. Accordingly, most baculovirus applications in mammalian cells so far have used transient expression strategies (10). It has been previously demonstrated, however, that antibiotic selection is a feasible strategy to select for random integration of expression cassettes transduced by baculoviruses into mammalian genomes (11, 12). In order to be able to efficiently achieve stable protein expression in human cell lines, we have constructed a BacMam vector with a bicistronic expression cassette including a resistance marker for puromycin selection (13). Use of a bicistronic vector linking the gene of interest and the selection marker via an internal ribosome entry site (IRES) predisposes almost all puromycin-resistant cells to express recombinant protein and thus eliminates the tedious work of screening single clones (14, 15). When using IRES elements to express two open reading frames (ORFs) from a single mRNA, the lower expression level of the downstream cistron is sometimes seen as a disadvantage, and as an alternative it was recently shown that up to four proteins can be stoichiometrically coexpressed from a lentivirus vector by linking them together via autonomous "self-cleaving" 2A peptides (16). For our approach, higher expression of the upstream cistron containing the gene of interest seems an advantage rather than a disadvantage; nevertheless, incorporating polycistronic, self-cleaving, peptide-based, expression cassettes into baculovirus backbones is an attractive option for future vector design. The baculovirus vector described here is based on the widely used Bac-to-Bac insect expression system from Invitrogen and represents a simple and efficient tool for transient or stable protein expression in mammalian cell lines. We provide a detailed protocol for construction of the bicistronic BacMam virus and its application to obtain stable protein expression in glioblastoma and hepatocellular carcinoma cells.

2. Materials

2.1. Cloning of Bicistronic Transfer Vectors (pFIP, pFIP-GFP)

1. Plasmids: pIRESneo3, pPUR, pEGFP-N1 (Clontech), pFastbac1 (Invitrogen) (see Note 1).
2. Restriction enzymes: *Age*I, *Hpa*I (New England Biolabs), *Xba*I, *Cfr*9I, *Bcu*I, *Eco*47III, *Eco*RI, and *Sma*I (Fermentas), plus the corresponding restriction enzyme buffers supplied by the manufacturers.
3. Calf intestinal alkaline phosphatase (CIAP, Fermentas).
4. LE agarose (Biozym).
5. TBE buffer [0.5×: 45 mM Tris–HCl, 45 mM boric acid, 1 mM ethylenediaminetetraacetic acid (EDTA)].
6. Vistra green loading buffer: 333 μL/mL 6× loading dye (Fermentas), 250 μL/mL 80% glycerol, 66.5 μL/mL 0.5 M EDTA, 0.5 μL/mL 10,000× stock Vistra green (GE Healthcare) (see Note 2).
7. 1-kb DNA ladder (Fermentas).
8. Wizard SV gel and PCR clean-up system (Promega).
9. T4 DNA ligase 5 U/μL (Fermentas).
10. LB-broth (Sigma).
11. LB agar plates containing 50 μg/mL ampicillin (Roche).
12. JM-109 competent *Escherichia coli* cells (Promega).
13. Wizard Plus SV Miniprep DNA Purification System (Promega).

2.2. Making Competent DH10Bac Cells

1. DH10Bac Max efficiency competent cells (Invitrogen).
2. SOB-broth (20% tryptone, 0.5% yeast extract, 0.1% NaCl, 2.5 mM KCl in H_2O, adjust pH to 7.0 with NaOH, autoclave, and store at 4°C. Just before use, add $MgCl_2$ to a final concentration of 10 mM).
3. Kanamycin sulfate (Sigma), tetracycline hydrochloride (Sigma).
4. Inoue transformation buffer (ITB): 55 mM $MnCl_2{\cdot}4H_2O$, 15 mM $CaCl_2{\cdot}2H_2O$, 250 mM KCl, 10 mM PIPES (stock: 0.5 M, pH 6.7).
5. Dimethyl sulfoxide (DMSO) (Amresco).

2.3. Transformation of DH10Bac with Transfer Vector (pFIP-GFP)

1. Competent DH10Bac cells (from Invitrogen or made competent as described in Subheading 3.3).
2. SOC-broth: 1 mL 1 M sterile glucose solution added to 49 mL SOB-broth.

3. LB-broth (Sigma).
4. Agar (Fluka).
5. Gentamicin sulfate salt (Sigma–Aldrich).
6. LB agar plates containing 50 μg/mL kanamycin, 7 μg/mL gentamicin, 10 μg/mL tetracycline, 100 μg/mL 5-bromo-4-chloro-3-indolyl beta-D-galacto-pyranoside [X-Gal, 1 μl/mL of 10% X-Gal in dimethyl formamide (DMF)], 40 μg/mL isopropyl-β-D-1-thiogalactopyranoside (IPTG, 2 μL/mL of 0.1 M IPTG in H_2O).

2.4. Identification of Recombinant Bacmids by Colony PCR

1. Go-Taq DNA polymerase and 5× reaction buffer (Promega), dNTPs (Fermentas).
2. Forward primer: M13for: GTTTTCCCAGTCACGAC.
3. Reverse primer: gent_rev: ATCAGCCGGACTCCGATTACC (see Note 3).
4. TBE buffer, 6× Vistra green loading buffer, 1-kb DNA ladder (Fermentas).

2.5. Alkaline Lysis Midiprep of Recombinant Bacmids

1. LB-broth with 50 μg/mL kanamycin and 7 μg/mL gentamicin.
2. Solution I (15 mM Tris–HCl, pH 8.0, 10 mM EDTA, 100 μg/mL RNase A; filter sterilize and store at 4°C).
3. Solution II (0.2 N NaOH, 1% SDS; filter sterilize).
4. 3 M potassium acetate, pH 5.5 (autoclave and store at 4°C).

2.6. Sf9 Insect Cell Culture and Generation of Virus Stocks

1. Insect express Sf9-S2 medium with L-glutamine (PAA Laboratories) supplemented with 10% fetal calf serum (FCS; PAA Laboratories) (see Note 4).
2. Cellfectin (Invitrogen) (see Note 5).

2.7. Determination of Virus Titer

1. High pure viral nucleic acid kit (Roche).
2. Fast SYBR green master mix (Applied Biosystems).
3. Microamp fast optical 96-well reaction plates and microamp optical adhesive film (Applied Biosystems).
4. Forward QPCR primer: CMV_QPCR_for: CCCATAGTAACGCCAATAGG (see Note 6).
5. Reverse QPCR primer: CMV_QPCR_rev: CGTAGATGTACTGCCAAGTAGG.

2.8. Transduction of Mammalian Cells, Selection of Stable Clones, and Identification of GFP-Expressing Cells

1. H52 (established as described in ref. 17) and MGC (kindly provided by T. Kurata) glioblastoma cells.
2. Hep3B hepatocellular carcinoma cells (American Type Culture Collection, ATCC).
3. Minimal essential medium (MEM) and RPMI-1640 medium (Sigma) supplemented with 10% FCS (PAA Laboratories).

4. 25-cm^2 tissue culture flasks with filter cap (Greiner Bio-One).
5. Phosphate-buffered saline (PBS; 137 mM NaCl, 2.7 mM KCl, 10 mM Na_2HPO_4, 2 mM KH_2PO_4 in bidistilled H_2O, sterilize by autoclaving, cool to 4°C before use).
6. Solution of 0.1% trypsin (BD Biosciences) and 0.01% EDTA.
7. Puromycin (PAA Laboratories, store at −20°C as 1 mg/mL solution in bidistilled H_2O).

3. Methods

3.1. Construction of a Transfer Vector with a Bicistronic Mammalian Expression Cassette (pFIP)

The mammalian baculovirus expression system described here is based on the Bac-to-Bac insect cell expression system from Invitrogen. This system uses a transfer plasmid (pFastbac1) into which the gene of interest is cloned. The following section describes the cloning of a bicistronic mammalian expression cassette containing a CMV promoter, an IRES sequence, and a puromycin resistance gene into pFastbac1 to yield pFIP (pFastbac IRES puro).

1. To construct pIRESpuro by replacing the neomycin resistance gene from pIRESneo3 with the puromycin resistance gene from pPUR (see Note 7), digest pIRESneo3 with the restriction enzymes *Xba*I and *Cfr*9I and pPUR with *Xba*I and *Age*I.
2. Dephosphorylate the generated DNA ends of the pIRESneo3 digest with CIAP (see Note 8).
3. Separate the fragments resulting from the two digestions by agarose gel electrophoresis in TBE buffer, visualize the bands with Vistra green or other DNA stain, and cut out the gel slices containing the 4.4-kbp pIRES vector backbone without the neomycin resistance gene and the 0.8-kbp fragment containing the puromycin resistance gene. Purify the DNA from the gel slices with the Wizard SV gel and PCR cleanup kit and determine the DNA concentration of the purified products spectrophotometrically (OD_{260}).
4. Ligate the purified pIRES vector backbone and the insert containing the puromycin resistance gene with T4 DNA ligase using 100–200 ng DNA. Use a three- to fivefold molar excess of insert over vector in a reaction volume of 10 µL.
5. Transform 3–5 µL of the ligation reaction into competent JM-109 (or similar *E. coli* strain, see Note 9) and plate on agar plates containing 50 µg/mL ampicillin.
6. Pick 5–10 of the resulting colonies, isolate plasmid DNA with the Wizard Plus SV Miniprep kit according to the instructions of the manufacturer, and identify recombinant pIRESpuro plasmids by a diagnostic restriction digest and agarose gel

electrophoresis. Digestion with *Sma*I should yield two fragments of 4.6 and 0.6 kbp.

7. From pIRESpuro, cut out the entire bicistronic expression cassette containing the CMV promoter, the MCS, a synthetic intron, the IRES sequence, and the puromycin resistance gene with *Bcu*I (or *Spe*I) and *Dra*I and purify the 2.5-kbp fragment as described above. In parallel, digest pFastbac1 with BcuI and HpaI (removing a 183-bp fragment from the MCS), dephosphorylate, and purify the 4.6-kbp vector backbone.
8. Ligate the expression cassette into the pFastbac1 vector backbone, transform into JM-109, and identify colonies containing the recombinant pFIP plasmid. Digestion of pFIP with *Eco*RI yields two fragments of 6.4 and 0.75 kbp.
9. The MCS of pFIP contains unique recognition sequences for the restriction endonucleases *Cla*I, *Nhe*I, *Eco*47III, *Afl*II, *Hpa*I, and *Not*I. Since *Eco*47III and *Hpa*I are blunt-end cutters, any blunt-ended restriction fragment or PCR product (as generated by most proofreading polymerases) can readily be cloned into pFIP. A vector map of pFIP is shown in Fig. 1a.

3.2. Cloning of GFP into pFIP

1. Digest pEGFP-N1 with *Eco*47III and *Dra*I and purify the 0.86-kbp fragment containing the GFP ORF as in Subheading 3.1 (see Note 10). Digest pFIP with *Eco*47III, dephosphorylate, and purify as described above.
2. Ligate, transform, and identify recombinant pFIP-GFP as in Subheading 3.1. Digestion of pFIP-GFP with *Eco*RI yields three fragments of 6.4, 0.87, and 0.75 kbp.

3.3. Preparation of Competent DH10Bac Cells

The *E. coli* strain DH10Bac contains the bacmid and a helper plasmid coding for the Tn7 transposase. The bacmid is a ~140-kbp hybrid DNA molecule consisting of most of the AcMNPV genome (without the polyhedrin gene), a plasmid (mini-F) origin of replication, and a Tn7 insertion site in an ORF encoding LacZα. The Tn7 transposase mediates the insertion of the expression cassette from pFIP into the bacmid. Highly efficient competent DH10Bac cells can be purchased from Invitrogen (Max efficiency DH10Bac). The following steps can be used to produce DH10Bac with lower transformation efficiency that we have nevertheless found sufficient for the subsequent transposition reaction.

1. Inoculate 2 mL SOB-broth containing 50 μg/mL kanamycin and 10 μg/mL tetracycline with 1–10 μL Max efficiency DH10Bac.
2. Incubate on a shaker at 37°C overnight.
3. Add the 2 mL overnight culture to 48 mL SOB-broth containing kanamycin and tetracycline at same concentrations as above and incubate on a shaker at 37°C until the OD_{600} has reached

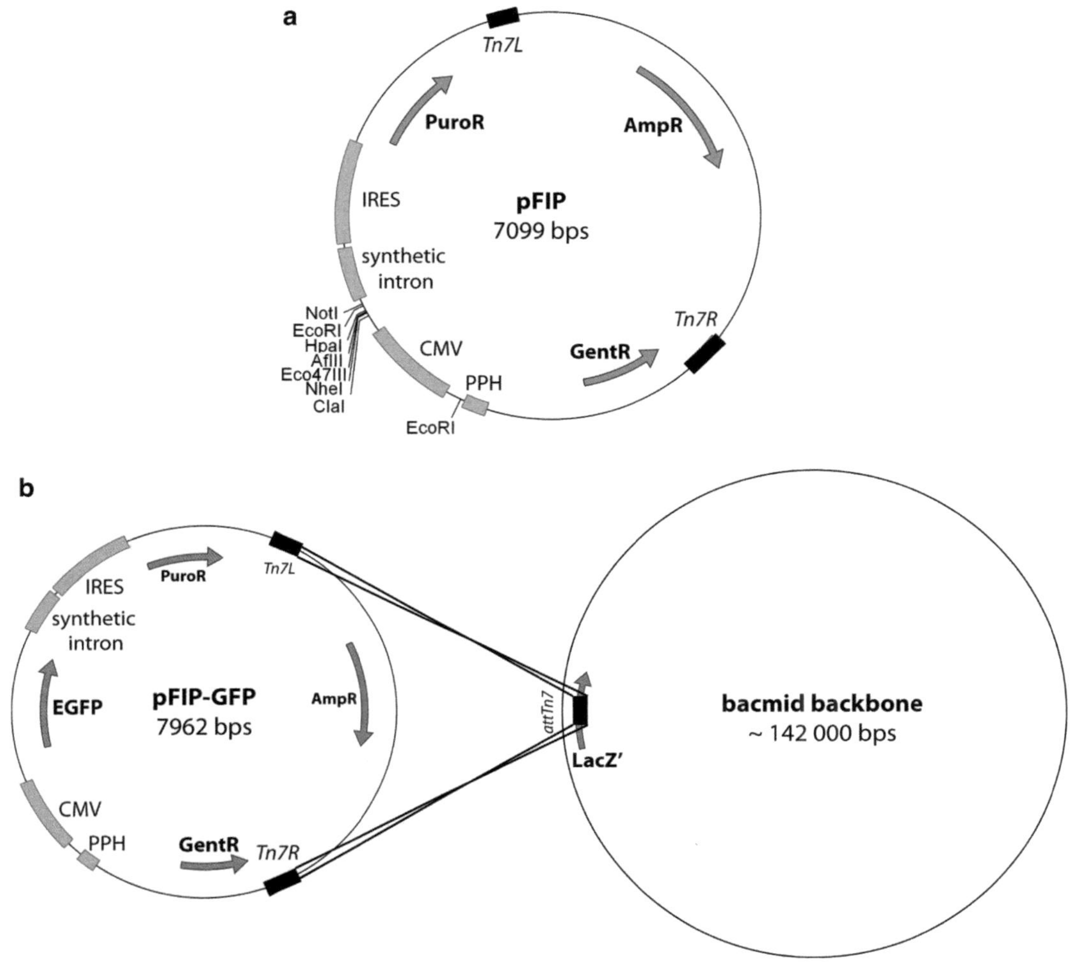

Fig. 1. Construction of baculoviruses with bicistronic expression cassettes for mammalian cells. (**a**) Annotated plasmid map of pFIP with positions of recognition sequences for single cutter restriction enzymes. *PPH* polyhedrin promoter, *CMV* CMV promoter, *IRES* internal ribosome entry site, *PuroR* puromycin resistance gene, *Tn7L* Tn7 transposon left arm, *AmpR* ampicillin resistance gene, *Tn7R* Tn7 transposon right arm, *GentR* gentamicin resistance gene. (**b**) Insertion of the expression cassette of pFIP-GFP into the bacmid by Tn7-mediated transposition.

0.55 (first measurement after 1 h 30 min, and then measure every 10–15 min).

4. Pellet cells at 2,500 × *g* and 4°C for 10 min.
5. Pour off supernatant, and vacuum aspirate remaining liquid.
6. Resuspend cells in 20 mL ice-cold ITB by swirling.
7. Pellet cells and discard supernatant as above and resuspend by swirling in 5 mL ITB.

8. Add 375 μL DMSO (dropwise), mix, and store on ice for 10 min.
9. Dispense aliquots (200–500 μL) into prechilled (5 min at −20°C) tubes.
10. Snap freeze each tube in liquid N_2 and store at −80°C.

3.4. Transformation of DH10Bac with pFIP-GFP

Competent DH10Bac cells are transformed with pFIP-GFP (or other gene of interest containing pFIP) resulting in transposition of the expression cassette and a gentamicin resistance marker into the bacmid. Successful recombinants are selected for their resistance to gentamicin and kanamycin. The protocol for transformation of pFIP constructs is similar to the one suggested by Invitrogen for pFastbac1 constructs. A graphic representation of the insertion of the expression cassette from pFIP-GFP into the bacmid is shown in Fig. 1b.

1. Thaw competent DH10Bac cells on ice, and transfer about 100 μL of the cell suspension for each construct to a chilled 1.5 mL tube.
2. Add 10–50 ng of pFIP-GFP plasmid DNA to the cell suspension, mix gently, and incubate on ice for 20–30 min.
3. Heat shock the cell suspension at 42°C in a heat block for 1 min.
4. Add 1 mL cold SOC medium and incubate the cell suspension at 37°C and 350 rpm for 4–5 h.
5. Harvest cells by centrifugation at 1,500 × *g* for 3 min at RT.
6. Pour off supernatant and resuspend cells in the remaining liquid (about 70 μL).
7. Plate cells on LB agar plates with kanamycin, tetracycline, gentamicin, IPTG, and X-Gal.
8. Incubate for 48 h at 37°C.

3.5. Identification and DNA Isolation of Recombinant Bacmid (FIP-Bac-GFP)

Typically, one to two dozen large colonies should be visible on the plate, some of them showing a blue color while others appear white. Integration of the expression cassette into the bacmid disrupts the ORF coding for the lacZα fragment, and the resulting bacterial colonies are no longer able to hydrolyse X-Gal to its blue metabolite and appear white (18) (see Note 11). Thus, these white colonies contain the desired recombinant bacmid.

1. To check correct transposition by PCR, set up one reaction mix (5 μL 5× GoTaq reaction buffer, 0.5 μL dNTPs (10 mM each), 0.5 μL M13for primer (20 μM), 0.5 μL gent_rev primer (20 μM), 0.6 U GoTaq DNA polymerase, nuclease-free water to 25 μL) per white colony you would like to test (five were usually sufficient in our hands) and one for a blue colony as

negative control. If available, also set up one PCR for a successfully created bacmid containing binding sites for the same primers as positive control.

2. Mark colonies on the back side of the Petri dish, and then transfer a bit of material into the appropriately labeled PCR tube.
3. Perform PCR with an initial denaturation of 95°C for 3 min, followed by 33 cycles with 30 s denaturation at 95°C, 30 s annealing at 60°C, and 1 min elongation at 72°C. For final elongation, use 72°C for 2 min.
4. Upon completion, add 6× Vistra green loading buffer to the PCRs and load onto a 1% agarose gel together with the DNA marker.
5. Depending on the dye used in the gel electrophoresis system, visualize nucleic acids either under UV light (e.g., ethidium bromide) or with a fluorescence scanning device (e.g., Vistra green).
6. Transformation of pFIP-GFP into DH10Bac results in formation of the recombinant bacmid FIP-Bac-GFP, which produces a PCR fragment of 0.8 kbp with primers M13for and gent_rev. Neither the transfer vector nor the unrecombined bacmid are amplified.
7. Once a successfully recombined bacmid has been identified, inoculate 100 mL of LB containing 50 μg/mL kanamycin, 7 μg/mL gentamicin, and 10 μg/mL tetracycline with bacteria from the corresponding colony on the Petri dish and incubate over night at 37°C and 200 rpm.
8. Mix 700 μL of the overnight culture with 300 μL 80% glycerol and store at −80°C as glycerol stock.
9. Pellet the remaining bacteria and extract the recombinant bacmid following a modified Midiprep procedure as described below.
10. Remove the supernatant by vacuum aspiration and resuspend the cell pellet in 10 mL of solution I. Gently vortex or pipet up and down to resuspend.
11. Add 10 mL of solution II and gently mix. Incubate at room temperature for 5 min.
12. Slowly add 10 ml of 3 M potassium acetate, pH 5.5, mixing gently during addition. A thick white precipitate of protein and *E. coli* genomic DNA forms. Place the sample on ice for 5–10 min.
13. Centrifuge for 10 min at 14,000 ×*g*.
14. Gently transfer the supernatant to a fresh tube containing 30 mL isopropanol. Do not transfer any white precipitate. Invert the tube a few times to mix and place on ice for 5–10 min or at −20°C overnight.
15. Centrifuge for 15 min at 14,000 ×*g* at room temperature.

16. Carefully remove the supernatant, taking care not to disturb the pellet. Add 10 mL of 70% ethanol. Invert the tube several times to wash the pellet.
17. Centrifuge for 10 min at 14,000 × *g* at room temperature.
18. Remove the supernatant, taking care not to lose the pellet. Air dry the pellet for 5–10 min at room temperature.
19. Dissolve the DNA pellet in 1 mL of 1× TE buffer, pH 8.0. To avoid shearing of the DNA, resuspend by gentle tapping or place at 4°C overnight. Do not vortex or pipet vigorously.
20. Store the DNA at 4°C (see Note 12).

3.6. Sf9 Insect Cell Culture

1. Sf9 cells are maintained in closed 25-cm² tissue culture flasks at 27°C in 5 mL insect express Sf9-S2 medium with 10% FCS.
2. For passaging, just remove the old growth medium and rinse the growth area of the flask several times with (the same) 5 mL of fresh medium. A considerable fraction of the cells detach without trypsinization.
3. Transfer 0.25–0.5 mL of the cell suspension to a fresh 25-cm² flask for maintenance and add 5 mL fresh medium.

3.7. Transfection of Sf9 Cells with FIP-Bac-GFP to Generate P1 Virus Stock

1. For transfection, transfer 2.5 mL of the cell suspension of Subheading 3.6 to a fresh 25-cm² flask and incubate for 1 h at 27°C (for cell attachment).
2. Meanwhile, dilute 15 μL bacmid DNA in 185 μL serum-free insect express Sf9-S2 medium.
3. Dilute 30 μL Cellfectin in 170 μL serum-free insect express Sf9-S2 medium.
4. Add Cellfectin dilution to DNA dilution.
5. Incubate for 45 min at RT.
6. Bring to a volume of 2 mL by adding 1.6 mL medium with 10% FCS.
7. Remove medium from the prepared 25-cm² flask and add transfection mixture.
8. Incubate at 27°C for 24 h and then change medium to fresh insect express Sf9-S2 with 10% FCS.
9. A further incubation of 5–6 days is necessary to get the first viral stock, referred to as P1 stock (see Note 13).
10. Harvest P1 virus stock by collecting cells and supernatant in a sterile 15 mL tube.
11. Centrifuge the P1 stock to pellet Sf9 cells and debris.
12. Transfer the supernatant containing the virus to a fresh tube and store at 4°C (see Note 14).

3.8. Virus Amplification to Generate P2 and P3 Stocks

1. To amplify the baculovirus, seed Sf9 cells in a 75-cm^2 flask and grow to about 50% confluence.
2. Add 5 mL P1 virus stock and incubate 5–7 days at 27°C. Occasional gentle agitation of the flask helps to distribute the virus.
3. Harvest and store the resulting P2 virus stock as described for the P1 stock (see Note 15).
4. To produce a P3 stock, transfer Sf9 cells grown in a 75-cm^2 flask to a 175-cm^2 flask and add fresh medium to a volume of 25 mL.
5. After attachment of the cells, add 25 mL P2 stock.
6. After another 5–7 days of incubation, harvest the P3 stock as described for the P1 stock.

3.9. Titer Determination of Virus Stocks

This section describes the use of quantitative real-time PCR (QPCR) for the determination of the virus titer in insect cell supernatants, as originally described by Lo and Chao (19). Alternatively, other methods, like plaque assays or endpoint assays (described in a step-by-step manner in ref. 20) or commercially available kits (described in ref. 21), can be used. In our experience, these methods, while delivering equally good results, are significantly more time consuming (several days as opposed to a few hours) and/or expensive. Derivation of the formula to calculate virus titers from the threshold cycle (Ct) values obtained by QPCR is shown in Fig. 2. We have used a serially diluted P3 stock of FIP-Bac-GFP, the titer of which was determined with the BaculoX rapid titer kit (Clontech) as described by the manufacturer to correlate Ct values to virus titer. QPCR was performed on an Applied Biosystems 7500 fast real-time PCR system with 7500 fast system SDS software. Other systems with similar specifications can of course also be used, but may require optimization of the conditions and adjustment of the parameters of the formula. Use of a Roche LightCycler instrument, for example, has been described by Lo and Chao (19).

1. Isolate viral DNA from 200 μL of cleared cell supernatant using high pure viral nucleic acid kit (Roche) according to the instructions of the manufacturer, eluting DNA in 50 μL of elution buffer. For a negative control, perform the isolation with the supernatant of insect cells not infected with a baculovirus (see Note 16).
2. Per sample to be tested, transfer 18 μL 2× Fast SYBR Green Master Mix and 1.5 μL of a 10 μM stock of the forward and reverse QPCR primer into an Eppendorf tube. Also prepare a reaction mix for the no-template control (NTC).
3. For each sample, transfer 7 μL of this preparation into three individual wells in the QPCR plate. Do the same for the NTC.

4. Add 5 μL of isolated viral DNA per well and mix by carefully pipetting up and down. For NTC, add 5 μL of water. Make sure to avoid or remove air bubbles.
5. Seal the plate with optical film.
6. Set up the QPCR system for the detection of SYBR green in fast 7500 mode.
7. Perform QPCR with initial denaturation at 95°C for 30 s, followed by 45 cycles of denaturation at 95°C for 3 s and annealing/extension at 60°C for 30 s.
8. Average the Ct values obtained in the triplicates from each sample. Use this value to calculate the virus titer (*c*) in plaque-forming units per mL (Pfus/mL) of the viral stocks by using the formula shown in Fig. 2c. P3 stocks of FIP-Bac-GFP contained between 0.5 and 3.0×10^8 Pfu/mL.

3.10. Transduction of Human Glioblastoma and Hepatocellular Carcinoma Cell Lines with FIP-Bac-GFP and Selection of Stable Clones

1. Seed $2.5–4 \times 10^5$ cells into 25-cm^2 flasks in appropriate growth medium (RPMI for H52 and Hep3B, MEM for MGC, see Note 17).
2. After 24 h, add P3 baculovirus stock at a multiplicity of infection (MOI = Pfu per cell) between 300 and 1,000 (see Note 18).
3. Change the growth medium 24–48 h after transduction and add puromycin at the appropriate concentrations (H52: 0.25 μg/mL; MGC: 0.6 μg/mL; Hep3B: 1 μg/mL, see Note 19).
4. Change selection medium every 3–5 days.
5. After 10 days to 2 weeks, selection is complete and the clone pools can be passaged.
6. From passage 5 on, it is sufficient to add puromycin to every third or fifth passage (see Note 20).

3.11. Flow Cytometry to Determine Protein Expression after Transduction with FIP-Bac-GFP

The effectiveness of transduction with FIP-Bac-GFP can be roughly assessed by fluorescence microscopy with GFP or FITC filters. To more accurately determine percentages of GFP-expressing cells in transiently transduced cultures or in stable clone pools after selection, flow cytometry can be used. In our hands, most glioblastoma and hepatocellular carcinoma cell lines were transduced with an efficiency of 80% and more. After puromycin selection, more than 90% of the cells showed GFP expression as shown for H52 (91%), MGC (99%), and Hep3B (99%) (see Note 21). Typical results of GFP detection by fluorescence microscopy and flow cytometry of stable clone pools of H52, MGC, and Hep3B cells are shown in Fig. 3. Once the selection process was completed, the percentage of GFP-expressing cells remained stable during prolonged passaging (>12 passages) and upon thawing cells after storage (>1 year) in liquid nitrogen.

a

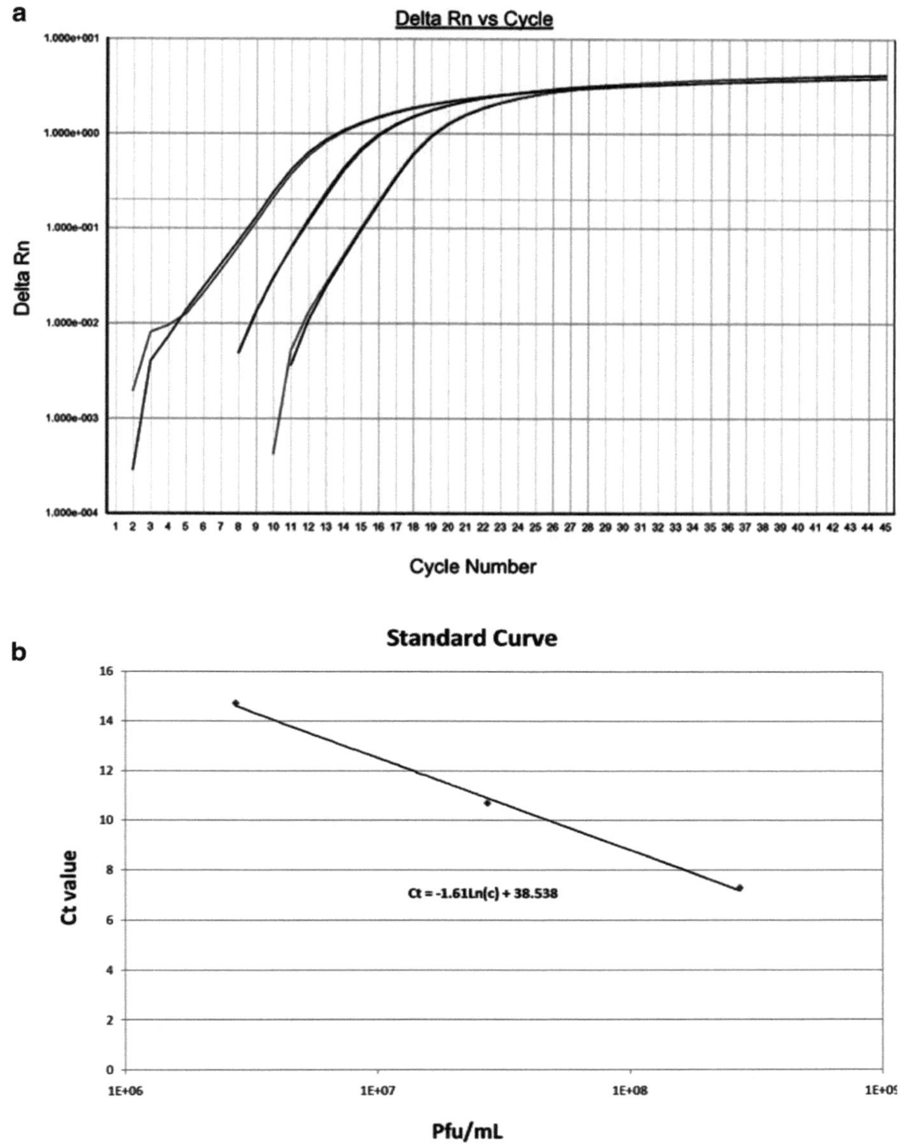

c

$$c = e^{-\left(\frac{Ct - 38.538}{1.6138}\right)}$$

Fig. 2. Titer determination of baculovirus stocks by QPCR. (**a**) Amplification plots of serial dilutions (10^{-1}, 10^{-2}) of a P3 stock of FIP-Bac-GFP of known titer. The Ct value is the cycle number at which an amplification plot reaches the set threshold level. (**b**) Standard curve obtained by plotting Ct values versus Pfu/mL. (**c**) Formula derived from the standard curve. *c* is the virus titer in Pfu/mL.

1. If transiently transduced cultures are measured, wash samples at least three times with PBS to remove remaining baculoviruses. If stable clone pools are measured, washing once with cold PBS is sufficient.

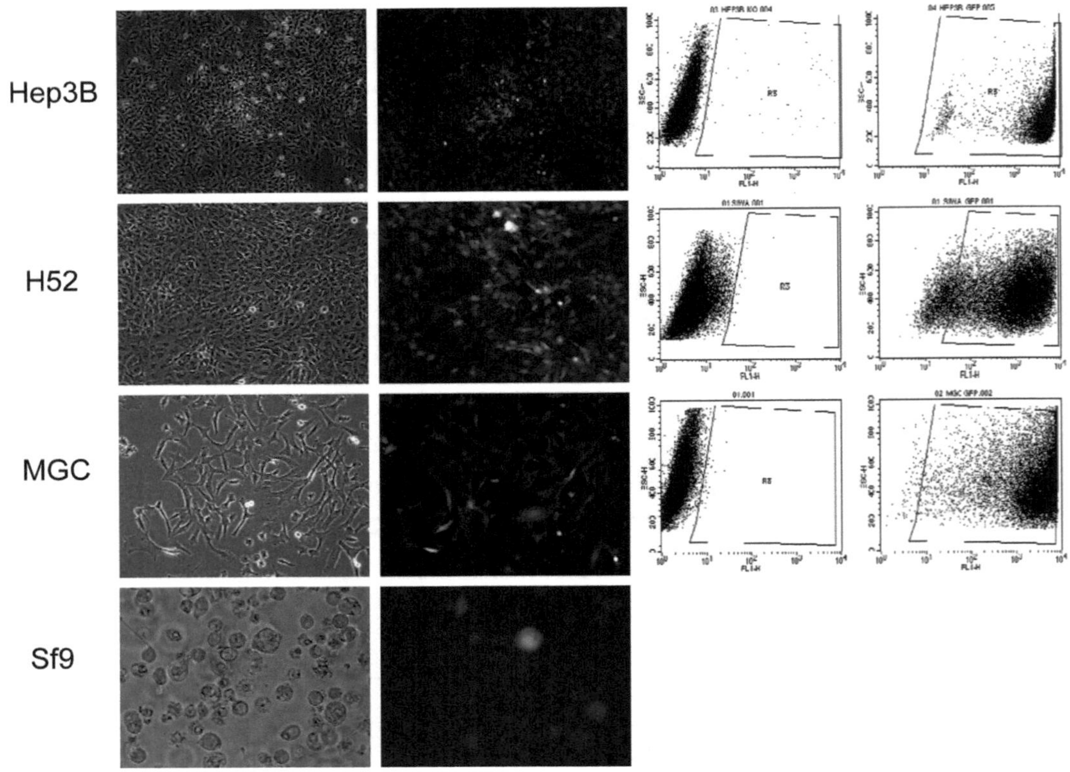

Fig. 3. Monitoring of GFP expression in clone pools of glioblastoma (H52, MGC) and hepatocellular carcinoma (Hep3B) cells and in Sf9 insect cells by fluorescence microscopy and flow cytometry. Panels from *left* to *right*: Micrographs of GFP-expressing cells in phase contrast, the same frames photographed with GFP filters, flow cytometry dot plots of GFP-negative control cells and of GFP-expressing clone pools.

2. For each cell line, use a nontransduced culture as negative control.
3. Trypsinize cells and resuspend in 5 mL medium containing 10% FCS to neutralize trypsin.
4. Centrifuge at 100–200 × *g* for 5 min, and discard supernatant.
5. Resuspend in 2 mL cold PBS, centrifuge at 100–200 × *g* for 5 min, discard supernatant, and resuspend in another 0.5 mL PBS.
6. Measure control sample (not expressing GFP) on a flow cytometry unit using GFP settings.
7. Set the gate for the negative controls to contain less than 1% false positives.
8. Measure percentages of GFP-positive cells.

4. Notes

1. The pFastbac1 plasmid, Max efficiency-competent DH10Bac cells, and Cellfectin transfection reagent for SF9 cells are all included in Invitrogen's Bac-to-Bac baculovirus expression kit, but can also be purchased separately.
2. If preferred, other dyes staining nucleic acids, like ethidium bromide, may similarly be used according to the suggestions of the manufacturer or standard protocols.
3. Alternatively, the primer combinations M13for × IRESrev (13) or M13for × M13rev (as described by Invitrogen in the Bac-to-Bac manual) can be used. While allowing to check the complete integration of the gene of interest, these primers generally require much longer elongation (and thus PCR) times or more expensive Long-PCR kits and might not always work in colony PCR.
4. Although Invitrogen recommends growing Sf9 cells in medium without FCS, cell growth was much better in serum-containing medium in our hands and virus production was not impaired.
5. While Cellfectin worked very well in our hands, other transfection reagents, like Fugene6 (Roche) and Turbofect (Fermentas), were equally effective.
6. The primers described here bind to the CMV promoter driving the expression of the gene of interest, allowing them to be used for the quantification of any piece of DNA containing this sequence. If you decide to use another promoter, design primers suitable for real-time PCR. Alternatively, primers binding directly to the bacmid backbone can be used as described by Lo and Chao (19). The disadvantage, however, is that wild-type viruses, if present, are amplified as well in the reaction.
7. The CMV promoter to drive expression and puromycin for selection of stable clones worked very well in our hands. Of course, the same principles as described here can also be applied to construct bicistronic expression vectors with different promoters or selection markers. For instance, if geneticin is preferred as selection marker, the entire expression cassette from pIRESneo3 can simply be cloned into pFastbac1.
8. Dephosphorylation may be omitted since *Xba*I and *Cfr*9I produce noncompatible DNA overhangs. This can, however, result in unwanted religation of copurified fragments, where only one of the enzymes has cut.
9. The *E. coli* strain JM-109 works very well for this purpose in our experience, but other *E. coli* strains, like DH5alpha or XL1-Blue, may be equally suitable. We would recommend to use only strains deficient in recombinase and endonuclease

activity (recA1, endA1) to avoid unwanted recombination events and degradation of plasmid DNA during preparation.

10. When cloning a gene of interest into pFIP, make sure that the ORF contains an ATG and optimally purines at −3 and +4 for good eukaryotic expression.
11. If it is not yet possible to distinguish between blue and white colonies, transfer 5-10 prospective white colonies to a fresh plate containing the same media additives using sterile toothpicks or pipette tips. Continue incubation at 37°C for another 24 h.
12. Do not store the purified bacmid DNA at −20°C as even a single freeze–thaw cycle may shear the DNA and strongly reduce transfection efficiency.
13. Virus infection and beginning cell lysis of the Sf9 cells should be visible microscopically 3–4 days after transfection depending on the transfection efficiency. Using the CMV promoter and GFP, we have observed weak but detectable transgene expression in the Sf9 cells by fluorescence microscopy (Fig. 3). Protein detection in Sf9 cells can, thus, be used to verify correct transgene expression before proceeding to further virus amplification. An aliquot of the P1 stock may also be used for an initial experiment in mammalian cells.
14. It is also possible to dispense P3 stocks into aliquots and store at −20°C alternatively to the storage at 4°C. Our experiments, though, showed no loss of activity when the virus stock was stored at 4°C for several months.
15. For the inexperienced user, it is advisable to quantify the amount of virus in the cell supernatant already at this stage to avoid using material that does not contain significant amounts of baculoviruses. In general, it is possible to verify successful amplification of the virus by the presence of a high number of lysed cells during the production of the P1 virus.
16. Kits by other manufacturers can probably also be used to give satisfactory results. Classical phenol/chloroform extraction, however, has been reported to not yield DNA of sufficient quality (19).
17. For infection of mammalian cells by baculoviruses, it is important that the cells are well-dispersed, as we have frequently observed lack of transduction in areas of the culture flasks where cells were closely packed.
18. Baculoviruses are able to transduce many cell lines with an efficiency of >80% when used at an MOI of 1,000. For selection of stable clones, however, lower MOIs and transient transduction rates are usually sufficient.
19. To determine the appropriate antibiotic concentration for selection, treat untransduced cells with a range of concentrations

(0.1–2 μg/mL puromycin is sufficient for most cell lines) and choose the lowest concentration killing all cells within a few days. Make sure that the culture plates are not confluent at the beginning of the selection phase as high cell densities may impair and slow down the selection process.

20. Some cell lines may, at varying frequencies, produce resistant clones which, despite the use of an IRES cassette, do not express GFP or another gene of interest. For those lines, isolation of single clones or cell sorting after the selection phase can eliminate resistant cells not expressing the gene of interest.
21. Similar results can be expected with other proteins but should be verified by immunological methods, like immunohistochemistry or Western analysis.

Acknowledgments

We thank Katharina Leopold for help with the plasmids and Irene Herbacek for performing the flow cytometry analyses.

References

1. Moscardi, F. (1999) Assessment of the application of baculoviruses for control of Lepidoptera. *Annu Rev Entomol* **44**, 257–89.
2. Luckow, V. A. (1993) Baculovirus systems for the expression of human gene products. *Curr Opin Biotechnol* **4**, 564–72.
3. Kost, T. A., Condreay, J. P., Jarvis, D. L. (2005) Baculovirus as versatile vectors for protein expression in insect and mammalian cells. *Nat Biotechnol* **23**, 567–75.
4. Boyce, F. M., Bucher, N. L. (1996) Baculovirus-mediated gene transfer into mammalian cells. *Proc Natl Acad Sci USA* **93**, 2348–52.
5. Hofmann, C., Sandig, V., Jennings, G., Rudolph, M., Schlag, P., Strauss, M. (1995) Efficient gene transfer into human hepatocytes by baculovirus vectors. *Proc Natl Acad Sci USA* **92**, 10099–103.
6. Kost, T. A., Condreay, J. P. (2002) Recombinant baculoviruses as mammalian cell gene-delivery vectors. *Trends Biotechnol* **20**, 173–80.
7. Kappes, J. C., Wu, X. (2001) Safety considerations in vector development. *Somat Cell Mol Genet* **26**, 147–58.
8. Losert, A., Mauritz, I., Erlach, N., Herbacek, I., Schulte-Hermann, R., Holzmann, K., *et al.* (2006) Monitoring viral decontamination procedures with green fluorescent protein-expressing adenovirus. *Anal Biochem* **355**, 310–2.
9. Volkman, L. E., Goldsmith, P. A. (1983) In Vitro Survey of Autographa californica Nuclear Polyhedrosis Virus Interaction with Nontarget Vertebrate Host Cells. *Appl Environ Microbiol* **45**, 1085–93.
10. Fornwald, J. A., Lu, Q., Wang, D., Ames, R. S. (2007) Gene expression in mammalian cells using BacMam, a modified baculovirus system. *Methods in molecular biology (Clifton, NJ)* **388**, 95–114.
11. Condreay, J. P., Witherspoon, S. M., Clay, W. C., Kost, T. A. (1999) Transient and stable gene expression in mammalian cells transduced with a recombinant baculovirus vector. *Proc Natl Acad Sci USA* **96**, 127–32.
12. Merrihew, R. V., Clay, W. C., Condreay, J. P., Witherspoon, S. M., Dallas, W. S., Kost, T. A. (2001) Chromosomal integration of transduced recombinant baculovirus DNA in mammalian cells. *J Virol* **75**, 903–9.
13. Lackner, A., Genta, K., Koppensteiner, H., Herbacek, I., Holzmann, K., Spiegl-Kreinecker, S., *et al.* (2008) A bicistronic baculovirus vector for transient and stable protein expression in mammalian cells. *Anal Biochem* **380**, 146–8.
14. Gurtu, V., Yan, G., Zhang, G. (1996) IRES bicistronic expression vectors for efficient creation of stable mammalian cell lines. *Biochem Biophys Res Commun* **229**, 295–8.

15. Rees, S., Coote, J., Stables, J., Goodson, S., Harris, S., Lee, M. G. (1996) Bicistronic vector for the creation of stable mammalian cell lines that predisposes all antibiotic-resistant cells to express recombinant protein. *Biotechniques* **20**, 102–4, 6, 8–10.
16. Carey, B. W., Markoulaki, S., Hanna, J., Saha, K., Gao, Q., Mitalipova, M., *et al.* (2009) Reprogramming of murine and human somatic cells using a single polycistronic vector. *Proc Natl Acad Sci USA* **106**, 157–62.
17. Spiegl-Kreinecker, S., Pirker, C., Marosi, C., Buchroithner, J., Pichler, J., Silye, R., *et al.* (2007) Dynamics of chemosensitivity and chromosomal instability in recurrent glioblastoma. *Br J Cancer* **96**, 960–9.
18. Barry, G. F. (1988) A broad-host-range shuttle system for gene insertion into the chromosomes of gram-negative bacteria. *Gene* **71**, 75–84.
19. Lo, H. R., Chao, Y. C. (2004) Rapid titer determination of baculovirus by quantitative real-time polymerase chain reaction. *Biotechnol Prog* **20**, 354–60.
20. Bernard, A., Payton, M., Radford, K. R. (2001) Protein expression in the baculovirus system. *Curr Protoc Neurosci* **Chapter 4**, Unit 4 19.
21. Kitts, P. A., Green, G. (1999) An immunological assay for determination of baculovirus titers in 48 hours. *Anal Biochem* **268**, 173–8.

Chapter 7

Using Matrix Attachment Regions to Improve Recombinant Protein Production

Niamh Harraghy, Montserrat Buceta, Alexandre Regamey, Pierre-Alain Girod, and Nicolas Mermod

Abstract

Chinese hamster ovary (CHO) cells are the system of choice for the production of complex molecules, such as monoclonal antibodies. Despite significant progress in improving the yield from these cells, the process to the selection, identification, and maintenance of high-producing cell lines remains cumbersome, time consuming, and often of uncertain outcome. Matrix attachment regions (MARs) are DNA sequences that help generate and maintain an open chromatin domain that is favourable to transcription and may also facilitate the integration of several copies of the transgene. By incorporating MARs into expression vectors, an increase in the proportion of high-producer cells as well as an increase in protein production are seen, thereby reducing the number of clones to be screened and time to production by as much as 9 months. In this chapter, we describe how MARs can be used to increase transgene expression and provide protocols for the transfection of CHO cells in suspension and detection of high-producing antibody cell clones.

Key words: Matrix attachment region, Recombinant protein production, Chinese hamster ovary cells, Antibody, IgG, ELISA

1. Introduction

From somewhat inauspicious beginnings in the 1980s, recombinant proteins have developed into a beneficial and profitable class of therapeutic products. The biopharmaceutical market was expected to approach or exceed $70 billion by the end of 2010 (1) and there is a growing need for clinical-grade proteins. Advances in genetic engineering techniques have opened the door to the expression of human molecules, with therapeutic antibodies representing the biggest group. However, antibodies are complex molecules that require post-translational modifications. Therefore, they cannot be produced in prokaryotic or simple eukaryotic systems. The most

James L. Hartley (ed.), *Protein Expression in Mammalian Cells: Methods and Protocols*, Methods in Molecular Biology, vol. 801, DOI 10.1007/978-1-61779-352-3_7,

extensively used system for the production of complex molecules, such as antibodies, is based on Chinese hamster ovary (CHO) cells. Although the productivity of CHO cells is less than that seen with prokaryotic systems, productivity from CHO cells has increased substantially over the last 20 years with modifications and improvements to vectors and cell culture conditions (2).

Following transfection, the gene of interest usually integrates randomly into the genome of the host cell. The vast majority of these recombination events occur in loci that are "silent", i.e. that are not transcribed to produce mRNAs, or alternatively they may disrupt the metabolism of the cell. Typically, hundreds to thousands of clones must be screened in order to isolate a few clones with an acceptable level of productivity. To increase the number of copies of the transgene, the dihydrofolate reductase (DHFR) gene amplification system has been used extensively (3). However, this system is time intensive, as each round of amplification takes weeks and several thousand clones must be screened because there is often significant heterogeneity in expression between cell clones. Moreover, the use of methotrexate, a mutagenic selection agent that can result in the amplification of up to thousands of copies of the transgene and induces chromosome breaks, often leads to genetic instability and unstable expression. Unstable expression has been linked to transgene copy loss. The selection pressure must, therefore, be maintained as removal quickly leads to a reduction in specific productivity by as much as 80% (4–6).

Other events, such as slow silencing effects and position effect variegation, are also known to contribute to clonal heterogeneity and expression instability ((7, 8), reviewed in ref. 9). The loss of transgene expression or poor expression is linked to changes in DNA methylation (10), propagation of heterochromatin, or insertion of the transgene in an unfavourable region (e.g. close to a negative regulatory element at the site of integration or in a heterochromatin domain). Recent work also suggested that (trans)genes expressed at very low levels are not necessarily permanently silent and may cycle between an active and an inactive state (7, 11, 12). Nonetheless, in general, high levels of gene expression require the insertion of the transgene in a transcriptionally favourable environment, such as in euchromatin.

DNA sequences that can recruit histone acetyltransferases and other chromatin-remodelling proteins that favour a chromatin configuration amenable to transcription should, therefore, be of great value for inclusion in recombinant protein expression vectors. By generating and sustaining an open chromatin configuration around the transgene, these sequences may allow stable, long-term expression of the recombinant protein and obviate the need for toxic substances for selection and maintenance. They may, thus, greatly improve the recombinant protein production process by reducing the time to production, as fewer clones would need to be screened. Fortunately, several such DNA elements have been reported (reviewed in ref. 13). One group of these DNA elements is called matrix attachment regions (MARs). These DNA sequences attach the chromatin to the nuclear matrix, and may thereby segregate the chromatin into distinct independent loops with an open chromatin configuration, which may facilitate transcription and protect genes within the loops from silencing.

The mechanisms by which MAR elements prevent gene silencing and increase transgene expression are not fully understood. Using a GFP reporter gene, we have shown that incorporation of a MAR in the vector reduces the number of silenced cells while increasing the fluorescence level of expressing cells (14, 15). We, and others, have found that MARs increase the number of integrated copies of the transgene, which may lead to a further increase in gene expression (16–18). We have also shown that these elements reduce the probability of the transgene switching from an active to an inactive state, thereby mediating a sustained expression of the transgene (12). Other advantageous features of MAR elements for recombinant protein production are that they reduce variegation (12) and transgene expression is maintained over long periods of time, even in the absence of selection.

About 500 MARs have been experimentally tested or predicted using bioinformatics (19), although in silico analyses suggest that there may be tens of thousands of MARs in the human genome (17). Only a few MARs have been extensively characterized, and the majority of these analyses have been performed in gene and cell therapy models (reviewed in ref. 20). In studies in CHO cells, with the view to improving protein production, four MARs have been documented to significantly increase the proportion of positive clones as well as to increase the level of protein production of the most productive clones (14, 15, 17, 21, 22). This was shown to greatly reduce the number of clones required for screening and hence the time to production. Expression of therapeutic proteins from the high-producing clones was shown to be very stable over at least 30 generations (23). MAR elements have been successfully used to improve the production of various therapeutic proteins, including EPO (22), HGF (22), TGF-β type II receptor (21), and anti-Rhesus D antibody (14).

In this chapter, we describe methods for using MARs to increase the production of recombinant proteins, such as antibodies, from CHO cells grown in suspension and for selecting high-producing clones. As MAR elements should function with all expression vectors, MAR-containing expression vectors may be generated by cloning MAR elements whose sequences are available in the GenBank databases. Alternatively, vectors containing MAR elements that have been optimized for high recombinant protein production may be purchased commercially (see Note 1). To assess the efficiency of the MAR, initial experiments should be performed with and without a MAR using GFP as a reporter gene. GFP expression levels from a pool of transfected cells are measured 2 weeks post selection by flow cytometry. Cells containing the MAR construct should exhibit a reduction in the number of low-expressing cells, with a concomitant increase in the fluorescence levels of the expressing cells (see Girod et al. 14 and Fig. 2 for examples). Once the construct and transfection conditions have been optimized, GFP can be replaced with the gene of interest. 48 h post transfection, selection of single high-producing clones commences using the limited dilution technique. Analyses of protein production can be performed 2 weeks after selection using the assay of choice (we provide protocols for screening for clones highly producing an antibody using fluorescence-activated cell sorting (FACS) or an IgG ELISA). High-expressing clones can then be expanded and assessed for stable, long-term protein production.

2. Materials

2.1. DNA Preparation

1. *Escherichia coli* strain containing the plasmid of interest.
2. Luria-Bertani broth (1 L) (Difco LB Broth, BD # 244620).
3. Antibiotic (e.g. ampicillin, AppliChem, #A0839,0010).
4. 14-mL tube for growing bacteria (Falcon 2059).
5. Unbaffled Erlenmeyer flask (250 mL).
6. Shaking incubator (e.g. 211DS, Labnet International Inc.).
7. Refrigerated centrifuge.
8. Maxiprep kit (e.g. JETStar 2.0, Genomed).
9. Sterile ultra-pure H_2O.
10. Restriction enzyme that cuts once or several times in the vector backbone, typically in the bacterial replicon or ampicillin resistance gene (e.g. *Pvu*I, NEB, # R0150L).
11. PicoGreen dsDNA AssayKit (Invitrogen # P11496) for quantitating the DNA.

2.2. Cell Culture

1. CHO-K1-S cells (Invitrogen).
2. Serum-free, chemically defined medium (e.g. CHO serum-free media, which is available from Lonza, Irvine Scientific, or other sources), supplemented with 4 mM L-glutamine.
3. Phosphate-buffered saline (PBS) (per litre, 1.15 g Na_2HPO_4; 8 g NaCl; 0.2 g KCl; 0.2 g KH_2PO_4). Adjust the pH to 7.4 with 1N HCl. Autoclave before use.
4. Antibiotics for use in selecting transfected CHO cells (according to the transfected selection plasmid, e.g. puromycin, Sigma #P9620; hygromycin HyClone #SV30070).

5. Shake flasks (such as Corning # 430421) or square bottles (Schott, 100 mL).
6. Centrifuge tubes (15 mL, Techno Plastic Products (TPP) #91015, 50 mL, SPL-Life Science Inc. #50050).
7. 96, 24, and/or 6-well tissue culture plates (TPP # 920969; 92024; 92006).
8. Humidified incubator 37°C, 5% CO_2.
9. Humidified incubator 37°C, 5% CO_2 with shaking platform (Kühner, shaker ISF-1-x, 120 rpm, 25 mm stroke or equivalent).
10. Inverted microscope.

2.3. Cell Transfection

1. 14-mL, polystyrene, round-bottom tubes (Becton-Dickinson #35-2057).
2. 1.5-mL Eppendorf tubes.
3. 50-mL spin tubes (Bioreactor TPP #87050).
4. Transfection reagent – Transfast (Promega #E2431).
5. OptiMEM medium (Invitrogen).
6. Vector (plasmid DNA containing MAR and transgene of interest).
7. Guava Express (Guava Technologies/Millipore) or other flow cytometry machine for GFP analysis.

2.4. Fluorometric Analysis of IgG-Producing Clones

1. Guava Express or equivalent flow cytometer.
2. 96 well cell culture plates pro-Bind™ (BD Falcon # 353910).
3. PBS.
4. Blocking solution (PBS + 0.5% BSA + 0.5% Triton). Make a fresh solution each time. BSA (Eurobio # GAUBSA G1-62); Triton 100X (Acros Organics # 215682500).
5. 37% paraformaldehyde (PAF).
6. Labelling antibody: Dylight™ 488-conjugated AffiniPure Goat Anti-Human IgG (Fc) Fragment Specific (Jackson ImmunoResearch Lab, Inc. #109-485-098). Dilute 1:400 in blocking solution. Make a fresh dilution each time.

2.5. IgG ELISA

1. 96-well microtiter plates (MaxiSorb, Nunc, #439454).
2. Incubator at 37°C.
3. ELISA washer.
4. ELISA reader (405 and 490 nm).
5. Goat Anti-human Kappa (Biosource/Invitrogen, # AHI0801). Store in 25-μL aliquots at −20°C.
6. Goat Anti-Human IgG conjugated with alkaline phosphatase (Biosource/Invitrogen, # AHI0305). Store in 25-μL aliquots at −20°C.

7. Human IgG1 whole molecule (Jackson Immuno Research, # 009-000-003). Adjust an aliquot to a final concentration of 80 ng/mL in blocking solution, for preparation of standards.
8. 20× PBS: (7.73 g NaH_2PO_4 H_2O; 25.63 g Na_2HPO_4; 170 g NaCl). Make up to 1 L with ultra-pure H_2O. Store at room temperature (RT).
9. PBS, pH 8: (50 mL 20× PBS; 2 mL 10% NaAzide). Make up to 1 L with ultra-pure H_2O. Adjust pH to 8 using NaOH. Store at RT.
10. Substrate buffer: (105.14 g diethanolamine 99% (=97 mL) mixed with 700 mL ultra-pure H_2O). Adjust pH to 9.8 using HCl. Add 500 μL 1 M $MgCl_2$ and 2 mL 10% NaAzide. Make up to 1 L with ultra-pure H_2O. Store at RT, protected from light.
11. Coating solution: Add 22 μL Capture Antibody (goat anti-human Kappa, Biosource/Invitrogen, # AHI0801) to 11 mL PBS pH 8.
12. Blocking solution: 5 g casein hydrolysate; 1 mL 10% NaAzide; 250 μL Tween 20. Complete to 500 mL with 1× PBS (out of the 20× PBS stock). Store at 2–8°C.
13. Detection solution: Add 22 μL Detection Antibody (goat anti-human IgG, conjugated with alkaline phosphatase, Biosource/Invitrogen, # AHI0305) to 11 mL blocking solution.
14. Washing solution: (100 mL 20× PBS; 2 mL 10% NaAzide; 200 μL Tween 20). Make up to 2 L.
15. Substrate solution: Add 16.5 mg NPP (Fluka, # 71768) to 11 mL substrate buffer. Make fresh each time.
16. 3 M NaOH.

3. Methods

3.1. Identification and Cloning of Matrix Attachment Regions

Although MARs do not display significant primary sequence homology, they do have a number of features in common. MARs have an AT-rich region, which may aid unwinding or destabilization of the DNA duplex or formation of curved DNA structures. MARs may also contain origins of replication and act as binding sites for chromatin-remodelling factors. Based on these features, a number of algorithms have been described and a comprehensive discussion of the various MAR prediction programs is found in Evans et al. (24). MAR predictor programs include, MAR-Finder (MAR-Wiz): http://www.genomecluster.secs.oakland.edu/marwiz (25) and SMARTest: http://www.genomatix.de/smartest (26). Vectors containing optimized MAR elements are also available commercially (see Note 1).

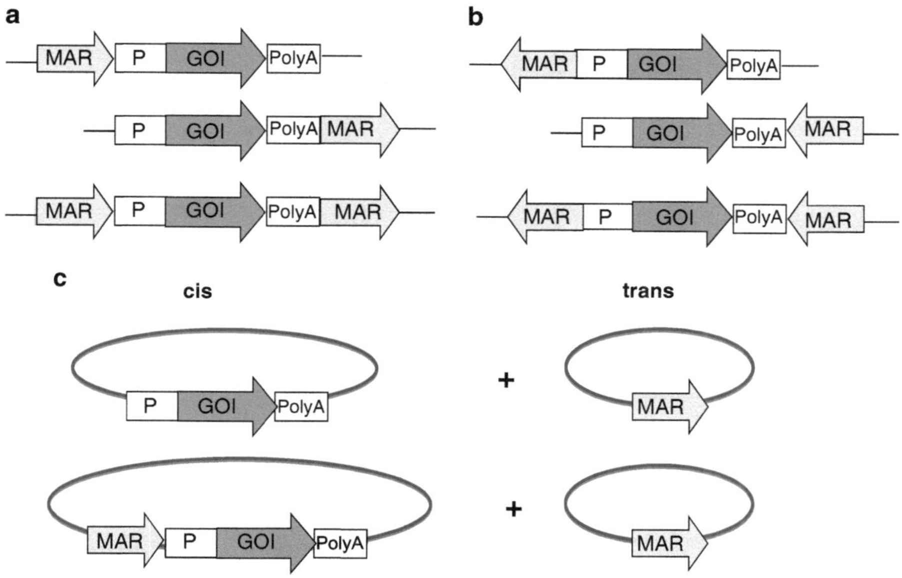

Fig. 1. Cloning of MARs in recombinant protein expression vectors. The MAR can be cloned either or both upstream or downstream of the promoter (P) and gene of interest (GOI) and usually in either orientation (**a** and **b**). The MAR can also be cloned on a separate plasmid in trans to be co-transfected with the expression vector, or alternatively the MAR can be cloned on both the plasmid containing the GOI and the plasmid transfected in trans (**c**).

Recombinant vectors containing MAR elements may have many configurations, which need to be determined for each MAR under consideration (see Fig. 1). In our constructs, the MAR element is not transcribed and so the vector becomes stably integrated into the chromosome. However, it has been shown that transcription through a transgene into the IFN-β MAR allows the plasmid to be retained in the cell as an episome for over 100 generations (27–29). This may, however, be related to specific properties of this vector and/or of this MAR (30).

1. As MAR elements range in size from 300 to 5,000 bp, they can be easily cloned in any expression vector.
2. The MAR should work with all promoters, although some optimization may be required (see Note 2).
3. The MAR can be cloned upstream or downstream of the expression cassette or both. The MAR can be cloned in either orientation. The MAR can also be co-transfected on a separate plasmid (15). The most effective configuration must be tested for each MAR and each condition examined. However, a common and favourable locus for introducing one MAR is upstream of the promoter and enhancer. When using potent MARs, this is sufficient to achieve high expression levels.

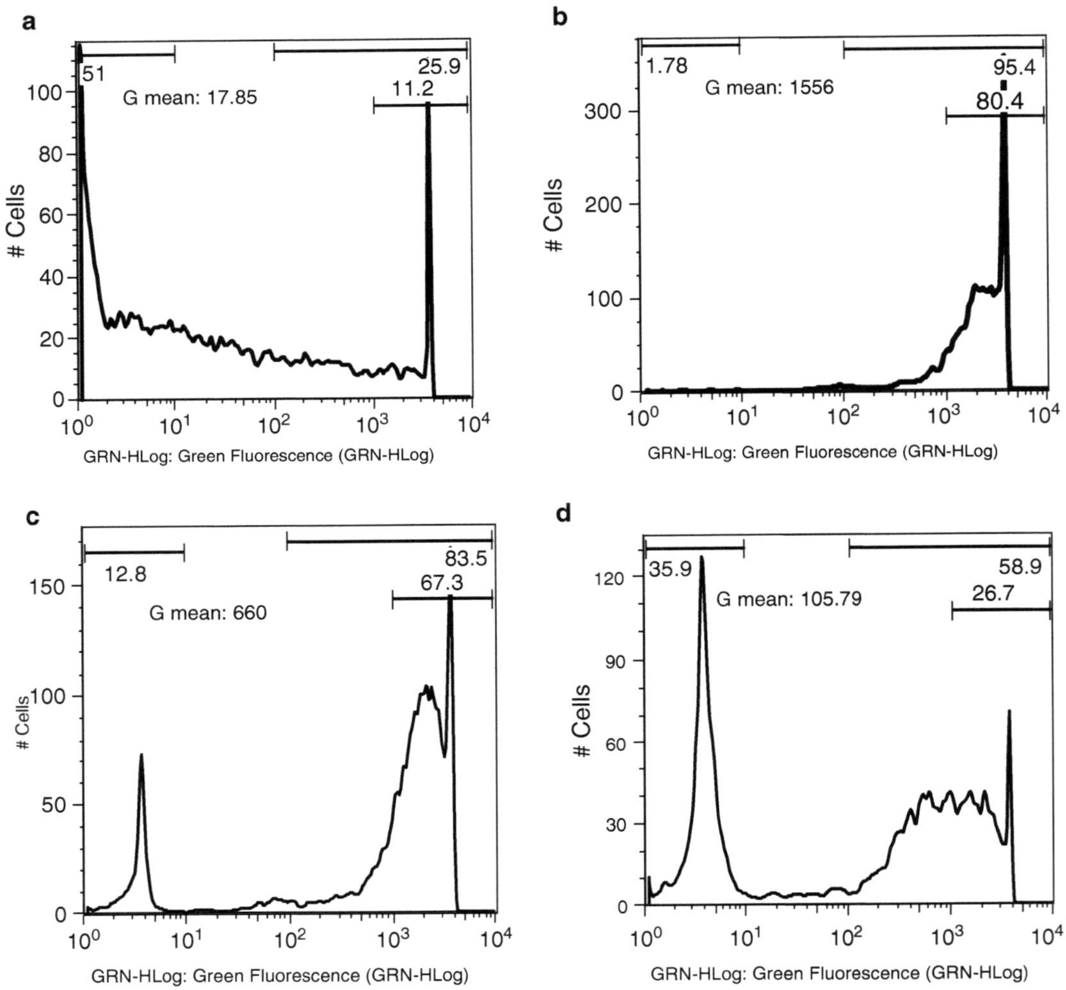

Fig. 2. Histograms of fluorescence (GRN-Hlog) vs. cell number produced by the Guava flow cytometer. Comparison of different MAR-containing constructs (**b–d**) and selection procedure (**a** vs. **b–d**). CHO-K1-S cells were analyzed after 1 week (**a**) or 1 month (**b–d**) of selection with puromycin after transfection. A more detailed description and interpretation of these FACS histograms is given in Note 7.

4. It is recommended that initial optimization experiments are performed using EGFP as a reporter gene rather than the gene of interest. It is important to assess that the selected MAR functions well with the promoter, vector, and cell line chosen for analysis. A flow cytometry analysis (e.g. see Fig. 2) allows the effectiveness of the MAR to be easily assessed. The GFP reporter construct can also be used as a control for transfection efficiency.
5. When performing co-transfections (e.g. with antibody heavy and light chains on separate vectors), the MAR is cloned in both vectors or alternatively it can be co-transfected on a separate plasmid, although this is less efficient (14, 15). However,

a combination of MAR in *cis* and *trans* can significantly augment transgene expression (14).

6. If the antibody heavy and light chain expression cassettes are co-transfected on separate plasmids, the optimal ratio of heavy to light chain must be determined empirically (see Note 3).
7. cDNA of the heavy and light chains is generally used. However, additional regulatory elements, such as introns and enhancers, may need to be incorporated into the vector for optimal performance.

3.2. Preparation of the DNA

1. A few isolated colonies of *E. coli*, transformed with the plasmid of interest, are used to inoculate 3 mL LB + antibiotic and grown for about 8 h with shaking at 37°C.
2. 1 mL of this culture is used to inoculate 150 mL LB + antibiotic in a 500-mL Erlenmeyer flask and grown with shaking (160–200 rpm) overnight at 37°C.
3. In the morning, the bacteria are pelleted by centrifugation (5,000 × *g* for 15 min).
4. Plasmid DNA is purified using a commercial Maxiprep kit according to the manufacturer's instructions.
5. The DNA pellet is dissolved in 250–300 μL of water. The final concentration of DNA should be in the range of 2–4 μg/μL.
6. 40 μL DNA is digested overnight in a final volume of 80 μL using a restriction enzyme that cuts once or several times in the portion of the plasmid not involved in expression or selection in CHO cells (typically, in the bacterial replicon or ampicillin resistance gene).
7. The DNA is used for transfection without any further purification. It is not necessary to separate and purify the digested fragments. However, the digested DNA is normally diluted 1:10–1:20 in PBS in order to obtain the correct concentration for transfection.
8. The digested DNA is quantitated using the PicoGreen dsDNA AssayKit according to the manufacturer's instructions (see Note 4).

3.3. Routine Cultivation of CHO-K1-S Cells

CHO-K1 cells grown in suspension (CHO-K1-S) were developed in response to the growing need for recombinant proteins for research or therapeutic purposes (31). Use of suspension rather than adherent CHO cells facilitates scale-up for production of recombinant proteins using bioreactors.

1. CHO-K1-S cells are cultivated in shake flasks in a serum-free medium (SFM) supplemented with 4 mM L-glutamine.
2. Cells are maintained under agitation (120 rpm, 25 mm stroke) at 37°C, 5% CO_2 in a humidified incubator.

3. Cells are counted three times a week (every other day) and adjusted to 2×10^5 cells/mL for a 2-day split or adjusted to 1.5×10^5 cells/mL for a 3-day split over the weekend.

3.3.1. Preparation of CHO-K1-S Cells for Transfection

Transfection is a method by which experimental DNA may be put into a cultured mammalian cell. Such experiments are usually performed using cloned DNA containing coding sequences and control regions (promoters, etc.) in order to test whether the DNA is expressed. Transfection is often carried out by mixing a cationic lipid with the experimental DNA to produce liposomes, which fuse with the cell plasma membrane and deposit their cargo inside. Genetic material (such as supercoiled or linearized plasmid DNA or siRNA constructs) or even proteins, such as antibodies, may be transfected. Several chemical or commercially available transfection reagents are available for transfection and work with various cell types. Care should be taken to use a reagent that works with cells grown in suspension in a chemically defined medium. In addition to lipofection, transfection can be carried out by electroporation, which typically involves opening transient pores or "holes" in the cell plasma membrane, to allow the uptake of genetic material.

1. The day prior to transfection, CHO-K1-S cells are passaged at a cell density of 3×10^5 cells/mL into fresh SFM.
2. The day of transfection, the cells are recounted and the total number of cells and the volume of the cell suspension to be used for the experiment are determined.
3. The required number of CHO-K1-S cells is centrifuged ($400 \times g$ for 5 min at 4°C), washed in 1 mL of 1× PBS, and resuspended gently in OptiMEM, at a cell density of 2×10^6 cells/mL. The cell suspension is stored in a 15-mL tube at 4°C in a refrigerator for 30 min.

3.3.2. Lipofection

1. In a polystyrene tube, combine OptiMEM, a total amount of 1 μg of linearized DNA (see Notes 5 and 6) and 3 μL TransFast transfection reagent to a final volume of 200 μL and mix gently by tapping the tube (do not mix by pipetting). TransFast reagent is added last.
2. Incubate the transfection mixture for 15 min at room temperature to allow DNA–lipid complexes to form.
3. Shortly before the end of the incubation time of the transfection mix, thoroughly resuspend the refrigerated cell suspension by pipetting up and down. Ensure that no cell clumps remain.
4. When the transfection mix incubation period is over, add 200 μL of the resuspended cells to the DNA–TransFast complexes (total volume is 400 μL). Mix well by pipetting up and down.

5. Transfer the 400 μL of the cell–DNA–TransFast mixture to one well of a 24-well plate.
6. Incubate the transfected cells for 1 h without agitation at 37°C, 5% CO_2 in a humidified incubator.
7. Add 1 mL of serum-free growth medium (warmed to room temperature and supplemented with L-glutamine) to each well of transfected cells.
8. Incubate the plate overnight without agitation at 37°C, 5% CO_2 in a humidified incubator.

3.4. Selection of a Pool of GFP-Expressing Cells for a Polyclonal Population Analysis

GFP is one of the most commonly used reporter genes for gene expression analysis. Major advantages of this reporter system include single cell analysis by FACS and that no substrates or cofactors are required for detection. This system allows the activity of the MAR to be easily assessed under different configurations, and the most effective configuration to be chosen for the recombinant protein expression vector. Although the GFP is retained within the cell (as opposed to being secreted like many recombinant proteins), and is relatively stable, we have found a good correlation between experiments performed initially with GFP and subsequent generation of cell lines with the gene of interest (14, 17). GFP can also be used as a marker of protein production in co-transfection experiments (32, 33), although the amount of DNA used must be adjusted so as not to interfere with recombinant protein production (34). The following protocol is our standard procedure for preparing CHO cells in suspension for analysis in a Guava or other FACS device.

1. 24 h post transfection, the cells are transferred to a 25- or 75-cm^2 flask containing 5 or 10 mL, respectively, of fresh medium containing the antibiotic for selection (such as puromycin, G418, etc.). The final concentrations of puromycin and G418 used are 2.5 and 300–500 μg/mL, respectively.
2. Grow the cells in the selection medium for 1–2 weeks, changing the medium once a week. A useful control consists of placing mock-transfected cells under similar culture conditions to confirm proper selection. Massive cell death occurs within 4 days when puromycin selection is used, whereas it takes approximately 7–10 days for selection in G418 to be complete.
3. The week following transfection, transfer the cells to spin tubes containing 5 mL of selection medium and culture the cells with agitation until the selection is completed (e.g. with no live cells remaining in the control).
4. After transfer of the cellular suspension to the spin tubes, the cells grow faster. It is, therefore, important that a proper cell density is maintained so that the cells are kept in the exponential

phase of growth. The cells probably need to be split twice per week. At this point, the number of cells are counted, the cellular viability is assessed, and the cells are split at 3×10^5 viable cells/mL.

5. Every week, remove some of the culture supernatant for a polyclonal flow cytometry population analysis.
6. On the day the FACS is performed, 20 μL of the cultured cells are transferred to a 96-well plate. To each well, 180 μL PBS is added. The cells are mixed by pipetting up and down.
7. Analyse GFP expression levels in the cells in PBS using the Guava System or other flow cytometry machine.
8. Examples of typical FACS profiles are shown in Fig. 2 (see also Note 7 for a detailed description on how to analyze these profiles).

3.4.1. Selection of Clones by Limiting Dilution: Identification of Single High-Producing Clones

1. 48 h post transfection, remove some supernatant for product formation analysis.
2. Prepare the cells for limiting dilution by diluting the cells to 20 cells/mL (i.e. 2 cells/well) in selection medium and prepare three dilutions such that 0.5 cells/well, 1 cell/well, and 2 cells/well are plated.
3. Transfer 100 μL of each dilution to the wells of a 96-well plate. Typically, about 1,000 clones (i.e. wells) are screened.
4. Allow the cell(s) to attach (1–2 days). Inspect the plates under the microscope and discard the dilution(s) that contain more than 1 cell/well.
5. Depending on cell growth (this differs depending on protein expression), cells are fed within 5–10 days by adding 100 μL of selection medium to each well. As a guide, the medium needs to be changed before it changes colour and/or when the colony takes up about 10–20% of the surface of the well.

3.5. Estimation of IgG Production by Fluorescence-Activated Cell Sorting

Once a sufficient number of cells have grown, it is necessary to screen for high-producing clones. This can be a tedious task as the recombinant protein is generally secreted; therefore, time-consuming and sometimes difficult protocols are often used for assaying product formation. Brezinsky et al. (35) hypothesized that it may be possible to detect the secreted protein while it is transiently associated with the cell surface and that this may correlate with the amount of protein being secreted from the cell. Based on this, they developed a protocol for the fluorescent labelling of cells and their detection by FACS. The protocol described below for the staining of recombinant CHO cells for surface antibody production is modified from Brezinsky et al. (35).

1. Cells ready for sorting are harvested and maintained at 0–4°C for all subsequent handling. For example, these cells may be clones derived from the limiting dilution selection described above containing different MAR constructs or different ratios of antibody constructs.
2. Use $1–2 \times 10^6$ viable cells for analysis.
3. Wash the cells with 500 μL PBS at 4°C twice and then centrifuge cells at $190 \times g$ for 10 min at 4°C.
4. Aspirate the supernatant carefully, taking care not to disturb the cell pellet.
5. Add 200 μL blocking solution per sample and transfer the cells to the wells of a 24-well plate and incubate the cells for 15 min at room temperature with orbital shaking at a speed high enough to keep the cells in suspension.
6. Centrifuge the cells in the 24-well plate at $190 \times g$ for 10 min at 4°C and carefully aspirate the supernatant.
7. Fix the cells with 200 μL 2% PAF in blocking solution for 15 min at room temperature with orbital shaking.
8. Centrifuge the cells in the 24-well plate at $190 \times g$ for 10 min at 4°C and carefully aspirate the supernatant.
9. Wash the cells in suspension for 15 min with PBS at 4°C with orbital shaking, twice.
10. Centrifuge the cells in the 24-well plate at $190 \times g$ for 10 min at 4°C and carefully aspirate the supernatant.
11. Add 200 μL of the diluted staining antibody (such as labelled anti-human IgG) in blocking solution and incubate the cells for 1 h at room temperature with orbital shaking in the dark.
12. Centrifuge the cells in the 24-well plate at $190 \times g$ for 10 min at 4°C and carefully aspirate the supernatant.
13. Wash the cells with 500 μL PBS at 4°C for 15 min with orbital shaking, three times.
14. Centrifuge the cells in the 24-well plate at $190 \times g$ for 10 min at 4°C and resuspend the cells in 500 μL PBS.
15. Transfer 100 μL to a 96-well plate and measure the fluorescence using the Guava Express or equivalent flow cytometer device.
16. An example of a FACS profile is shown in Fig. 3.

3.6. Quantification of IgG Production by ELISA

1. Coat the 96-well microtitre plate 1 day before with 100 μL of coating solution and incubate overnight at 2–8°C (see Note 8).
2. Prepare dilutions of the cell culture supernatant samples in blocking solution (approximate dilutions are 1:500 for transient; 1:4,000 to 1:30,000 for stable).

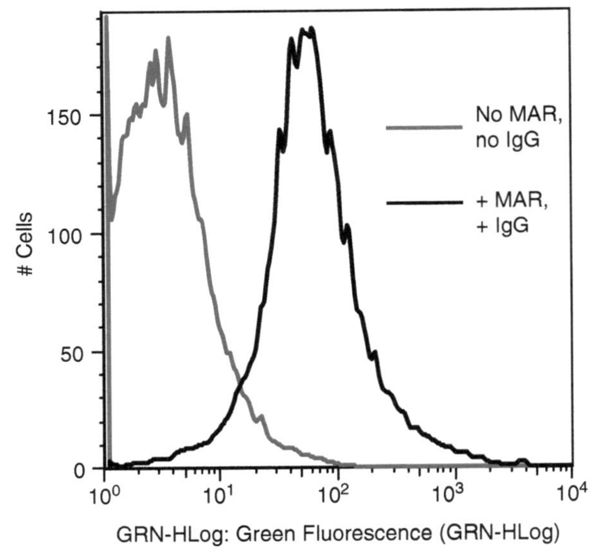

Fig. 3. Estimation of relative IgG production of clones by flow cytometry using a modified Brezinsky protocol (Subheading 3.5). The population of the cells lacking an IgG construct (negative control) showed levels of fluorescence within the first decade of the log scale, whereas the cells containing the MAR-IgG construct displayed fluorescence levels falling within the second-to-third decade.

3. Wash the IgG-coated plate three times with 300 μL of washing solution. Ensure that any remaining liquid is removed by blotting the plate on tissue paper.
4. Add 100 μL of blocking solution to each well.
5. Dilute the stock IgG molecule to 80 ng/mL in blocking solution. The two standard series are prepared in a separate plate with seven successive 1:2 dilutions of the IgG molecule (80 ng/mL), giving a standard range of 0.3125–40 ng/mL. 100 μL of each dilution is then transferred in one step to well A3 to A10 and B3 to B10, respectively, using a multi-channel pipette.
6. To wells C1 to H12, add 100 μL of the (diluted) samples in duplicate. Mix the samples by gently pipetting.
7. Incubate for 30 min at 37°C.
8. Wash the plate three times with 300 μL of washing solution.
9. Add 100 μL of detection solution.
10. Incubate for 30 min at 37°C.
11. Wash the plate four times with 300 μL of washing solution.
12. Add 100 μL of substrate solution to each well.
13. Cover the plate and incubate for 15 min in the dark at room temperature.
14. Stop the reaction by adding 50 μL of 3 M NaOH to each well.

15. Read the absorbance with the ELISA reader within 30 min (405 nm; reference 490 nm).
16. A polynomial standard curve is used for calculation. Software for generating the standard curve and performing the calculations usually accompanies the plate reader.

4. Notes

1. Vectors containing MAR elements, which have been optimized for high-level, stable protein production in eukaryotic cells, are commercially available from Selexis SA (http://www.selexis.com). MAR elements may also be found in the GenBank database, e.g. L22754 (human β-globin MAR); M83137 (human interferon-β MAR); X98408 (chicken lysozyme MAR subfragment).
2. We have successfully tested a number of human MARs with the CMV, SV40, EF1α, and other promoters. However, the promoter/MAR combination may need optimization. For instance, higher long-term productivities were obtained with the SV40 promoter when compared with the CMV promoter (14), but higher yields are obtained from strong natural or synthetic promoters. The protocol given here is optimized for strong MARs and promoters of human origin.
3. Due to a faster synthesis rate of the light chain (36) and higher degradation rates of heavy chain mRNA and polypeptide chains (37), the optimal ratio of heavy to light chain genes must be determined empirically. The reader is referred to Schlatter et al. (38) for further information.
4. It has been reported that other methods for DNA quantitation, such as the spectrophotometric method, may over-estimate the amount of DNA in the sample compared to the PicoGreen Assay (39). The parameters given here for lipofection of CHO-K1-S have been optimized based on DNA quantitation by the PicoGreen Assay. If alternative methods for DNA quantitation are to be used, they should be compared to those obtained by PicoGreen and the protocols for lipofection adjusted, if necessary. Similarly, if alternative methods of transfection will be used, the final amount of DNA for optimal transfection will have to be determined.
5. The total amount of linearized DNA to be transfected refers to the sum of all the plasmids to be co-transfected.
6. The final amount of DNA used for transfection must be optimized for every host cell line and when alternative media are used. The best ratio of heavy to light chain for antibody expression

must be determined empirically by transient transfections, with supernatants analyzed after 3–4 days of incubation.

7. Data shown in the FACS histograms (Fig. 2), which represents polyclonal populations of different constructs (a–d) performed either 1 week (a) or 1 month (b–d) after transfection, can be analyzed in two ways – the overall mean fluorescence of the population (G mean) or the percentage of cells giving fluorescence within a particular range. We classify the cells as follows: M1 cells are those found in the first decade (fluorescence from 10^0 to 10^1) and represent low or non-expressing cells; M2 cells are found in the third and fourth decades and are classified as medium- to high-expressing cells; M3 cells are high-expressing cells and are found in the fourth decade. Panel A shows a representative FACS profile of GFP-expressing cells 1 week after puromycin selection. The overall mean fluorescence is low and there is a high percentage of low and non-expressing cells. Panel B shows a construct following 1 month of selection. This construct gives a high G mean fluorescence with the vast majority of cells expressing GFP to a high level. Due to a number of very high-expressing cells whose fluorescence saturates the detector, the FACS profile ends in a straight line. Under these circumstances, the voltage would need to be adjusted to view the higher expressing cells. However, this may come at the expense of being unable to view the profile of the low-expressing cells, (e.g. the M1 cells seen in Panel C). Panel C shows a reduced G mean fluorescence when compared with B due to a higher percentage of M1 cells. Note the three peaks in this profile, one to the left, representing the low-expressing cells and two to the right, representing high- and very high-expressing cells. The construct shown in D has a higher percentage of M1 cells compared to the other constructs, which results in a decrease in the G mean. The percentage of M2 and M3 cells is also reduced.

   The fluorescence pattern of Panel A indicates incomplete selection of the polyclonal population of stably transfected cells. The profiles shown in panels C and D can result from the use of a less-efficient construct and/or from an insufficient or ineffective selection pressure. The profile in panel B is, therefore, most likely to be indicative of an efficient MAR and vector elements combination, and of adequate transfection and selection processes.

8. Plates can be stored for 5 days at 2–8°C.

References

1. Walsh, G. (2006) Biopharmaceutical benchmarks 2006. *Nat Biotechnol*, **24**, 769–776.
2. Wurm, F.M. (2004) Production of recombinant protein therapeutics in cultivated mammalian cells. *Nat Biotechnol*, **22**, 1393–1398.
3. Kaufman, R.J. (1990) Selection and coamplification of heterologous genes in mammalian cells. *Methods Enzymol*, **185**, 537–566.
4. Kim, N., Byun, T. and Lee, G. (2001) Key determinants in the occurrence of clonal variation in

humanized antibody expression of CHO cells during dihydrofolate reductase mediated gene amplification. *Biotechnol Prog*, **17**, 69–75.

5. Kim, S., Kim, N., Ryu, C., Hong, H. and Lee, G. (1998) Characterization of chimeric antibody producing CHO cells in the course of dihydrofolate reductase-mediated gene amplification and their stability in the absence of selective pressure. *Biotechnol Bioeng*, **58**, 73–84.
6. Chusainow, J., Yang, Y.S., Yeo, J.H., Toh, P.C., Asvadi, P., Wong, N.S. and Yap, M.G. (2009) A study of monoclonal antibody-producing CHO cell lines: what makes a stable high producer? *Biotechnol Bioeng*, **102**, 1182–1196.
7. Pilbrough, W., Munro, T.P. and Gray, P. (2009) Intraclonal protein expression heterogeneity in recombinant CHO cells. *PLoS One*, **4**, e8432.
8. Raj, A., Peskin, C., Tranchina, D., Vargas, D. and Tyagi, S. (2006) Stochastic mRNA synthesis in mammalian cells. *PLoS Biol*, **4**, e309.
9. Gorman, C., Arope, S., Grandjean, M., Girod, P. and Mermod, N. (2009) Use of MAR elements to increase the production of recombinant proteins. *Cell Engineering*, **6**, 1–32.
10. Yang, Y., Mariati, Chusainow, J. and Yap, M.G. (2010) DNA methylation contributes to loss in productivity of monoclonal antibody-producing CHO cell lines. *J Biotechnol*, **147**, 180–185.
11. Ferrai, C., Xie, S.Q., Luraghi, P., Munari, D., Ramirez, F., Branco, M.R., Pombo, A. and Crippa, M.P. (2010) Poised transcription factories prime silent uPA gene prior to activation. *PLoS Biol*, **8**, e1000270.
12. Galbete, J.L., Buceta, M. and Mermod, N. (2009) MAR elements regulate the probability of epigenetic switching between active and inactive gene expression. *Mol Biosyst*, **5**, 143–150.
13. Kwaks, T.H. and Otte, A.P. (2006) Employing epigenetics to augment the expression of therapeutic proteins in mammalian cells. *Trends Biotechnol*, **24**, 137–142.
14. Girod, P.A., Zahn-Zabal, M. and Mermod, N. (2005) Use of the chicken lysozyme 5′ matrix attachment region to generate high producer CHO cell lines. *Biotechnol Bioeng*, **91**, 1–11.
15. Zahn-Zabal, M., Kobr, M., Girod, P.A., Imhof, M., Chatellard, P., de Jesus, M., Wurm, F. and Mermod, N. (2001) Development of stable cell lines for production or regulated expression using matrix attachment regions. *J Biotechnol*, **87**, 29–42.
16. Phi-Van, L., von Kries, J.P., Ostertag, W. and Stratling, W.H. (1990) The chicken lysozyme 5′ matrix attachment region increases transcription from a heterologous promoter in heterologous cells and dampens position effects on the expression of transfected genes. *Mol Cell Biol*, **10**, 2302–2307.
17. Girod, P.A., Nguyen, D.Q., Calabrese, D., Puttini, S., Grandjean, M., Martinet, D., Regamey, A., Saugy, D., Beckmann, J.S., Bucher, P. *et al.* (2007) Genome-wide prediction of matrix attachment regions that increase gene expression in mammalian cells. *Nat Methods*, **4**, 747–753.
18. Dang, Q., Auten, J. and Plavec, I. (2000) Human beta interferon scaffold attachment region inhibits de novo methylation and confers long-term, copy number-dependent expression to a retroviral vector. *J Virol*, **74**, 2671–2678.
19. Liebich, I., Bode, J., Frisch, M. and Wingender, E. (2002) S/MARt DB: a database on scaffold/matrix attached regions. *Nucleic Acids Res*, **30**, 372–374.
20. Harraghy, N., Gaussin, A. and Mermod, N. (2008) Sustained transgene expression using MAR elements. *Curr Gene Ther*, **8**, 353–366.
21. Kim, J.-M., Kim, J.-S., Park, D.-H., Kang, H., Yoon, J., Baek, K. and Yoon, Y. (2004) Improved recombinant gene expression in CHO cells using matrix attachment regions. *J Biotechnol*, **107**, 95–105.
22. Kim, J., Yoon, Y., Hwang, H.-Y., Park, J., Yu, S., Lee, J., Baek, K. and Yoon, J. (2005) Efficient selection of stable Chinese hamster ovary (CHO) cell lines for expression of recombinant proteins by using human interferon b SAR element. *Biotechnol Prog*, **21**, 933–937.
23. Varghese, J., Alves, W., Brill, B., Wallace, M., Calabrese, D., Regamey, A. and Girod, P. (2008) Rapid development of high-performance, stable mammalian cell lines for improved clinical development. *Bioprocess J*, 7, 30–36.
24. Evans, K., Ott, S., Hansen, A., Koentges, G. and Wernisch, L. (2007) A comparative study of S/MAR prediction tools. *BMC Bioinformatics*, **8**, 71.
25. Singh, G.B., Kramer, J.A. and Krawetz, S.A. (1997) Mathematical model to predict regions of chromatin attachment to the nuclear matrix. *Nucleic Acids Res*, **25**, 1419–1425.
26. Frisch, M., Frech, K., Klingenhoff, A., Cartharius, K., Liebich, I. and Werner, T. (2002) In silico prediction of scaffold/matrix attachment regions in large genomic sequences. *Genome Res*, **12**, 349–354.
27. Jenke, A.C., Stehle, I.M., Herrmann, F., Eisenberger, T., Baiker, A., Bode, J., Fackelmayer, F.O. and Lipps, H.J. (2004) Nuclear scaffold/matrix attached region modules linked to a transcription unit are sufficient for replication and maintenance of a mammalian episome. *Proc Natl Acad Sci USA*, **101**, 11322–11327.

28. Piechaczek, C., Fetzer, C., Baiker, A., Bode, J. and Lipps, H.J. (1999) A vector based on the SV40 origin of replication and chromosomal S/MARs replicates episomally in CHO cells. *Nucleic Acids Res*, **27**, 426–428.
29. Stehle, I.M., Postberg, J., Rupprecht, S., Cremer, T., Jackson, D.A. and Lipps, H.J. (2007) Establishment and mitotic stability of an extra-chromosomal mammalian replicon. *BMC Cell Biol*, **8**, 33.
30. Giannakopoulos, A., Stavrou, E.F., Zarkadis, I., Zoumbos, N., Thrasher, A.J. and Athanassiadou, A. (2009) The functional role of S/MARs in episomal vectors as defined by the stress-induced destabilization profile of the vector sequences. *J Mol Biol*, **387**, 1239–1249.
31. Rosser, M.P., Xia, W., Hartsell, S., McCaman, M., Zhu, Y., Wang, S., Harvey, S., Bringmann, P. and Cobb, R.R. (2005) Transient transfection of CHO-K1-S using serum-free medium in suspension: a rapid mammalian protein expression system. *Protein Expr Purif*, **40**, 237–243.
32. Albano, C.R., Randers-Eichhorn, L., Bentley, W.E. and Rao, G. (1998) Green fluorescent protein as a real time quantitative reporter of heterologous protein production. *Biotechnol Prog*, **14**, 351–354.
33. Meng, Y.G., Liang, J., Wong, W.L. and Chisholm, V. (2000) Green fluorescent protein as a second selectable marker for selection of high producing clones from transfected CHO cells. *Gene*, **242**, 201–207.
34. Pick, H.M., Meissner, P., Preuss, A.K., Tromba, P., Vogel, H. and Wurm, F.M. (2002) Balancing GFP reporter plasmid quantity in large-scale transient transfections for recombinant anti-human Rhesus-D IgG1 synthesis. *Biotechnol Bioeng*, **79**, 595–601.
35. Brezinsky, S.C., Chiang, G.G., Szilvasi, A., Mohan, S., Shapiro, R.I., MacLean, A., Sisk, W. and Thill, G. (2003) A simple method for enriching populations of transfected CHO cells for cells of higher specific productivity. *J Immunol Methods*, **277**, 141–155.
36. Bergman, L.W., Harris, E. and Kuehl, W.M. (1981) Glycosylation causes an apparent block in translation of immunoglobulin heavy chain. *J Biol Chem*, **256**, 701–706.
37. Bibila, T. and Flickinger, M.C. (1991) A structured model for monoclonal antibody synthesis in exponentially growing and stationary phase hybridoma cells. *Biotechnol Bioeng*, **37**, 210–226.
38. Schlatter, S., Stansfield, S.H., Dinnis, D.M., Racher, A.J., Birch, J.R. and James, D.C. (2005) On the optimal ratio of heavy to light chain genes for efficient recombinant antibody production by CHO cells. *Biotechnol Prog*, **21**, 122–133.
39. English, C., Merson, S. and Keer, J. (2006) Use of elemental analysis to determine comparative performance of established DNA quantification methods. *Anal Chem*, **78**, 4630–4633.

Chapter 8

Controlling Apoptosis to Optimize Yields of Proteins from Mammalian Cells

Matthew P. Zustiak, Haimanti Dorai, Michael J. Betenbaugh, and Tina M. Sauerwald

Abstract

Apoptosis is the foremost method of cell death in bioreactors and can be caused by nutrient limitation, toxin accumulation, and growth factor withdrawal. By delaying the onset of this form of programmed cell death, one can achieve longer sustained viabilities in culture, thereby increasing product yield. Described here is a genetic-based, step-by-step method to generate an apoptosis-resistant cell line. This cell line, then, can be used as a platform for biotherapeutic protein production. The key steps include antiapoptotic transgene selection and transfection followed by clonal isolation and screening. With the proper screening methods, one can obtain a robust cell line that resists the harsh conditions of late-stage and/or high-density culture.

Key words: Apoptosis, Transfection, Selection, Screening, Flow cytometry, bcl-X_L, CHO, Monoclonal antibody, Mammalian cell culture

1. Introduction

Mammalian cells are used to produce protein-based therapeutics, including monoclonal antibodies, cytokines, vaccines, and fusion proteins. Over the past 20 years, volumetric productivity of therapeutics has improved 100-fold (1). This major improvement has been accomplished primarily through (1) the adaptation of production cell lines to animal protein-free medium; (2) optimization of culture medium and improved feed strategies; (3) design of expression vectors, including that of the selection markers; and (4) the use of robotics to screen very large numbers of colonies. The new era of improvement is likely to come from genetic-based alterations of the host cells themselves. Among the many challenges

James L. Hartley (ed.), *Protein Expression in Mammalian Cells: Methods and Protocols*, Methods in Molecular Biology, vol. 801,
DOI 10.1007/978-1-61779-352-3_8,

faced during fed-batch processes are the buildup of toxins, growth factor withdrawal, and nutrient and oxygen deprivation. All of these insults together can potentially lead to cell death, which ultimately limits product yield. The predominant method of death in bioreactors is through apoptosis (2). Once the cell initiates this pathway, a signaling cascade takes place ultimately ending in death. One way to prolong cell life in bioreactors is to prevent or delay the cell from entering this pathway. This may be done through improved feed strategies and media development, but another more precise method is to engineer the cell line to be genetically resistant to apoptosis. In fact, antiapoptosis engineering has been applied successfully to increase culture longevity and increase product yield in biotechnologically relevant cell lines (3, 4).

There are several proteins known to have antiapoptotic effects when overexpressed in mammalian cells, which resulted in increased culture viability and product titer. Many of these come from the Bcl-2 family of proteins, perhaps the most famous being Bcl-X_L (3). Other examples include Aven and E1B-19K, where Nivitchanyong et al. (4) showed the beneficial effects of a combination of two antiapoptosis genes in a perfusion system. In this study, it was revealed that substantially decreased specific perfusion rates without the typically accompanying apoptosis were achievable without any decrease in cell viability. This lower perfusion rate could reduce the cost for media during production; however, the specific productivity was observed to decrease with the lower perfusion rates (4). Therefore, caution must be used and a balance must be reached between reduced cost and reduced productivity when applying these methods. Accordingly, overexpression of these genes is not a guarantee of increased productivity (5).

In order to generate cell lines overexpressing an antiapoptotic gene, the same steps are followed as if producing a cell line overexpressing a biotherapeutic product. First, one should decide on a host cell line that can withstand regulatory hurdles, namely, the host should be capable of growth in chemically defined animal protein-free (CD-APF) medium and have a well-documented origin. This could be a cell line that is already producing the product of interest or if creating a generic cell line resistant to apoptosis, it should be one that later can be used in a production setting. Once the cell line is chosen, one needs to decide which antiapoptotic gene to incorporate. Next, a vector must be chosen to allow for transfection of the host cell with the antiapoptotic gene. If the cell line chosen already contains the desired protein product, then a vector with a different selection marker will need to be chosen. In addition, a strong promoter would be useful to ensure sufficient expression of your antiapoptotic gene. The next steps are common to biotechnology and include transfection followed by selection, and then plating and screening for the clones with the desired characteristics. A second screening step, which includes an apoptotic insult, should be performed to identify clones that are resistant to apoptosis. Once individual clones are selected, analysis should be performed to ensure that both product and antiapoptotic genes are being expressed. This can be done by Western blot using appropriate antibodies. For production cell lines containing antiapoptotic genes, monitoring the product titer over multiple passages is indicative of stability. Following confirmation of gene expression, growth studies with apoptotic insults should be performed to ensure apoptotic resistance, improved culture viability, and ultimately productivity.

2. Materials

2.1. Transfection

1. An antiapoptosis gene, such as bcl-X_L, cloned into a mammalian expression plasmid containing a chemical selection marker, such as pcDNA3.1/Neo (Invitrogen, Carlsbad, CA) (see Notes 1 and 2).
2. A suspension-adapted mammalian cell line grown in chemically defined animal protein-free medium (see Note 3). The suspension-adapted Chinese hamster ovary cell line CHO-K1 is widely used in academia and industry and is used herein.
3. CD-APF medium, such as CD CHO (Invitrogen, Cat. #10743-011), supplemented with any additives necessary for growth, such as L-glutamine (Invitrogen, Cat. #25030-081) (see Note 4).
4. CEDEX Cell Counting System (Innovatis, Bielefeld, Germany) and related equipment or a hemacytometer and trypan blue.
5. Sterile conical tubes, 15 and 50 mL (BD Falcon, Cat. #35-2097, 35-2098, San Jose, CA).
6. Dulbecco's Phosphate Buffered Saline (D-PBS) without calcium or magnesium (Invitrogen, Cat. #14190-144).
7. Electroporation cuvettes BTX #640 or equivalent (Harvard Apparatus, Holliston, MA).
8. Ice bucket and ice.
9. Serum-containing medium to support CHO-K1 cell growth during recovery from transfection, such as Advanced DMEM+F12 (Invitrogen, Cat. #12634-010) supplemented with 10% FBS (Thermo Scientific, Cat. #SH30071.03, Waltham, MA) and 4 mM L-glutamine.
10. Electroporation system, such as the BTX ECM 630 Exponential Decay Wave Electro Cell Manipulator or equivalent (Harvard Apparatus).
11. Sterile, vented cap, 75 cm^2 tissue culture flasks (T75) (Corning, Cat. #430641, Corning, NY).

2.2. Selection

1. Serum-containing medium to support CHO-K1 cell growth, such as Advanced DMEM+F12 supplemented with 10% FBS and 4 mM L-glutamine.
2. Geneticin selection reagent (Invitrogen, Cat. #10131-035).
3. CEDEX Cell Counting System (Innovatis) and related equipment or a hemacytometer and trypan blue.
4. Sterile, vented cap, 125 or 250 mL Erlenmeyer flasks (shake flask) (Corning, Cat. #431143, 431144).

2.3. Plating

1. 96-well sterile tissue culture plates (Costar, Cat. #3595, Corning, NY).
2. 24-well sterile tissue culture plates (Costar, Cat. #3526, Corning, NY).
3. Serum-containing medium to support CHO-K1 cell growth, such as Advanced DMEM+F12 supplemented with 10% FBS and 4 mM L-glutamine.
4. Geneticin selection reagent (Invitrogen, Cat. #10131-035).
5. Inverted brightfield tissue culture microscope with 5×, 10×, and 20× objectives (Zeiss Axiovert 25 or equivalent, Carl Zeiss, Thornwood, CA).

2.4. Screening

1. Flow cytometer (Guava EasyCyte Plus or equivalent, Guava Technologies, Hayward, CA).
2. Guava Nexin Reagents Kit 4500-0450 or equivalent (Guava Technologies).
3. 96-well round bottom plates (BD Falcon, Cat.# 353910).

3. Methods

Generation of stable cell lines for the expression of an antiapoptotic gene and a protein product is a multistep procedure. This procedure has been outlined in Fig. 1.

3.1. Transfection

Because of transfection variability, perform the following electroporation in duplicate at a minimum.

1. Add 15 μg to 30 μg plasmid DNA into a sterile, disposable, electroporation cuvette (see Note 2).
2. Perform a cell count and aliquot 1×10^7 CHO-K1 cells, growing in exponential phase in CD-APF medium, into a sterile conical tube.
3. Centrifuge the cells for 5 min at $300 \times g$ and 4°C.

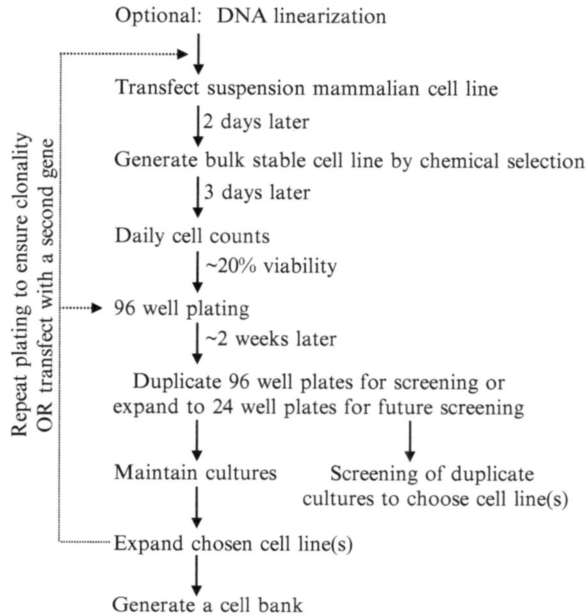

Fig. 1. Cell line generation process.

4. Decant the supernatant and resuspend in 10 mL sterile 1× D-PBS to wash the cells.
5. Centrifuge the cells for 5 min at 300×*g* and 4°C.
6. Decant the supernatant and resuspend the cells in sterile 1× D-PBS so that the final concentration is 1×10^7 cells/mL.
7. Transfer 1 mL of the culture to the electroporation cuvette containing the DNA, place the cap on the cuvette, and invert several times to mix the DNA and cells.
8. Incubate the cuvette on ice for 10–15 min.
9. While the cuvette is on ice, transfer 9 mL of cold serum-containing medium to a 15-mL sterile conical tube on ice.
10. Dry off the cuvette and place it in the electroporator. Shock the cells at 1,200 μF and 200 V (see Note 5).
11. Place 1 mL of the electroporated cells in the 9 mL cold medium, cap and invert to mix several times, and then return to ice for 5–10 min.
12. While on ice, remove a sample of the culture for a cell count.
13. Using the entire volume of cells from step 11, seed a T75 flask at 0.2×10^6 viable cells/mL using fresh media to dilute the cells as necessary but not to exceed 50 mL.
14. Incubate the culture for 48 h at 37°C and 5% CO_2.

3.2. Selection

If transient transfection instead of a stable transfection is desired, Subheadings 3.2 and 3.3 should be omitted. Proceed with Subheading 3.4 (see Note 6).

1. Replace the transfection medium with fresh serum-containing medium containing selection reagent (see Note 7). For the neomycin selectable marker, the corresponding selection reagent is Geneticin, also called G418.
2. Three days post selection, expand the culture to a shake flask at 0.2×10^6 viable cells/mL in medium containing selection reagent (see Note 8).
3. Four days post selection, begin taking daily cell counts. When the culture viability has dropped to approximately 20%, proceed to Subheading 3.3. Useful data may be derived by continuing daily cell counts to the point, where the culture has rebounded and the viability is improving.

3.3. Plating

If a bulk stable culture is desired instead of a clonal stable cell line, Subheading 3.3 should be omitted and proceed with Subheading 3.4 (see Note 9).

1. Seed each well of at least thirty 96-well plates at 100 viable cells/well in serum-containing medium with selection reagent (see Note 10).
2. Approximately 2 weeks later, expand the wells containing actively growing cultures to 24-well plates in fresh serum-containing selection medium. Visual detection of each well serves as a brute force approach for determining the cultures to expand; however, high-throughput robotics capable of plate expansion, such as the Hamilton StarPlus (Hamilton Company, Reno, NV) or Tecan Cellerity (Tecan, Männedorf, Switzerland) allow for more rapid expansion of a very large number of clones.
3. As the 24-well cultures become confluent, the cultures need to be duplicated or expanded for maintenance. Duplication into new 96-well or 24-well plates or expansion into 6-wells and then tissue culture flasks and shake flasks is dictated by the method of screening used.

3.4. Screening

In order to increase the probability of isolating stable cell lines with the desired antiapoptosis characteristics, a great number of independent colonies (transfectomas) need to be screened. Therefore, it is desirable to use a screening technique that can be performed in a plate format. For some of these tests, an inducer of apoptosis is needed (see Note 11).

Cells undergoing apoptosis are identifiable by a number of characteristics, including changes in mitochondrial membrane potential, activation of caspase proteases, DNA fragmentation in the nucleus, and transport of phosphatidylserine (PS) to the membrane surface (see Note 12). One of the most common tests for apoptosis resistance involves the latter, i.e., detection of externalized phosphatidylserine by flow cytometry. In order to test for apoptosis inhibition in this manner, one can compare the degree of apoptosis initiation and progression in each clone before and after exposure to an apoptotic insult. Using flow cytometry and an Annexin V binding assay, one can detect early signs of apoptosis. Annexin V-PE binds to PS, which is a membrane phospholipid normally located on the internal surface of the cell membrane but translocates to the outer surface early in apoptosis. A second dye called 7-amino-actinomycin D (7-AAD), that is excluded in healthy cells but enter cells whose membrane has lost structural integrity, can simultaneously be used to determine cells entering late-stage apoptosis. With these two dye systems, one can monitor populations of cells as they enter early-stage apoptosis and transition into late-stage apoptosis and cell death.

1. Seed 96-well plates such that the concentration is 2×10^5–1×10^6 cells/mL (2×10^4–1×10^5 cells in 100 μL) (see Note 13). Add the same apoptotic insult to each well and incubate the plates for a set amount of time dependent upon insult used (see Note 11).
2. Bring a vial of nexin to room temperature and add 100 μL of this reagent to each well.
3. Incubate samples at room temperature in the dark for 20 min.
4. Run each 96-well plate on the flow cytometer.
5. Analyze results: Nonapoptotic cells are negative for both Annexin V and 7-AAD, early-stage apoptotic cells are positive for Annexin-V and negative for 7-AAD while late-stage apoptotic and dead cells stain positive for both.

Once the number of cultures has been narrowed down to a manageable range, further tests should be performed. These include Western blots to verify expression of the antiapoptotic gene(s) and growth profiles with and without exposure to apoptotic insults. Additionally, the final candidate cell lines should be tested for stability of expression of the antiapoptotic gene. For this purpose, the selected cell lines should be monitored over multiple passages, preferably excluding the selection agent used during the screening process (see Note 14). From these screening methods, an apoptotic-resistant host cell line should be identified for use in the generation of a manufacturing cell line. Generate a frozen cell bank.

3.5. Generation of a Manufacturing Cell Line

Once an apoptosis-resistant cell line has been identified, this cell line can be used now as a host for the generation of a protein of interest, such as a monoclonal antibody or growth factor. In order to generate a cell line for the manufacture of a protein product, similar protocols

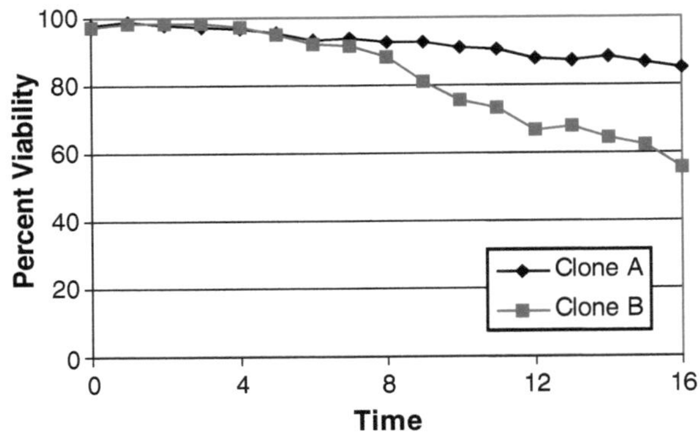

Fig. 2. Percent viability versus time for a fed-batch bioreactor run. Clone A is a suspension-adapted CHO cell line expressing a monoclonal antibody and an antiapoptotic gene. Clone B is a suspension-adapted CHO cell line expressing a monoclonal antibody only.

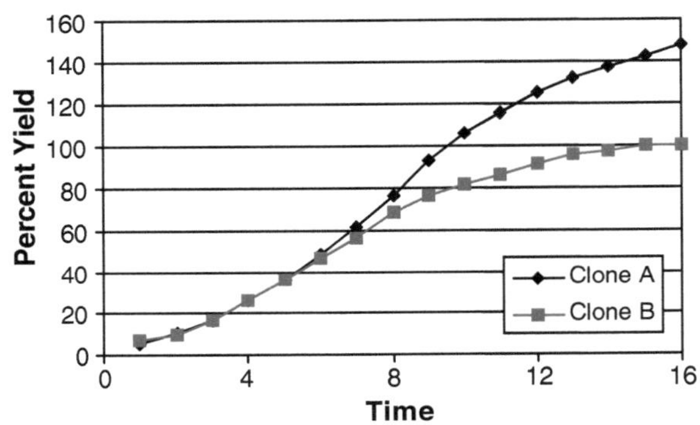

Fig. 3. Percent yield of monoclonal antibody production versus time for a fed-batch bioreactor run. Clone A is a suspension-adapted CHO cell line expressing a monoclonal antibody and an antiapoptotic gene. Clone B is a suspension-adapted CHO cell line expressing a monoclonal antibody only. The day 16 yield for clone B was set as 100% to allow for determination of the increase in yield of clone A.

used to generate the antiapoptosis host cell line are followed for transfection, screening, and plating (see Note 15). Once clones have been identified, additional screening consists of methods specific for the protein product, such as an ELISA or Western blot, followed by growth profiles and a stability study (see Note 14).

Expressing an antiapoptosis gene along with the gene coding for the product of interest may have great benefit. Increased culture viability is observed which may lead to increased product yield. An example of a mammalian cell line expressing a monoclonal antibody along with an antiapoptosis gene is shown in Figs. 2 and 3. In Fig. 2, at day 16, clone A, which expresses the antiapoptosis gene and a monoclonal antibody, has a viability 30% greater than that of clone B expressing only the monoclonal antibody. In Fig. 3, clone A has an end of production yield of almost 150% of clone B (see Note 16).

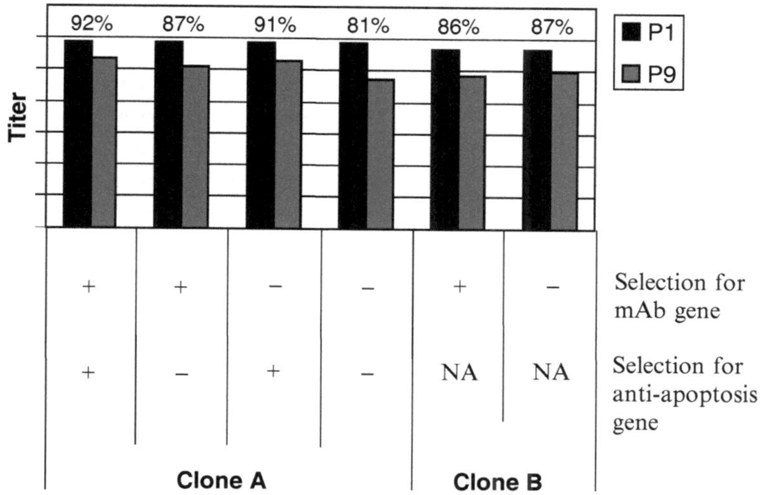

Fig. 4. Nine passage shake flask stability study comparing clone A to clone B. Clone A is a suspension-adapted CHO cell line expressing a monoclonal antibody (mAb) and an anti-apoptotic gene. Clone B is a suspension-adapted CHO cell line expressing a monoclonal antibody only.

When a nine passage stability study is performed on these same two clones in the presence and absence of selection, no discrepancies in the modest drops in titer are observed with time (Fig. 4). Both clones, regardless of the selection pressure, maintain monoclonal antibody production at the same levels within the error of the experimental measurement. Therefore, the presence of an anti-apoptosis gene does not affect the stability of the cell line producing a monoclonal antibody.

Not just product titer, but product quality may improve when manufactured in apoptosis-resistant cell lines. This is particularly true if serum-free medium is being employed. Absence of serum enhances a subset of the cellular proteolytic repertoire, which would otherwise be adsorbed by serum proteins nonspecifically (6). For products that are susceptible to proteolytic degradation in cell culture, use of apoptotic-resistant hosts may prove to be advantageous. With higher viabilities observed with antiapoptotic cell lines, less cell lysis occurs resulting in a lower amount of proteases in the culture medium resulting in less product degradation.

4. Notes

1. There are many antiapoptotic genes available for use. The most common genes include Bcl-2, Bcl-X_L, and E1B19K that inhibit the upstream portion of the mitochondrial pathway as well as

caspase inhibitor genes, including XIAP, that suppress apoptosis downstream in the apoptotic pathway. Additionally, there are numerous mammalian selection markers available. These include chemical inhibitors, such as neomycin, hygromycin, Zeocin, and blasticidin, as well as metabolic pathway inhibitors, such as the glutamine synthetase and dhfr^{-} systems. If it is desired to overexpress two antiapoptosis genes in a cell line, both genes may be placed on one plasmid or a dual transfection may be performed. It is advised to have each gene under its own selectable marker to ensure incorporation into the genome. When an upstream mitochondrial pathway inhibitor is paired with a downstream caspase inhibitor, the culture's viability may be greatly extended compared to either gene expressed individually (7).

2. Prior to transfection, the plasmid may be linearized by restriction enzyme digestion in the bacterial selection gene assuming that the bacterial selection is not the mammalian selection, as with Zeocin. In this case, linearization should occur elsewhere in the plasmid backbone so as not to disturb the transcription of the gene of interest. This step will be useful if insertion site of the plasmid into the genome will be examined for regulatory filings at a later time.
3. There are several commonly used suspension cell lines available. These include CHO, baby hamster kidney (BHK), NS0, and human embryonic kidney (HEK 293). Other less common mammalian cell lines may work also. If it is desired to use adherent cells, the protocols used will differ from those described herein.
4. The cell line used dictates the medium and there are commercially available media for many of the common cell lines. However, in industry, proprietary custom-made medium is typically used.
5. Electroporation conditions should be optimized. Both square wave and exponential wave electroporators may be utilized and optimization of the shock conditions should be performed for each machine and the cells used. Square wave electroporation, such as with the ECM 830 Square Wave Electroporator (Harvard Apparatus), allows for higher viabilities upon transfection, whereas the exponential wave electroporator allows for more flexibility in optimization. Lipid-based transfection methods may be used also. Examples include Lipofectamine and Lipofectamine 2000 by Invitrogen and FuGENE HD by Roche (Indianapolis, IN). For the highest transfection efficiency, it is recommended to follow the manufacturer's protocols for these commercially available products.

6. In some instances, it is desired to perform a transient transfection as opposed to a stable transfection. The benefits to performing a transient transfection include testing the plasmid to verify expression of the antiapoptosis gene and determining the effects of the gene of interest in a rapid manner. For this approach, the HEK 293 cell line is an ideal choice. However, if a cell line is needed that continuously expresses the gene of interest, Subheadings 3.2 and 3.3 should be performed.
7. Prior to selection, a dose-response curve should be performed on the untransfected host cell line using a range of selection reagent concentrations. This allows the user to determine the optimal concentration of selection reagent for the cell type used. For CHO-K1 cells, approximately 400 μg/mL Geneticin works well. Briefly, seed five flasks of untransfected cells at 2×10^5 cells/mL in medium containing your selection agent ranging from 50 to 1,000 μg/mL. Passage these flasks as needed (every 3–5 days) in medium containing the same concentration of selection agent. The concentration that kills the untransfected cells in 10–14 days is the one that should be used for selection after transfection.
8. Some suspension-adapted cell lines adhere to the tissue culture surface post transfection and selection. If the cells are not easily removed by gentle smacking of the flask on a flat surface and repeat pipetting, the use of Trypsin–EDTA or TrypLE (Invitrogen, Cat. #12563029) will be needed to disassociate the cells from the plastic surface. Briefly, after removing the growth medium containing the suspension cells from the flask into a sterile conical tube, wash the cells with 1× PBS at 1–2 mL/25 cm^2. Remove the PBS and add 1 mL/25 cm^2 Trypsin–EDTA, and ensure coverage of all adhered cells by rocking the flask back and forth to distribute the Trypsin evenly. Incubate the cells at room temperature for 2–10 min until the cells are released from the surface. Triterate the cells several times to disperse any clumps, and then place the cells containing Trypsin into the culture in the sterile conical tube. Continue with cell counting using an aliquot from the culture in the conical tube, and then perform routine passaging. The use of TrypLE is similar and manufacturer's instructions are included with the product.
9. In some instances, it is desired to obtain a bulk stable culture as opposed to a stable clonal cell line. This allows for the detection of trends and behaviors resulting from the gene of interest without being confounded by clonal isolation effects.
10. The seeding density may need to be optimized for the cell type and chemical selection used. The goal is to obtain cultures that derive from 1 cell/well, but if the seeding density is that low, the cultures may not survive. Therefore, plate at a cell

concentration that allows for the growth of colonies and establish the cell line in growth medium. Then too, ensure clonality of clones derived from plating at greater than 1 cell/well; a second round of 96-well plating may be performed at a lower seeding density, such as 1 cell/well. The cells may have a greater chance at survival in the subcloning stage at the lower seeding density because their growth in selection medium has been firmly established. At a minimum, plate at least thirty 96-well plates. The greater number of plates, the greater the probability of isolating clones with the desired antiapoptosis characteristics of high viability and culture longevity. However, if robotics are not available for high-throughput culture maintenance, this step may become burdensome to the scientist.

11. There are many methods to induce apoptosis in cell culture, including nutrient deprivation and chemical exposure. Some commercially available chemicals used to induce apoptosis in cell culture include staurosporine, etoposide, thapsigargin, and cisplatin. Because the rate of onset of apoptosis varies for each chemical and cell line, dosing concentrations vary dependent on cell line used. Therefore, a dose-response curve should be performed on the untransfected host cell line prior to use as a screening tool (see Notes 7 and 12).
12. There are numerous commercially available kits that allow for the detection of apoptosis in cell culture. These include various caspase assays and DNA fragmentation assays. Some of these kits allow for adaptation for high-throughput analysis using cultures in a 96-well plate format. Others require the use of cell lysates or supernatants that add additional time to the screening procedure. Methods for more cytometric and biochemical techniques have been described by Ishaque and Al-Rubeai (8).
13. Cell concentrations among all culture samples must be similar. In addition, both positive and negative controls should be run allowing for effective discrimination of noninduced, early-, and late-stage apoptosis.
14. A shake flask study should be set up for the chosen cell line with and without selection. The study should last for a duration equivalent to the number of generations observed from thaw of a cell bank through to the completion of the bioreactor run. Approximately once a week or whenever deemed necessary, a sample should be removed for Western blot analysis. At the end of the study, all samples should be run on a Western blot in chronological order and the bands analyzed by densitometry to determine if a change in expression of the transfected anti-apoptosis gene has occurred. If the cell line also has been transfected with a gene encoding for a product of interest, cultures should be split periodically maintaining one culture in normal passaging while setting aside the other to overgrow. Once the

overgrown cultures are at approximately 20% viability or lower, remove samples to obtain the protein titers over time. Ideally, the newly generated cell line is more resistant to apoptosis and the time it takes to reach 20% viability increases compared to the non-antiapoptotic gene-expressing cell line. A stable cell line demonstrates stable expression with time.

15. A new dosing curve is required (see Note 7) once an antiapoptosis cell line is established as this cell line is more resistant to death by many selection markers as compared to the parent cell line.

16. Other examples displaying the benefits of using antiapoptosis genes in conjunction with the expression of a protein product exist in literature. In Figueroa et al., an inducible system used in combination with fed-batch bioreactor operation led to an approximately 50% increase in monoclonal antibody production for a CHO-E1819K+Aven cell line when compared to the parental CHO line (9). When E1B19K+Aven was expressed in combination with recombinant Factor VIII in a perfusion bioreactor system, specific productivity was increased for the antiapoptotic BHK line compared to the parental BHK line for all perfusion rates studied (4).

References

1. DePalma, A. (2008) Strengthening mammalian cell culture. *Genetic Engineering & Biotechnol. News*. **28**.
2. Arden, N. and Betenbaugh, M.J. (2004) Life and death in mammalian cell culture: strategies for apoptosis inhibition. *Trends Biotechnol.* **22**, 174–180.
3. Chiang, G.G. and Sisk, W.P. (2005) Bcl-x(L) mediates increased production of humanized monoclonal antibodies in Chinese hamster ovary cells. *Biotechnol. Bioeng.* **91**, 779–792.
4. Nivitchanyong, T., Martinez, A., Ishaque, A., Murphy, J.E., Konstantinov, K., Betenbaugh, M., and Thrift, J. (2007) Anti-apoptotic genes Aven and E1B-19 K enhance performance of BHK cells engineered to express recombinant factor VIII in batch and low perfusion cell culture. *Biotechnol. Bioeng.* **98**, 825–841.
5. Kim, Y.G. and Lee, G.M. (2009) Bcl-xL overexpression does not enhance specific erythropoietin productivity of recombinant CHO cells grown at 33 degrees C and 37 degrees C. *Biotechnol. Prog.* **25**, 252–256.
6. Dorai, H., Nemeth, J., Cammaart, E., Wang, Y., Tang, Q., Magill, A., Lewis, M.J., Raju, T.S., Picha, K., O'Neil, K., Ganguly, S., and Moore, G. (2009) Development of mammalian production cell lines expressing CNTO 736, a glucagon like peptide-1-MIME TIBODY™: Factors that influence productivity and product quality. *Biotechnol. Bioeng.* **103**, 162–176.
7. Sauerwald, T.M., Figueroa Jr., B., Hardwick, J.M., Betenbaugh, M.J. and Oyler, G.A. (2006) Combining caspase and mitochondrial dysfunction inhibitors of apoptosis to limit cell death in mammalian cell culture. *Biotechnol. Bioeng.* **94**, 362–372.
8. Ishaque, A., and M. Al-Rubeai. (2007). Measurement of apoptosis in cell culture, in *Methods in Biotechnology, Animal Cell Biotechnology* (Ralf Portner, ed.) Human Press, Totowa.
9. Figueroa Jr., B., Ailor, E., Osborne, S., Hardwick, J.M., Reff, M., and Betenbaugh, M.J. (2006) Enhanced cell culture performance using inducible anti-apoptotic genes E1B-19 K and Aven in the production of a monoclonal antibody with Chinese hamster ovary cells. *Biotechnol. Bioeng.* **97**, 877–892.

Chapter 9

Post-transcriptional Regulatory Elements for Enhancing Transient Gene Expression Levels in Mammalian Cells

Mariati, Steven C.L. Ho, Miranda G.S. Yap, and Yuansheng Yang

Abstract

Low yield from transient gene expression in mammalian cells limits its application to areas where large amount of proteins are needed. One effective approach to enhance transient gene expression levels is to use post-transcriptional regulatory elements (PTREs). We have evaluated the effect of five PTREs on the transient gene expression of three proteins in two cell lines. Most of the elements increased expression but exhibited cell-specific and gene-specific effects. The tripartite leader sequence of human adenovirus mRNA linked with a major late promoter enhancer gave the most universal and highest enhancement of gene expression levels. It increased the expression of all three proteins in HEK293 cells and two proteins in CHO K1 cells by 3.6- to 7.6-fold. Combinations of multiple PTREs increased protein expression as much as 10.5-fold.

Key words: Post-transcription, Mammalian cells, Transient gene expression, Untranslated region, Intron, WPRE

1. Introduction

Transient gene expression in mammalian cells allows for rapid generation of recombinant proteins for applications in biochemical, biophysical, and preclinical studies (1–4). However, low productivity remains a major obstacle for the application of this technology to areas where large amounts of proteins are needed (5). One effective approach for enhancing transient gene expression levels is optimizing the expression vector to increase the protein expression level per gene copy (5–7). Gene expression in mammalian cells is regulated by a cascade of events that includes transcription, RNA processing, mRNA transport, mRNA degradation, translation, post-translational modification, and secretion (for secreted proteins) (7–9). Besides using strong promoters and enhancers to increase

James L. Hartley (ed.), *Protein Expression in Mammalian Cells: Methods and Protocols*, Methods in Molecular Biology, vol. 801,
DOI 10.1007/978-1-61779-352-3_9, © Springer Science+Business Media, LLC 2012

transcription, the protein expression level per gene copy can also be increased by improving the efficiency of downstream events by using post-transcriptional regulatory elements (PTREs), such as introns, polyadenylation signals, untranslated regions (UTRs) of mRNA, and signal peptides (10–16).

PTREs that have been reported to enhance gene expression levels include: (1) UTRs derived from either eukaryotic or viral mRNA, such as the 5′ UTR of the human heat shock protein 70 mRNA (Hsp70) (17), the 163-bp splice variant derived from the 5′ UTR of vascular endothelial growth factor mRNA (SP163) (18), and the tripartite leader sequence (5′ UTR of human adenovirus mRNA) linked with a major late promoter enhancer (TM) (19–21), (2) introns, such as the first intron of human cytomegalovirus immediate early gene (Intron A) (22–25), and (3) PTREs derived from hepadnaviruses, such as the woodchuck hepatitis virus (WPRE) (10, 26–30). Although the exact mechanisms of these elements have not been elucidated, it has been proposed that UTRs contribute to mRNA stability and translation, introns enhance polyadenylation, mRNA stability and export, and WPRE stimulates RNA processing, mRNA export, and translation (17–30).

In this work, we evaluated the effect of five PTREs, Hsp70, SP163, TM, Intron A, and WPRE, on the transient expression of firefly luciferase (Fluc), human interferon γ (IFN-γ), and anti-HER2 monoclonal antibody (Trastuzamab) in HEK293 and CHO K1 cells. The protocols can be adapted to evaluate the effect of other PTREs on the transient expression of other proteins in different cell lines.

2. Materials

2.1. Plasmid Construction and Purification

1. pcDNA3.1(+) (Invitrogen, Carlsbad, CA, USA), store at −20°C.
2. Vectors used for cloning of PTREs are listed in Table 1, store at −20°C.
3. QIAquick gel extraction and PCR purification kits (QIAGEN, Duesseldorf, Germany).
4. *Escherichia coli* competent cells (DH5α) (Invitrogen).
5. PureYield™ Plasmid Midiprep System (Promega, Madison, WI, USA).

2.2. Transient Transfection

1. CHO K1 (CCL-61) and HEK293 (CRL-1573) (ATCC, Manassas, VA, USA).
2. Dulbecco's modified Eagle's medium (DMEM) + GlutaMax™ (Invitrogen), store at 4°C.

Table 1
Vectors used for cloning of PTREs

Vector	Source	PTRE in vector	Accession number	PCR primer (5′–3′)
–	From Dr. Vivinus (17)	Hsp70 5′ UTR[a]	M11717	F: AACGGCTAGCCTGAGGAG R: GCGGTTCCCTGCTCTCTG
pcDNA4/ HisMax	Invitrogen	SP163 5′ UTR	–	F: AGCGCAGAGGCTTGGGGC R: GGTTTCGGAGGCCGTCCG
pTT3	From Dr. Durocher (6)	TM Adeno 5′ UTR – enhancer	–	F: CAGATCCTCACTCTCTTC R: TTGGACCTGGGAGTGGAC
gWiz	Genlantis	CMV Intron A	M60321	F: TCAGATCGCCTGGAGACG R: CTGCAGAAAAGACCCATG
pSF91EGFP	From Dr. Schambach (28)	Woodchuck PRE	J02442	F: AGCATCTTACCGCCATTT R: GAAAGGACGTCAGCTTCC

[a]Hsp70 was synthesized by overlapping PCR based on the sequence provided by Dr. Vivinus

3. Fetal bovine serum (FBS, Sigma, St. Louis, USA), store at −20°C.
4. 0.5% trypsin-EDTA (Invitrogen).
5. FuGENE 6 (Roche, Indianapolis, IN, USA), store at −20°C for up to 2 years.
6. pRL-CMV (Promega), store at −20°C.

2.3. Luciferase Assay

Dual-Glo™ luciferase assay system (Promega): (1) 10 mL Dual-Glo™ luciferase buffer, store at 4°C; (2) 1 Vial Dual-Glo™ luciferase substrate (lyophilized), store at −20°C; (3) 10 mL Dual-Glo™ Stop & Glo® buffer, store at 4°C; (4) 100 μL Dual-Glo™ Stop & Glo® substrate, store at −20°C.

3. Methods

High variation in transfection efficiency can make the subtle changes in transient gene expression levels resulting from added DNA elements hard to detect. To obtain reliable and reproducible results, it is important to use identical culture conditions and cell numbers across different transfections, utilize plasmids of high quality, and co-transfect an internal control vector to normalize the transfection efficiency. The method of using Fluc as the reporter gene and renilla luciferase (Rluc) as the internal control to evaluate the strength of five PTREs (Table 1), either alone or in combination, is presented (see Note 1). Transfection is carried out in 96-well plate. The Fluc and Rluc activities are measured by using Dual-Glo™ luciferase assay system. Dual-luciferase method is recommended for the initial evaluation of PTREs because of its quickness, simplicity, and sensitivity.

The effect of PTREs may vary with the cell lines and genes to be expressed. It is advisable to evaluate the benefit of a PTRE for a gene of interest in the host cell lines despite its positive effects in other cells or for other genes. The procedures for evaluating PTREs on the transient gene expression of another two proteins, IFN-γ and trastuzamab, are described. Rluc is not co-transfected because its expression is inhibited by the expression of IFN-γ and trastuzamab (see Note 2). Without internal control, transfections should be repeated at least one more time using independently prepared plasmids to confirm that the observed effect is not due to variation in plasmid quality. Moreover, transfection in six-well plate is more favorable than in smaller sized well plates since the number of cells used in transfection can be accurately determined. The concentration of IFN-γ and trastuzamab in the supernatant, which are secreted proteins, is quantified by using standard enzyme-linked immunosorbent assay (ELISA) (protocols not shown). The ELISA assay allows accurate quantification of protein concentration. Appropriate methods to measure gene expression, such as Western blot, GFP fusions, or activity assays, can be chosen depending on the specific experiment and purpose.

3.1. Plasmid Construction and Purification

1. Clone reporter genes or the gene of interest into a standard mammalian expression vector, such as pcDNA3.1(+), using restriction enzymes and ligase according to standard techniques (Fig. 1) (see Notes 3–5).
2. Amplify PTREs to be tested from corresponding templates using PCR primers with 3′ termini shown in Table 1, and restriction sites in the 5′ termini according to your cloning scheme (see Note 6).
3. Verify the sizes and good yield of PCR products by gel electrophoresis, before extracting and purifying PCR products from the gel using QIAquick gel extraction kit.
4. Digest PCR products and corresponding plasmids with restriction enzymes to give compatible ends. Dephosphorylate the digested plasmid ends to reduce occurrences of self-ligation. After verifying complete digestion of plasmids by gel electrophoresis, purify the digested plasmids and PCR products using QIAquick gel extraction and PCR purification kits, respectively, and then ligate according to standard techniques (see Note 7).

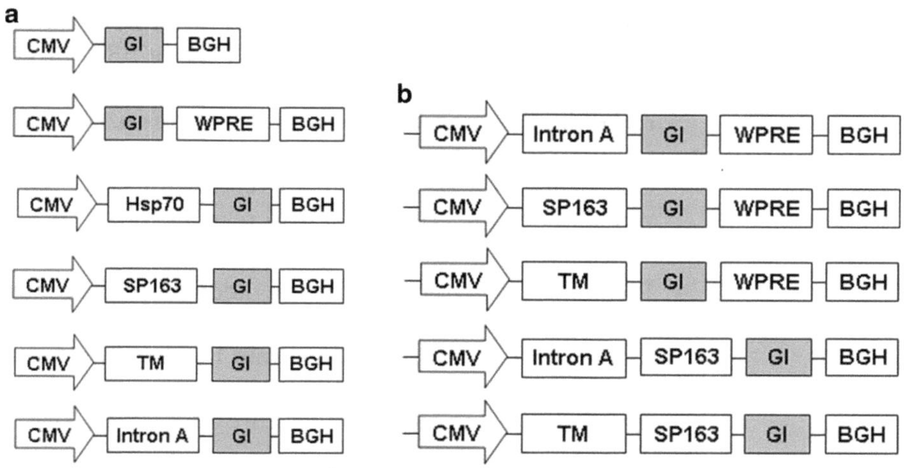

Fig. 1. Schematic representation of vectors. (**a**) Vectors containing single post-transcriptional element. (**b**) Vectors containing two post-transcriptional elements. *CMV* human cytomegalovirus immediate early gene promoter, *BGH* bovine growth hormone polyadenylation signal, *Intron A* first intron of human cytomegalovirus immediate early gene, *Hsp70* 5′ untranslated region (UTR) of human heat shock protein 70 mRNA, *SP163* 163-bp splice variant derived from the 5′ UTR of vascular endothelial growth factor mRNA, *TM* tripartite leader sequence (the 5′ UTR of human adenovirus mRNA) linked with a major late promoter enhancer, *WPRE* post-transcriptional regulatory element from woodchuck hepatitis virus, *GI* gene of interest. Reproduced from (Protein Expression and Purification 2010) with permission from ELSEVIER (31).

5. Transform each ligation into a standard cloning strain and grow in 2 mL culture. Extract plasmids and confirm all sequences that have been PCR amplified (see Note 8).
6. Expand the strain which contains the verified plasmid in 50 mL culture, extract plasmids and purify using PureYield™ Plasmid Midiprep System. Determine concentration and purity at wavelength of 260 nm and 280 nm on a spectrophotometric reader (see Note 9).

3.2. Transient Transfection

1. Maintain HEK293 and CHO K1 cells in DMEM + GlutaMax™ supplemented with 10% FBS in a humidified 37°C incubator with 5% CO_2. Conduct subculture every 3–4 days by detaching the cells with 0.05% trypsin-EDTA, counting by trypan blue exclusion method, and diluting in fresh medium to 2×10^5 cells/mL in T-flasks.
2. One day before the transfection experiment, detach cells at exponential phase (~1×10^6 cells/mL) from T-flasks, dilute in fresh medium to $2–3 \times 10^5$ cells/mL, seed 100 μL to each well of a tissue culture treated 96-well plate and 2 mL to each well of a tissue culture treated six-well plate, and culture in the incubator until transfection (see Notes 10 and 11).
3. Calculate the amount of FuGENE 6 transfection reagent, plasmid DNA, and serum-free DMEM + GlutaMax™ needed to perform the desired transfections (see Note 12). The recipe for

one transfection of luciferase vectors in 96-well plate is 0.2 μg of the appropriate Fluc vector, 0.02 μg of pRL-CMV vector, and 0.6 μL of FuGENE 6 mixed in 50 μL of medium. The recipe for one transfection of IFN-γ or trastuzamab vectors in six-well plate is 2 μg of the appropriate IFN-γ, or 1 μg of light chain and 1 μg of heavy chain vectors for expressing trastuzamab, and 6 μL of FuGENE 6 reagent mixed in 100 μL of medium.

4. Prepare transfection cocktails by adding specified amount of serum-free DMEM + GlutaMax™ into sterile tubes, followed by FuGENE 6 reagent and plasmid DNA, gently tap the tube to mix the contents (avoid vigorous vortexing), and incubate for 15–45 min at room temperature (see Note 13).
5. Add the transfection cocktail of 50 μL to each culture in 96-well plates or 100 μL to each culture in six-well plates, with gentle swirling to ensure even distribution around the well.
6. Return the plates into the incubator until the measurement of luciferase activities at 24 h post-transfection or collection of supernatant for the analysis of IFN-γ and trastuzamab concentration at 48 h post-transfection (see Note 14).

3.3. Luciferase Assay

1. Prepare all materials required for the assay: (1) Dual-Glo™ luciferase reagent, prepared by dissolving the Dual-Glo™ substrate in the Dual-Glo™ luciferase buffer; (2) Dual-Glo™ Stop&Glo® reagent, prepare only the amount needed for the experiment by diluting the Dual-Glo™ Stop&Glo® substrate 1:100 into an appropriate volume of Dual-Glo™ Stop&Glo® buffer (see Note 15).
2. Measure Fluc activity: (1) At 24 h post-transfection, add 75 μL of Dual-Glo™ luciferase reagent into each culture in 96-well plates and mix; (2) Incubate at room temperature for at least 10 min but not more than 2 h; (3) Measure the firefly luminescence on a Tecan plate reader.
3. Measure Rluc activity: (1) After the measurement of Fluc activity, add 75 μL of Dual-Glo™ Stop&Glo® reagent into each well and mix; (2) Incubate at least 10 min but not more than 2 h at room temperature; (3) Measure the renilla luminescence on the Tecan plate reader.
4. Calculate the ratio of Fluc to Rluc reading from each transfection. Normalize this ratio to the ratio of cultures transfected with the control vector which does not contain any PTREs. Examples of results produced for individual and combination of elements are shown in Figs. 2–4, respectively.

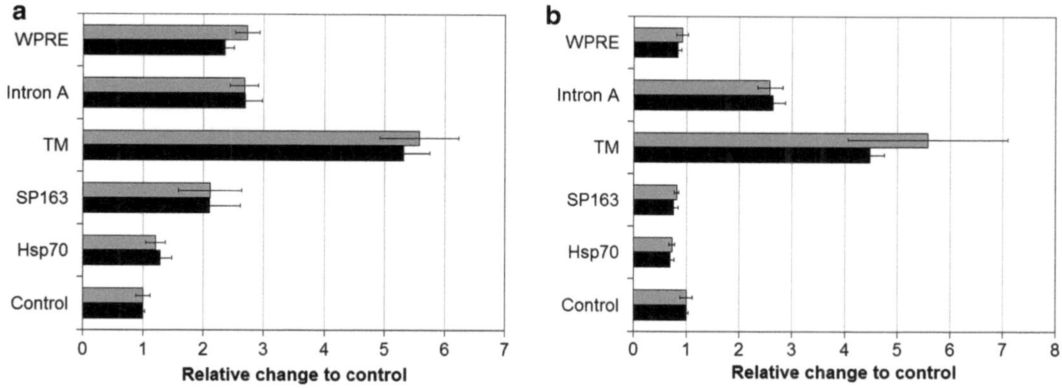

Fig. 2. Effects of individual post-transcriptional elements on transient luciferase expression in HEK293 (**a**) and CHO K1 (**b**) cells. Results represent the ratio of Fluc to Rluc activities normalized to the control vector. Each point represents the average and standard deviation of measurements from four transfections, and that *black* and *gray bars* represent separate experiments done with independently prepared plasmids at different times. Reproduced from (Protein Expression and Purification 2010) with permission from ELSEVIER (31).

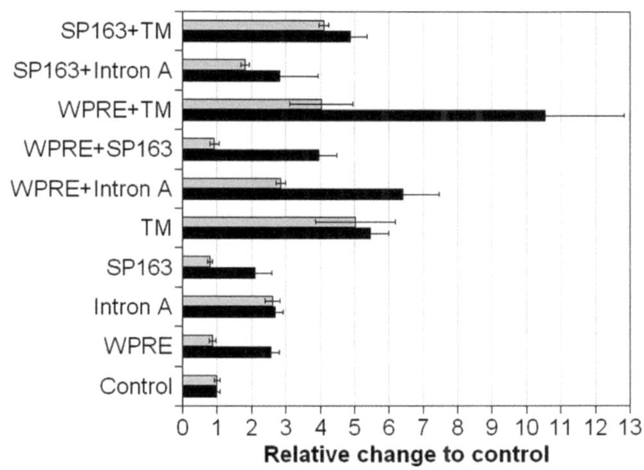

Fig. 3. Combinatorial effects of different post-transcriptional elements on transient luciferase expression. Results represent the ratio of Fluc to Rluc activities normalized to the control vector. Each point represents the average and standard deviation of measurements from eight transfections carried out in two independent experiments. *Black* and *grey bars* represent HEK293 and CHO K1 cells, respectively. Reproduced from (Protein Expression and Purification 2010) with permission from ELSEVIER (31).

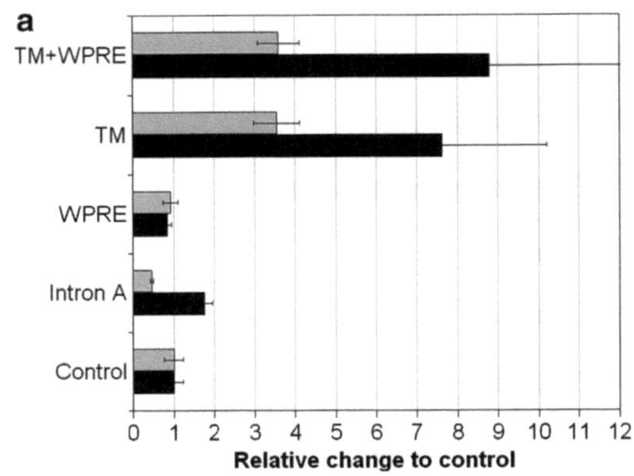

Fig. 4. Effect of post-transcriptional elements on transient expression of IFN-γ (**a**) and trastuzamab. (**b**) In HEK293 cells, the control vector lacking post-transcriptional elements gave 0.9 μg/mL of IFN-γ and 1.0 μg/mL of trastuzamab. In CHO K1 cells, the control vector gave 7.0 μg/mL of IFN-γ and 8.2 μg/mL of trastuzamab. Results in the figures represent the changes in protein concentration from each vector normalized to the control vector. The concentration of IFN-γ and trastuzamab was measured using ELISA. Each point represents the average and standard deviation of measurements from four transfections carried out in two independent experiments. *Black* and *grey bars* represent HEK293 and CHO K1 cells, respectively. Reproduced from (Protein Expression and Purification 2010) with permission from ELSEVIER (31).

4. Notes

1. Variation in transfection efficiency due to plasmid quality could be greater than 30%. Co-transfection with another reporter plasmid as internal control is highly recommended to normalize the transfection efficiency variation.
2. In transfections which increased the expression of IFN-γ and Trastuzamab, such as from Intron A and TM, the expression of the co-transfected internal control, Rluc, was much lower than that in cultures transfected with the control vector (~10% of that in control), suggesting its expression was inhibited.
3. The same plasmid backbone should be used for all comparisons so that changes in gene expression across different vectors are specifically a result of adding PTREs.
4. Because PTREs will be cloned into these plasmids in subsequent steps, it is important to have the complete cloning scheme for each plasmid planned ahead of time so that all MCS restriction sites are available and unique for your particular gene of interest.
5. The location and orientation of PTREs in the vector are designed based on literatures. Placement of an element in different locations may not provide positive effect. For instance, WPRE only functions when it is placed between the reporter gene and polyadenylation signal in the sense orientation (30).

6. Extra bases need to be added to the ends of the PCR product for good cleavage for some restriction enzymes. Flanking the restriction sites with six bases will ensure efficient cleavage for most enzymes.
7. It is recommended to determine the concentration of digested vector and insert by gel electrophoresis before ligation. Using a ratio of digested vector and insert at a ratio of 1:3 is optimal for successful ligation.
8. It is critical to confirm all sequences that have been PCR amplified so that the wrong conclusions are not drawn about how best to increase the expression of your genes.
9. Plasmid quality is critical for transfection efficiency. It is recommended to use only preparations with A_{260}/A_{280} between 1.8 and 2.0 for transfection.
10. Cells at exponential phase are preferable for higher transfection efficiency. Do not use cells after 30 passages for transfection.
11. The appropriate seeding density depends on the growth rate and size of culture plates. Cells that are 50–80% confluent are preferable on the day of transfection.
12. The ratio of FuGENE 6 (μL):DNA (μg) is critical for transfection efficiency. For a specific cell line, it would be advisable to test a broader range of ratios for optimal transfection efficiency.
13. Chemical residues in plastic vials can significantly decrease the biological activity of the FuGENE 6 transfection reagent. Always store it in the original polypropylene tubes and do not aliquot into new plastic vials. When preparing transfection cocktails, it is compulsory to add the serum-free medium first before addition of FuGENE 6 reagent.
14. Transient gene expression usually peaks at 24–48 h post-transfection. The time of collecting samples can be extended for higher yield of secreted proteins.
15. Dual-Glo™ assay reagents are stable at room temperature for a few hours. Prepare the Dual-Glo™ luciferase reagent on the day it is used. The leftovers can be stored at −70°C for up to 1 month. Do not thaw the reconstituted Dual-Glo™ luciferase reagent at temperatures above 25°C. For Dual-Glo™ Stop&Glo® reagent, prepare only the amount required on the day it is to be used.

Acknowledgments

This work was supported by the Biomedical Research Council/Science and Engineering Research Council of A*STAR (Agency for Science, Technology and Research), Singapore.

References

1. Baldi, L., Hacker, D. L., Adam, M., and Wurm, F. M. (2007) Recombinant protein production by large-scale transient gene expression in mammalian cells: state of the art and future perspectives, *Biotechnology Letters* *29*, 677–684.
2. Baldi, L., Muller, N., Picasso, S., Jacquet, R., Girard, P., Thanh, H. P., Derow, E., and Wurm, F. M. (2005) Transient gene expression in suspension HEK-293 cells: Application to large-scale protein production, *Biotechnology Progress* *21*, 148–153.
3. Carpentier, E., Paris, S., Kamen, A. A., and Durocher, Y. (2007) Limiting factors governing protein expression following polyethylenimine-mediated gene transfer in HEK293-EBNA1 cells, *Journal Of Biotechnology* *128*, 268–280.
4. Wurm, F., and Bernard, A. (1999) Large-scale transient expression in mammalian cells for recombinant protein production, *Current Opinion In Biotechnology* *10*, 156–159.
5. Backliwal, G., Hildinger, M., Chenuet, S., Wulhfard, S., De Jesus, M., and Wurm, F. M. (2008) Rational vector design and multi-pathway modulation of HEK 293E cells yield recombinant antibody titers exceeding 1 g/l by transient transfection under serum-free conditions, *Nucleic Acids Research* *36*.
6. Durocher, Y., Perret, S., and Kamen, A. (2002) High-level and high-throughput recombinant protein production by transient transfection of suspension-growing human 293-EBNA1 cells, *Nucleic Acids Research* *30*.
7. Kaufman, R. J. (2000) Overview of vector design for mammalian gene expression, *Molecular Biotechnology* *16*, 151–160.
8. Barnes, L. M., Bentley, C. M., and Dickson, A. J. (2003) Stability of protein production from recombinant mammalian cells, *Biotechnology And Bioengineering* *81*, 631–639.
9. Kim, C. H., Oh, Y., and Lee, T. H. (1997) Codon optimization for high-level expression of human erythropoietin (EPO) in mammalian cells, *Gene* *199*, 293–301.
10. Brun, S., Faucon-Biguet, N., and Mallet, J. (2003) Optimization of transgene expression at the posttranscriptional level in neural cells: Implications for gene therapy, *Molecular Therapy* *7*, 782–789.
11. Knappskog, S., Ravneberg, H., Gjerdrum, C., Trosse, C., Stern, B., and Pryme, I. F. (2007) The level of synthesis and secretion of Gaussia princeps luciferase in transfected CHO cells is heavily dependent on the choice of signal peptide, *Journal Of Biotechnology* *128*, 705–715.
12. Liao, M. M. L., and Sunstrom, N. A. (2006) A transient expression vector for recombinant protein production in Chinese hamster ovary cells, *Journal Of Chemical Technology And Biotechnology* *81*, 82–88.
13. Luo, M. J., and Reed, R. (1999) Splicing is required for rapid and efficient mRNA export in metazoans, *Proceedings Of The National Academy Of Sciences Of The United States Of America* *96*, 14937–14942.
14. Makrides, S. C. (1999) Components of vectors for gene transfer and expression in mammalian cells, *Protein Expression And Purification* *17*, 183–202.
15. Zhang, L., Leng, Q. X., and Mixson, A. J. (2005) Alteration in the IL-2 signal peptide affects secretion of proteins in vitro and in vivo, *Journal Of Gene Medicine* *7*, 354–365.
16. Zhao, J., Hyman, L., and Moore, C. (1999) Formation of mRNA 3′ ends in eukaryotes: Mechanism, regulation, and interrelationships with other steps in mRNA synthesis, *Microbiology And Molecular Biology Reviews* *63*, 405–45.
17. Vivinus, S., Baulande, S., van Zanten, M., Campbell, F., Topley, P., Ellis, J. H., Dessen, P., and Coste, H. (2001) An element within the 5′ untranslated region of human Hsp70 mRNA which acts as a general enhancer of mRNA translation, *European Journal Of Biochemistry* *268*, 1908–1917.
18. Stein, I., Itin, A., Einat, P., Skaliter, R., Grossman, Z., and Keshet, E. (1998) Translation of vascular endothelial growth factor mRNA by internal ribosome entry: Implications for translation under hypoxia, *Molecular And Cellular Biology* *18*, 3112–3119.
19. Logan, J. S., T. (1984) Adenovirus tripartite leader sequence enhances translation of mRNAs late after infection, *Proc. Natl. Acad. Sci. USA* **81**, 3655–3659.
20. Massie, B., Couture, F., Lamoureux, L., Mosser, D. D., Guilbault, C., Jolicoeur, P., Belanger, F., and Langelier, Y. (1998) Inducible overexpression of a toxic protein by an adenovirus vector with a tetracycline-regulatable expression cassette, *Journal Of Virology* *72*, 2289–2296.
21. Massie, B., Mosser, D. D., Koutroumanis, M., Vitte-Mony, I., Lamoureux, L., Couture, F., Paquet, L., Guilbault, C., Dionne, J., Chahla, D., Jolicoeur, P., and Langelier, Y. (1998) New adenovirus vectors for protein production and gene transfer, *Cytotechnology* *28*, 53–64.
22. Chapman, B. S., Thayer, R.M., Vincent, K.A., and Haigwood, N.L. (1991) Effect of intron A

from human cytomegalovirus (Towne) immediately-early gene on heterologous expression in mammalian cells, *Nucleic Acids Research* ***19***, 3979–3986.

23. Simari, R. D., Yang, Z. Y., Ling, X., Stephan, D., Perkins, N. D., Nabel, G. J., and Nabel, E. G. (1998) Requirements for enhanced transgene expression by untranslated sequences from the human cytomegalovirus immediate-early gene, *Molecular Medicine* *4*, 700–706.
24. Xia, W., Bringmann, P., McClary, J., Jones, P. P., Manzana, W., Zhu, Y., Wang, S. J., Liu, Y., Harvey, S., Madlansacay, M. R., McLean, K., Rosser, M. P., MacRobbie, J., Olsen, C. L., and Cobb, R. R. (2006) High levels of protein expression using different mammalian CMV promoters in several cell lines, *Protein Expression And Purification* ***45***, 115–124.
25. Xu, Z. L., Mizuguchi, H., Ishii-Watabe, A., Uchida, E., Mayumi, T., and Hayakawa, T. (2002) Strength evaluation of transcriptional regulatory elements for transgene expression by adenovirus vector, *Journal Of Controlled Release* ***81***, 155–163.
26. Donello, J. E., Loeb, J. E., and Hope, T. J. (1998) Woodchuck hepatitis virus contains a tripartite posttranscriptional regulatory element, *Journal Of Virology* *72*, 5085–5092.
27. Mahonen, A. J., Airenne, K. J., Purola, S., Peltomaa, E., Kaikkonen, M. U., Riekkinen, M. S., Heikura, T., Kinnunen, K., Roschier, M. M., Wirth, T., and Yla-Herttuala, S. (2007) Post-transcriptional regulatory element boosts baculovirus-mediated gene expression in vertebrate cells, *Journal Of Biotechnology* ***131***, 1–8.
28. Schambach, A., Wodrich, H., Hildinger, M., Bohne, J., Krausslich, H. G., and Baum, C. (2000) Context dependence of different modules for posttranscriptional enhancement of gene expression from retroviral vectors, *Molecular Therapy* *2*, 435–445.
29. Xu, Z. L., Mizuguchi, H., Mayumi, T., and Hayakawa, T. (2003) Woodchuck hepatitis virus post-transcriptional regulation element enhances transgene expression from adenovirus vectors, *Biochimica Et Biophysica Acta-General Subjects* ***1621***, 266–271.
30. Zufferey, R., Donello, J. E., Trono, D., and Hope, T. J. (1999) Woodchuck hepatitis virus posttranscriptional regulatory element enhances expression of transgenes delivered by retroviral vectors, *Journal Of Virology* *73*, 2886–2892.
31. Mariati, Ho, S. C. L., Yap, M. G. S., and Yang, Y. S. (2010) Evaluating post-transcriptional regulatory elements for enhancing transient gene expression levels in CHO K1 and HEK293 cells, *Protein Expression And Purification* ***69***, 9–15.

Chapter 10

Converting Monoclonal Antibodies into Fab Fragments for Transient Expression in Mammalian Cells

Joanne E. Nettleship, Aleksandra Flanagan, Nahid Rahman-Huq, Rebecca Hamer, and Raymond J. Owens

Abstract

In this chapter, protocols are described for converting mouse monoclonal antibodies into recombinant Fabs for transient expression in mammalian cells. Variable region genes are cloned by reverse transcription: PCR using either sequence specific or mixed 5′ primers that hybridise to the first framework sequence of the mouse light and heavy chains and 3′ primers that bind to the heavy- and light-chain constant regions. The amplified sequences are inserted into mammalian cell expression vectors by In-Fusion™ cloning. This method allows vector and amplified DNA sequences to be seamlessly joined in a ligation-independent reaction. Transient co-expression of light-chain and heavy-chain genes in HEK 293T cells enables production of recombinant Fabs for functional and structural studies.

Key words: HEK293T, PCR cloning, Fabs

1. Introduction

Monoclonal antibodies have been generated to a variety of antigens, including human proteins using mouse hybridoma technology. For some applications, for example as co-crystallisation chaperones in structural biology (1, 2) or as targeting agents in vivo (3, 4), only the antibody-binding Fab fragment is required. Fabs are ideal reagents for these purposes being relatively small, stable, and monovalent binding proteins. Traditionally, Fabs have been produced by proteolysis from mouse IgGs, though this requires careful optimisation and is very time-consuming (1). A way round this is to clone the antibody genes from the parent hybridoma and re-express them in the format of a Fab fragment; i.e. light chain linked by a disulphide bridge to the Fd fragment of the heavy chain.

James L. Hartley (ed.), *Protein Expression in Mammalian Cells: Methods and Protocols*, Methods in Molecular Biology, vol. 801, DOI 10.1007/978-1-61779-352-3_10, © Springer Science+Business Media, LLC 2012

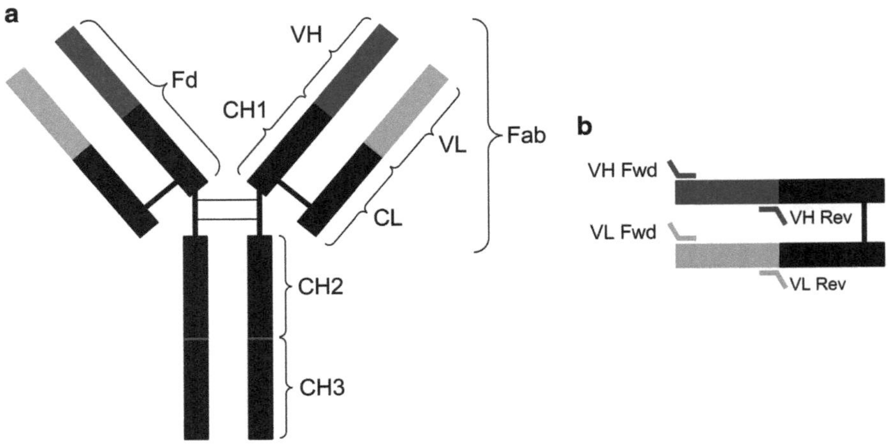

Fig. 1. (**a**) Schematic diagram of a typical antibody showing the Fab fragment; the Fd fragment of the heavy chain; the heavy-chain variable region (VH) shown in *dark grey*; the heavy-chain constant regions CH1, CH2, and CH3 in *black*; the light-chain variable region (VL) in *light grey*; and the light-chain constant region (CL) in *black*. (**b**) Schematic diagram to show the position of the PCR primers for amplification of the heavy and light variable domain genes (Fwd indicates a forward primer and Rev a reverse primer).

The Fd fragment comprises the heavy-chain variable (VH) domain together with the first domain of the heavy-chain constant region (CH1) (Fig. 1a). Typically, the heavy- and light-chain genes are amplified by PCR from B cell cDNA using forward primers that hybridise to either the relatively conserved first framework (5–7) or to the signal sequences (8, 9) of the variable regions in combination with reverse primers that bind to the constant regions. Expression of the cloned heavy-chain Fd and light-chain genes in *Escherichia coli* offers a rapid way of producing the recombinant Fabs. However, successful expression and secretion from these cells is affected by the sequence of the antibody variable region (10, 11). By contrast, mammalian cells contain the appropriate chaperones that should enable the expression of Fab fragments irrespective of their variable region sequence (12, 13). The time-consuming process of establishing mammalian stable cell lines for protein production can be avoided by using large-scale transient expression (14, 15), which can produce yields of Fabs in the milligram range suitable for functional and structural studies (16).

In this chapter, protocols for the cloning and transient expression of recombinant Fab fragments in mammalian cells are described. A list of PCR primers is given for the amplification of heavy- and light-chain variable region genes from purified hybridoma mRNA. The 5′ primers are designed to amplify as much of the diversity of the variable regions as possible. PCR-amplified genes are inserted by In-Fusion™ cloning into expression vectors which supply signal sequences and the appropriate constant regions. The heavy-chain vector also adds a C-terminal hexahistidine tag for detection and purification. Resulting clones are sequenced to confirm authenticity and plasmids used to co-transfect HEK cells at small scale to test for expression which is detected by sandwich enzyme-linked immunosorbent assay (ELISA). Large-scale transient co-expression is used to produce recombinant Fabs for subsequent functional and structural studies.

2. Materials

2.1. Enzymes and Buffers

1. RNeasy® Mini Kit (Qiagen, Crawley, UK) or similar product for extraction and purification of total RNA from animal cells.
2. OneStep RT-PCR Kit consisting of RT-PCR Enzyme Mix (comprising Omniscript Reverse Transcriptase, Sensiscript Reverse Transcriptase, and HotStarTaq DNA Polymerase), 5× reaction buffer (Tris–HCl, pH 8.7, KCl, $(NH_4)_2SO_4$, 12.5 mM $MgCl_2$, and DTT), and dNTPs (10 mM) (Qiagen, Crawley, UK).
3. AccuScript® High Fidelity 1st Strand cDNA Synthesis Kit consisting of AccuScript® High Fidelity reverse transcriptase, oligo(dT), and random hexamer primers reaction buffer (0.5 M Tris–HCl, pH 8.3, 0.75 M KCl, 0.03 M $MgCl_2$), dNTPs (100 mM), DTT (10 mM), and RNase inhibitor (Stratagene; Agilent Technologies Ltd, Stockport, UK).
4. Biomix Red (BioLine, UK) or similar product (see Note 1).
5. In-Fusion™ enzyme available lyophilised with the buffer components in microtube format (8 or 96) which conveniently can be stored at room temperature. The enzyme and buffers can also be purchased separately in liquid form (Clontech, Oxford, UK).

2.2. Gel Electrophoresis and Purification of DNA

1. Tris–borate–EDTA (TBE) running buffer (10×): 108 g Tris base, 55 g boric acid, 9.3 g EDTA dissolved in 1 l water; store at room temperature.
2. DNA gel loading dye: 0.25% (w/v) Bromophenol blue in 30% (v/v) glycerol/TE: store at room temperature.
3. Visualisation of DNA: SYBRSafe™ (Invitrogen, Paisley, UK).
4. Nucleospin® Extract II kit (Macherey-Nagel GmbH, Düren, Germany).
5. Wizard® kit (Promega, Madison Wisconsin, USA).

2.3. Cloning-Grade E. coli

Chemically competent cells with an efficiency of at least 10^8 cfu/μg circular plasmid DNA are required for In-Fusion™ cloning, for example, TAM1 cells from Active Motif (Rixensart, Belgium), OmniMax2 cells (Invitrogen, Paisley, UK) and Fusion-Blue™ Competent Cells (Clontech).

2.4. Plastic-Ware in Multi-well Format

1. Thermo-Fast 96-well skirted PCR plates sealed with a clear film (e.g. ABGene, Epsom, UK).
2. 96-Deep well plates sealed with gas permeable film (e.g. ABGene).
3. 96-Well microtitre plates (e.g. Greiner Bio-One, Frickenhausen, Germany) and foil seals (e.g. ABGene).
4. 24-Well tissue culture plates with lids (e.g. Corning, Lowell, MA, USA).
5. 96-Well plates for ELISA (e.g. Nunc Maxisorp™ plates Sigma-Aldrich, Gillingham, UK).

2.5. Cell Culture

1. HEK 293T cells (ATCC no. CRL-1573 – LGC) Promochem, Teddington, UK.
2. Dulbecco's modified Eagles medium (DMEM) supplemented with L-glutamine (1:100) and non-essential amino acids (1:100) plus 10% foetal calf serum (FCS) (Invitrogen, Paisley, UK).
3. Roswell Park Memorial Institute (RPMI) 1640 supplemented with 15% FCS and L-glutamine (1:100) (Invitrogen, Paisley, UK).
4. Trypsin–ethylenedinitrilotetraacetic acid (trypsin–EDTA) (Sigma-Aldrich Ltd., Dorset, UK).
5. Phosphate-buffered saline (PBS): 0.01 M Phosphate buffer, 0.0027 M potassium chloride, 0.137 M sodium chloride, pH 7.4.
6. Polystyrene roller bottles with expanded surface (2,125 cm^2) (Cat. no.: 681070 Greiner Bio-One, Stonehouse, UK).

2.6. Small and Large-Scale Cell Transfection

1. Plasmid DNA with an A_{260}/A_{280} ratio of greater than 1.8.
2. GeneJuice™ (Novagen, Nottingham, UK) or similar transfection reagent.
3. Polyethylenimine (PEI) (25 kDa branched PEI, Sigma-Aldrich, Gillingham, UK). Prepare as a stock solution of 100 mg/ml in water. Then dilute to 1 mg/ml, neutralise with HCl, filter-sterilise and store frozen in aliquots.

2.7. Enzyme-Linked Immunosorbent Assay

1. Goat anti-(mouse light chain) polyclonal antibody (Cat. no.: AP200; Millipore Ltd., Livingston, UK).
2. PBST; phosphate buffered saline containing 0.05% Tween-20.
3. HisProbe™-HRP (horseradish peroxidase) (Pierce Protein Research Products: Thermo Fisher Scientific: Perbio Science, Cramlington, UK).
4. TMB (3,3′,5,5′-tetramethylbenzidine) substrate kit (Pierce Protein Research Products).
5. Stop solution: 2 M Sulphuric acid.

2.8. Large-Scale Fab Purification

1. 5 ml HisTrap column (GE Healthcare, Little Chalfont, UK).
2. HiLoad 16/60 Superdex 75 column (GE Healthcare, Little Chalfont, UK).
3. Nickel Wash Buffers (filtered and de-gassed): 50 mM Tris–HCl, pH 7.5, containing 150 mM NaCl and either 20 or 50 mM imidazole.
4. Nickel Elution Buffer (filtered and de-gassed): 50 mM Tris–HCl, pH 7.5, 150 mM NaCl, 500 mM imidazole.
5. Size Exclusion Buffer (filtered and de-gassed): 150 mM NaCl and 10 mM HEPES, pH 7.5.
6. Simply Blue Safe Stain (Invitrogen, Paisley, UK).

3. Methods

3.1. Primer Design

A number of comprehensive primer sets have been designed for the PCR amplification of the variable regions of rearranged mouse immunoglobulins. In general, these prime in the first framework of the light- and heavy-chain genes and are synthesised as mixtures of oligonucleotides. An alternative to designing "universal" primers is to first sequence the variable domains of the target antibody. This is usually carried out by cloning the respective heavy- and light-chain genes by 5′ rapid amplification of cDNA ends (5′ RACE (17)) and then sequencing of the variable regions. Sequence-specific PCR primers may then be designed for further sub-cloning of the variable domains. In both cases the framework primers are combined with sequence-specific primers that hybridise to the 5′ end of the constant domain of the light chain and first constant domain of the heavy chain (Fig. 1b). Using generic primers would be the preferred option for cloning a large number of antibody genes, though there is a cost overhead in synthesising large numbers of PCR primers. To clone one or just a few variable domains using gene specific primers following 5′ RACE and sequencing would be advisable, since amplification of the required sequences would be guaranteed.

3.1.1. Forward Primers

The primer templates defined by Essono et al. (5) are used as the basis of the forward primers. Comprehensive coverage of mouse heavy- and light-chain variable region sequences requires the synthesis of relatively large numbers of primers. If these are synthesised as redundant mixtures then equal representation of a specific sequence cannot be guaranteed due to differences in the efficiencies of base coupling reactions during synthesis. Therefore, to reduce the complexity of these primer sets we have written a Perl script to filter the primer sequences according to degeneracy and

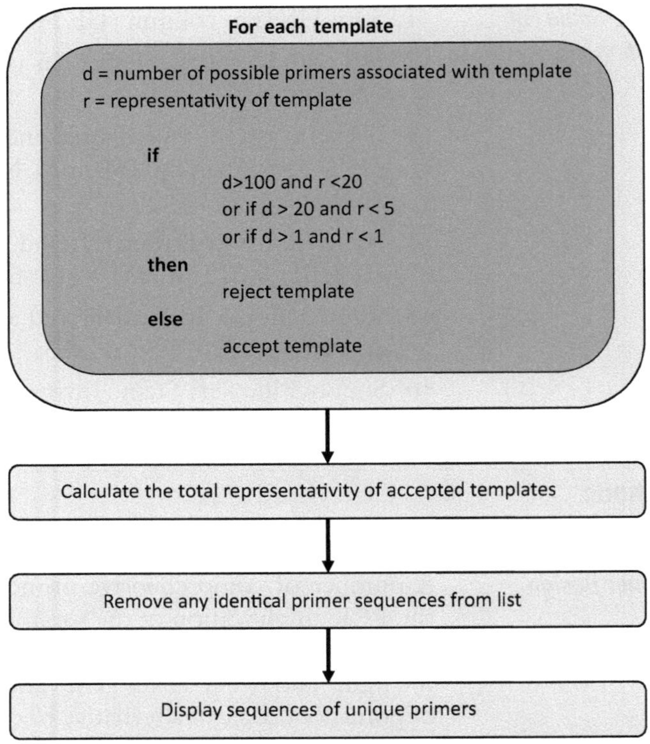

Fig. 2. Schematic representation of programme to filter heavy- and light-chain sequences according to degeneracy and representation amongst the sequences of mouse antibodies compiled by Essono et al. (5). The programme can be downloaded from http://www.oppf.ox.ac.uk/OPPF/public/supp/antibodyPrimers/antibodies.pl.

representation in mouse variable domain genes. The programme is shown schematically in Fig. 2 and an example of a filtered list of 5′ primers is given in the Appendix. By altering the parameters, the number of primers can be reduced but at the expense of the number of variable domain sequences that would be covered.

The 5′ regions of homology required for In-Fusion™ cloning are generated by adding 15 bp extensions to the forward primers (Table 1). The sequences of these extensions match the ends of the recipient vectors exposed by linearisation of the vector at the position into which the PCR product is to be inserted (Fig. 3). Typically, oligonucleotide primers are approximately 35-bp long (including the extension and gene-specific region) and purification of the primers is not necessary.

3.1.2. Reverse Primers

The reverse primers for amplifying the heavy- and light-chain V-region genes are given in Table 1. The primers anneal to the 5′ ends of the mouse Cγ1 heavy-chain domain and Cκ light chain, respectively, which are resident in the expression vectors (Figs. 1 and 3). The Cγ1 primer matches the sequence of IgG1 but has

Table 1
Sequences of the extensions to the forward primers and complete reverse primers for cloning VH and VL sequences by In-Fusion™

Primer/extension	Sequence	Vector
VH Fwd extension	TGGGTTGCGTAGCT	pOPINVH
VL Fwd extension	TGGGTTGCGTAGCT	pOPINVL
VH Rev primer	GGGTGTCGTTTTGGC	pOPINVH
VL Rev primers	TGCAGCATCAGCCCG	pOPINVL

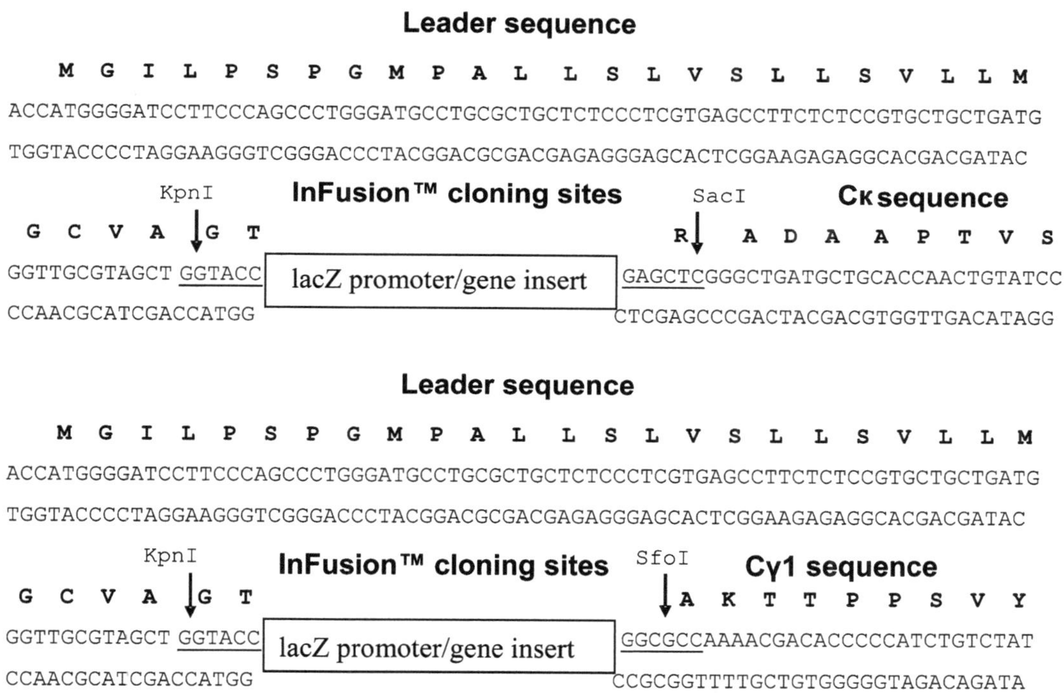

Fig. 3. Diagram showing the DNA and corresponding amino acid sequences flanking the In-Fusion™ cloning sites for the vectors (**a**) pPOINVL and (**b**) pOPINVH. The positions of the restriction enzymes used to linearise the plasmids are indicated by *vertical arrows*.

only one or two mismatches with the sequences of IgG2b and IgG2a heavy chains, respectively, so should also amplify antibodies with these isotypes.

3.2. PCR Amplification of Variable Domain Genes

The V-region genes of the target mouse monoclonal antibody are cloned by reverse transcription coupled to the polymerase chain reaction (RT-PCR) from total RNA extracted from the hybridoma cells producing the antibody (Subheading 3.2.1). VH and VL genes are amplified using sets of forward primers that hybridise to the first framework sequence and reverse primers that bind to the beginning of the light- and heavy-chain constant regions, respectively (Subheading 3.3.2). If the DNA sequences of the variable region genes are already known the RT-PCR can be conveniently carried out in a single tube reaction using specific forward framework primers in combination with the reverse constant domain primers (Subheading 3.2.3).

3.2.1. Preparation of Total RNA from Hybridoma Cells

1. For each monoclonal antibody, grow a 50 ml culture of the hybridoma cells in RPMI 1640 medium in 37°C incubator at 5% CO_2. After 4 days, harvest the cells by centrifugation at $800 \times g$ for 10 min (total of approximately 10^7 cells).
2. Extract total RNA from approximately 10^7 hybridoma cells using RNeasy kit (Qiagen) according to the manufacturer's protocol (Handbook 04/2006: purification of Total RNA from Animal Cells using Spin Technology's RNeasy Mini). Briefly, homogenise cells in 600 μl of RLT buffer by passing the lysate through 0.6 mm needle fitted to 1 ml syringe (5–10 times).
3. Mix lysate with 600 μl 70% ethanol and apply onto an RNeasy spin column.
4. Wash column with 700 μl RW1, twice with 700 μl RPE and elute total RNA with 2× 30 μl with RNase-free water. The yield of total RNA is approximately 12–30 μg.

3.2.2. Option 1: Two-Step RT-PCR

1. First strand cDNA is synthesised using reverse transcriptase primed with oligo dT (12–18).
2. Mix 50 mM Tris–HCl (pH 8.3), 40 mM KCl, 6 mM $MgCl_2$, 500 μM of each dNTP, 5 mM DTT, 1 μg oligo dT (12–18), 5 μg RNA, and 40 U Superscript™ reverse transcriptase (Invitrogen) in a 40 μl volume and incubate 45 min at 42°C.
3. Amplify VH and VL region sequences in separate reactions by mixing the following components in a total volume of 50 μl:

BioMixRed® Taq polymerase mix	25 μl
5′ Primers (10 μM each primer; see Note 2)	3.0 μl
3′ Primer (10 μM)	3.0 μl
Water (ddH_2O)	Up to 50 μl

4. Perform amplification using the three-step cycling parameters: denaturation (1 min, 94°C), annealing (1 min, 55°C), extension (1 min, 72°C), for 30 cycles followed by a final extension for 10 min at 72°C.

5. When thermal cycling is complete, analyse PCR products by adding 5 μl of the reaction to DNA loading dye and running on a 1.25% TBE agarose gel.

3.2.3. Option 2: One-Step RT-PCR

1. If the nucleotide sequences of the variable regions of the target antibody are known then the genes can be amplified by one step RT-PCR using total RNA as a template.
2. Set up one 50 μl reaction each for the heavy and light chains according to the Qiagen One-step RT-PCR kit protocol:

5× Qiagen OneStep RT-PCR buffer	10 μl
dNTP mix (10 mM of each dNTP)	2 μl
5′ Primer (10 μM)	3 μl
3′ Primer (10 μM)	3 μl
Qiagen OneStep RT-PCR enzyme mix	2 μl
Template RNA (1 mg/ml)	2 μl
RNase-free water	30 μl

3. Perform reverse transcription followed by PCR by incubating for 30 min at 50°C followed by 15 min at 95°C to activate the HotStartTaq polymerase. Then carry out 3-step cycling: denaturation (45 s, 94°C), annealing (45 s, 50°C), extension (1 min, 72°C), 30 cycles followed by a final extension for 10 min at 72°C.
4. Analyse PCR products on a 1.25% TBE agarose gel. A band of approximately 250 bp is expected for both VH and VL amplicons.

3.3. Construction of Expression Vectors

Two expression vectors, namely pOPINVL and pOPINVH, have been constructed containing a resident signal sequence from human μ-phosphatase (18) and either κ light chain or Cγ1 constant genes, respectively (Fig. 2). The heavy-chain expression vector also adds a His-tag to the C-terminus of the heavy-chain fragment. Since antibody heavy chains are only secreted in association with light chains, only assembled Fabs would be detected by an anti-His tag antibody or purified by metal affinity chromatography (19, 20). Purified PCR-amplified V-region genes (Subheading 3.3.1) are inserted into the expression vectors by In-Fusion™ cloning (Subheading 3.3.2). This cloning method enables the signal sequence, V-region, and constant domain sequences to be precisely joined in a ligation-independent reaction. The expression vectors pOPINVH and pOPINVL are available from the corresponding author on request, and the vector sequences have been deposited in GenBank (EU733645 and EU733646).

3.3.1. Purification of PCR Products and Preparation of Vectors

1. Purify PCR products separated by preparative gel electrophoresis using Nucleospin® Extract II kit or similar product. Excise the electrophoretically separated PCR products from the agarose gel and place each gel slice into an individual tube. Add two volumes of NT buffer to one volume of gel (i.e. 200 μl per 100 mg gel slice).
2. Incubate at 50°C for 10 min (or until gel is completely dissolved) and apply dissolved gel to a Nucleospin® Extract II spin column and centrifuge for 1 min at 11,000 × *g*. Wash by centrifugation with 0.6 ml of NT3 wash buffer. Re-spin column to remove any residual wash buffer.
3. Place spin column into a clean tube and elute DNA by adding 25 μl of elution buffer (EB). Collect purified PCR product by centrifugation.
4. Digest vector DNA prepared using QIAPrep (Qiagen) spin column (or similar DNA purification method) with appropriate restriction enzyme(s) using standard conditions. For example, incubate 5 μg pOPINVH plasmid DNA with 50 U *Kpn*I and 50 U *Sfo*I and pOPINVL with *Kpn*I/*Sac*I in a total reaction volume of 100 μl for 2 h at 37°C.
5. Purify linearised vector by preparative gel electrophoresis (see above) or by a spin column (e.g. Nucleospin® Extract II kit). If the plasmid DNA has been prepared without the use of an alcohol precipitation step the spin column method is sufficient, otherwise gel purification is recommended.

3.3.2. Infusion Cloning

1. Combine 10–100 ng of each purified insert and 100 ng of the appropriate linearised and purified vector (these can be prepared and stored in advance) in a total volume of 10 μl of either elution buffer (above) or H_2O (see Note 3).
2. Add this to a well of the dry-down In-Fusion™ plate. Mix contents briefly by pipetting up and down, taking care that the lyophilised enzyme/buffer pellet is re-suspended. Cover the In-Fusion plate with a self-adhesive foil plate seal.
3. Incubate the plate for 30 min at 42°C in either a thermal cycler or water bath.
4. Dilute immediately with 40 μl TE and either transform into *E. coli* either immediately or freeze the reaction for use later.
5. Thaw competent *E. coli* on ice add 50 μl of cells to 5 μl of the diluted In-Fusion reaction and incubate on ice for 30 min.
6. Heat-shock the cells for 30 s at 42°C and return the cells to ice for 2 min.
7. Add 450 μl of Luria Broth (LB) per tube, transfer to 37°C incubator and incubate for 1 h.

8. Plate on LB Agar supplemented with the appropriate antibiotic for the vector, X-Gal and IPTG (see Note 3). Plates are prepared by the addition of 1 ml of molten LB agar (plus appropriate supplements) to each well of the 24-well plates. Plate 10 μl of cells/well, shake plates laterally/orbitally by hand to distribute the culture and allow at least 10–15 min for the plates to dry before inverting.
9. Following overnight incubation at 37°C, wells should contain predominantly white colonies. Any blue colonies are derived from inefficiently linearised parental plasmid and are non-recombinant (see Note 4). Pick a minimum of two colonies for small-scale of DNA preparation in multiwell plates, for example using the Wizard® kit (see Note 5).
10. PCR screen the plasmid mini-preps using the PCR protocol described in Subheading 3.2.2 replacing the forward primers with a T7 forward primer (5′ TAATACGACTCACTATAGGG 3′); 25 μl reactions can be used for screening. Amplicons should be about 260 bp larger than the PCR products from Subheading 3.2, above.
11. It is critical to sequence PCR verified pOPINVH and pOPINVL expression clones to obtain a consensus sequence for VH and VL genes of the parent monoclonal antibody particularly if a *Taq* polymerase without proofreading activity has been used in combination with a mixture of 5′ framework primers (see Note 6). A problem that is frequently encountered when amplifying variable region genes is the expression of non-functional light- and heavy-chain genes by myeloma fusion partners derived from the original MOPC21 tumour (e.g. X63-Aq8, NS-1, P3X63Ag8.653) (see Note 7).

3.4. Production of Fabs in HEK 293T Cells

The expression of the cloned Fabs can be rapidly assessed by transient co-transfection of the pOPINVH and VL plasmids in mammalian cells (e.g. HEK 293T cells). The DNA prepared by the procedure described in Subheading 3.3.2, step 9 is of sufficient quantity and quality for directly transfecting into HEK 293T cells. The amount of secreted protein can be assessed by sandwich ELISA within a week of cloning the gene-of-interest. A protocol for the parallelised small-scale co-expression screening of plasmids using HEK 293T cells is given in Subheadings 3.4.1 and 3.4.2. Transient production of Fabs in HEK 293T cells can be readily scaled up to one litre cell culture for subsequent purification of protein by metal chelate affinity chromatography in combination with gel filtration as described in Subheadings 3.4.3 and 3.4.4. This protocol gives yields of Fabs of 2–4 mg/l cell culture depending upon the antibody (16). A semi-automated version of this general method has been described (21).

3.4.1. Co-transfection of Cells in 24-Well Plates

1. Seed the HEK293T cells into a 24-well plate at a density to give 75–80% confluency after 24 h (typically ~1.5×10^5 cells/ml).
2. Mix 2 μl of 1.33 mg/ml GeneJuice™ with 60 μl of DMEM supplemented with 1× non-essential amino acids and 1 mM glutamine in a V-well micro-titre plate.
3. Add ~0.5 μg each of the pOPINVH and pOPINVL plasmid DNAs, mix thoroughly and incubate for 30 min at room temperature.
4. Carefully aspirate the media from the cell layer in the 24-well plate and discard.
5. Make up the DNA/GeneJuice™ cocktail to 1 ml with DMEM supplemented with 2% FCS, 1× non-essential amino acids and 1 mM glutamine and add to the plated cells.
6. Incubate the cells at 37°C in a 5% CO_2/95% air environment for 4 days.
7. Harvest cell media, centrifuge for 10 min at 12,000×*g* to remove any cells, and test for Fab expression by ELISA.

3.4.2. ELISA

1. Prepare ELISA plates by adding 200 μl of anti-mouse Light-Chain antibody (5 μg/ml) in PBS at pH 7.5 to each well of a Nunc Maxisorp™ plate and incubate overnight at 4°C.
2. Wash wells 3× with 200 μl PBST.
3. Block wells by adding 200 μl PBS containing BSA (10 mg/ml) and incubating at room temperature for 4 h (or 4°C overnight).
4. Wash 3× with 200 μl PBST.
5. To generate a standard curve, serially dilute a purified mouse Fab of known concentration in PBS containing BSA (1 mg/ml), add 200 μl dilute protein to the ELISA plate, and incubate for 1 h at room temperature.
6. Add conditioned media from small-scale transient expression experiment to the ELISA plate and incubate for 1 h at room temperature.
7. Wash 3× with 200 μl PBST.
8. Add 100 μl HisProbe solution to each well and incubate at room temperature for 15 min.
9. Wash 4× with 200 μl PBST.
10. Mix equal volumes of TMB solution and peroxide solution. Add 100 μl per well and incubate for 15–30 min until the desired colour develops.
11. Read absorbance of wells at 630 nm in a spectrophotometer.
12. Add 100 μl stop solution.

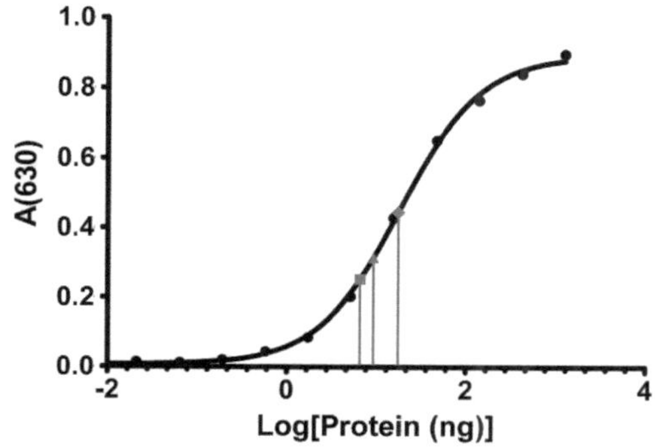

Fig. 4. Quantitation of Fabs by ELISA. Sandwich ELISA results showing a standard curve (*black circles*) generated using known concentrations of a reference Fab (FABOX117 (16)) and 1:10 dilutions of media containing FABOX108 (*grey square*), FABOX117 (*grey triangle*), and FABOX119 (*grey diamond*) from expression screening in HEK 293T cells. These results show yields of 0.67 μg/ml for FABOX108; 0.93 μg/ml for FABOX117; and 1.75 μg/ml for FABOX119.

13. Read absorbance at 450 nm.
14. Plot standard curve and estimate the Fab concentration in the conditioned media by reference to the standard (Fig. 4).

3.4.3. Large-Scale Transient Co-expression

1. For a 1 l culture, 4× 175-cm^2 flasks containing fully confluent HEK 293T cells are required.
2. Remove cells by trypsinisation and add all the cells (typically ~7.5×10^5 cells/ml) from one flask into one roller bottle with 250 ml of DMEM containing 1× non-essential amino acids and 1 mM glutamine and 2% FCS.
3. Incubate the cells for 4 days at 37°C with the bottles rolling at 10 rpm in a suitable incubator (e.g. Wheaton Science Products, NJ, USA).
4. Replace media in each roller bottle with 200 ml of DMEM supplemented with 1× non-essential amino acids and 1 mM glutamine and 2% FCS. Incubate the cells at 37°C in the roller incubator for 2 h.
5. Pipette 1 mg of each pOPINVH and pOPINVL plasmid (see Note 6) into 200 ml of DMEM containing 1× non-essential amino acids and 1 mM glutamine in a sterile conical flask. Mix and then add 3.5 ml of 1 mg/ml polyethyleneimine (PEI) to form the transfection cocktail.
6. Mix thoroughly and incubate at room temperature for 30 min.
7. Add 50 ml of the transfection cocktail to each of the roller bottles.
8. Incubate the cells for 4–5 days at 37°C with the bottles rolling at 10 rpm.

9. To harvest the secreted protein, remove the media from the cells and centrifuge at 5,000 × *g* for 30 min to remove any non-attached cells (see Note 8).
10. Dialyse the media twice against five volumes of PBS (see Note 9) and filter sterilise through a 0.2-μm membrane using a bottle-top filter. The conditioned media may be stored at 4°C prior to purification.

3.4.4. Purification

1. Supplement the harvested media with 10 mM Tris–HCl, pH 8.0 and 10 mM imidazole and load onto a 5 ml HisTrap column (GE Healthcare) at 5 ml/min flow rate using the external pump of an Akta system (Purifier or FPLC).
2. Wash the HisTrap™ column with five column volumes of wash buffer containing 20 mM imidazole followed by five column volumes of wash buffer with 50 mM imidazole. The Fab fragment is eluted with elution buffer containing 500 mM imidazole (see Note 10). Concentrate and buffer exchange the eluted protein into 150 mM NaCl, 10 mM HEPES, pH 7.5 using an Amicon Ultra 10-kDa concentrator.
3. Apply the concentrated protein onto a Superdex 75 HiLoad 16/60 column equilibrated with 150 mM NaCl and 10 mM HEPES, pH 7.5 and collect fractions. The Fab fragment elutes at approximately 60–62 ml and is well separated from any bovine serum albumin, carried over from the tissue culture media. Depending upon the Fab, yields of 2–4 mg/l purified are typically obtained by this protocol.
4. Analyse purified protein by SDS-PAGE under reducing and non-reducing conditions to confirm assembly of Fabs (Fig. 5).

4. Notes

1. Although a variety of DNA polymerases have been used to amplify V-region genes, Taq polymerases with no added proof-reading activity are preferred if the primers contain a large number of degenerate codons primers (7). BioMix™ Red is a complete ready-to-use 2× reaction mix containing an ultra-stable *Taq* DNA polymerase. It contains an additional inert red dye that permits easy visualisation and direct loading onto a gel.
2. Individual 5′ primers are pooled so as to give a final concentration of 10 μM for each primer; for example, mixing ten primers each at 100 μM gives a final concentration of 10 μM for each primer.

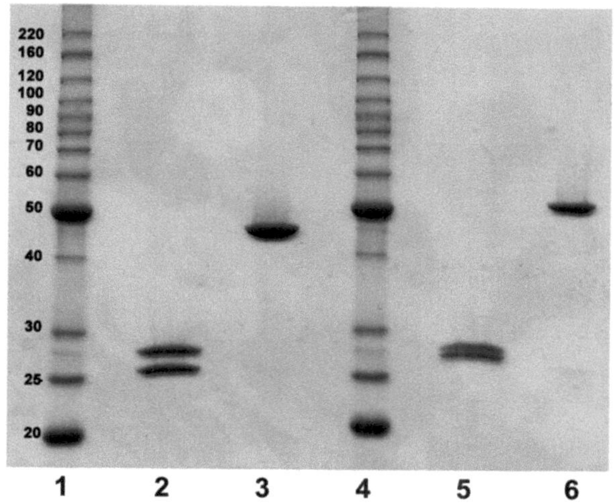

Fig. 5. Purified recombinant Fabs. The light-chain and Fd heavy-chain genes of two mouse monoclonal antibodies (3H5 and 8A1) raised to the EIII protein of Dengue virus were co-transfected into HEK 293T cells. Recombinant Fabs were purified from the culture supernatant by Ni-NTA affinity chromatography coupled to size exclusion chromatography. Purified Fabs (3H5: *lanes 2, 3* 8A1 *lanes 4, 5*) were analysed by SDS-PAGE under reducing (*lanes 2* and *5*) and non-reducing conditions (*lanes 3* and *6*). *Lanes 1* and *4* are molecular weight markers numbered in kD.

3. Do not use buffers containing chelating agents such as EDTA, as they inhibit the Mg^{2+}-dependent activity of the DNA polymerases to be used.
4. The pOPIN vectors contain the LacZ promoter gene between the restriction sites used to linearise the vectors for In-Fusion™ cloning, which serves as a quality for preparing the vectors by double restriction digestion. Therefore, by plating transformants onto agar containing IPTG/X-gal, any blue colonies obtained correspond to inadequately linearised vector.
5. In our experience, picking two clones gives an average cloning efficiency for cloning 96 PCR products in parallel of approximately 90%; picking a further two clones can improve this to approximately 95%.
6. The use of *Taq* polymerase, which lacks proofreading function, results in an error rate of approximately 1/1,000 bases, which means that for the most part error-free versions of variable region sequences are obtained. Errors in amplified VH and VL sequences can be readily identified by sequencing several clones.
7. The single aberrant light sequence is characterised by mutation of the invariant cysteine at residue 23 to tyrosine and a

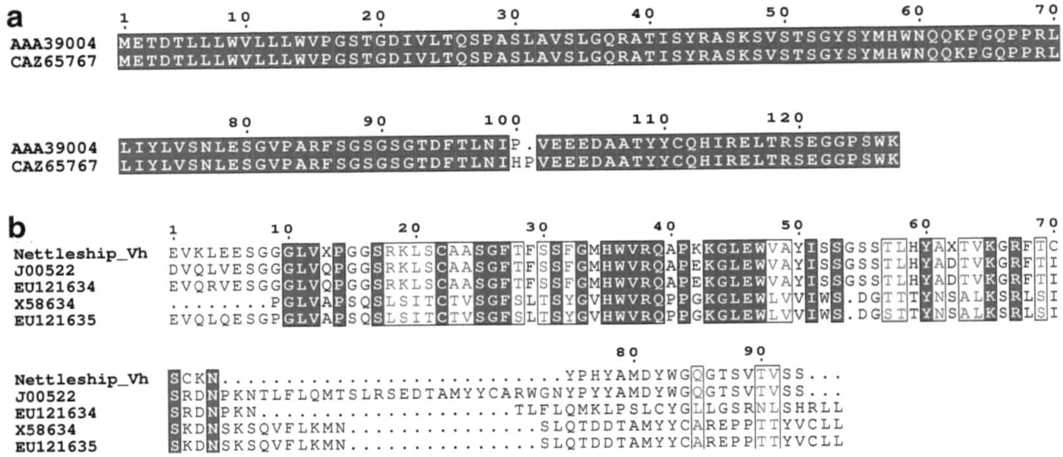

Fig. 6. Sequences of aberrant antibody variable domains. (**a**) sequence of the non-functional VL gene produced by MOPC21 derived myeloma fusion partners (22, 25); (**b**) multiple sequence alignment using CLUSTALW of the variable domains of non-functional heavy-chain chains produced by MOPC21 derived myeloma fusion partners. Nettleship VH was cloned by the authors, EU121635 and EU121634 were reported by Irani et al. (26), and X58634 was reported by Kutemeier et al. (27). The numbers refer to the GenBank accession identifiers.

4-bp deletion at the junction of the variable and joining (V/J) domains (22, 23). The sequence can, therefore, be readily recognised amongst the cloned light-chain variable regions. A method to suppress amplification of the sequence using a peptide nucleic acid (PNA)-based analogue specific for the complementarity determining region 3 of the aberrant light chain has been reported (24). In the case of VH genes, there are a number of related non-functional sequences that have been cloned from hybridomas which show different internal deletions (Fig. 6), making it difficult to apply a PNA-mediated PCR clamping to suppress their amplification.

8. Good quality DNA can be obtained using a standard mega plasmid preparation kit such as the PureLink HiPure Plasmid Megaprep Kit (Invitrogen, Paisley, UK).
9. Fresh media can be added to the harvested roller bottles and the cells are returned to the roller bottle incubator and harvested after a further 4–5 days. Generally, the yield from the second harvest is approximately 75% that of the first one.
10. Dialysis removes small molecular media components, which can strip the Ni-NTA affinity matrix and hence reduce binding of the His-tagged Fabs.

Acknowledgements

The Oxford Protein Production Facility is supported by grants from the Medical Research Council, UK, the Biotechnology and Biological Sciences Research Council, UK. AF is the recipient of a Wellcome Trust Research Studentship.

Appendix

The nucleotide sequences of all of the known mouse antibody heavy (VH) and light-chain (VL) variable domains are grouped into a series of templates (5). Variation within each template is shown by the wobble codons. For each template, the number of different variants and hence the number of different primers required to amplify the sequences in the template is given (d). The occurrence of any given template in mouse antibody sequences is shown as a percentage (%rep). For both VH and VL sequences, an example is given of using a filter to limit the total number of primers needed to amplify most of the sequences. Thus, by eliminating VH and VL sequences with high variability and low representation, the total number of unique primers required can be reduced to 43 unique primers for VH and 116 unique primers for VL, which cover 87 and 83% of known mouse VH and VL sequences, respectively. The Infusion cloning tags shown in Table 1 are added to the 5′ end of each primer. The primers are synthesised individually to ensure equal representation of each sequence in the final pooled primer sets.

VH sequences ordered by representation

		d	%rep
Template:	SAG GTC CAG CTG CAG CAG YYT GG	8	28.6
Template:	GAG GTT CAG CTG CAG CAG TCT GK	2	16.1
Template:	GAD GTG MAM TTK SWG SWG TCT GG	384	12.5
Template:	CAG GTR CAG CTG AAG SAG TCA GG	4	10.7
Template:	GAK GTG CAG CTT CAG CAG TCR GG	4	8.9
Template:	GAV GTG AWG CTG GTG GAG TCT GR	12	7.1
Template:	GAA GTG CAG CTG TTG GAG ACT GG	1	3.6
Template:	CAG GTT CAC CTA CAA CAG TCT GG	1	3.5
Template:	CAG ATG CAG CTT CAG GAG TCA GG	1	1.8
Template:	GAA GTG AAG CTT GAG GAG TCT GG	1	1.8
Template:	CAG GTT ACT CTG AAA GAG TCT GG	1	1.8
Template:	GAG GTG AAG CTG RTG GAR TCT GR	8	1.8
Template:	GAG GTG CAG CTT GTT GAG WCT GG	2	1.8

Code for wobbles: R=A/G, Y=C/T, M=A/C, K=G/T, S=C/G, W=A/T).

```
If all templates are used:
Total rep: 100
total d: 429

The following restrictions were then put on the data:  Reject the
template sequence if:
(d>100) AND (rep<20) or
(d>20) AND (rep<5) or
(d>1) AND (rep<1)

This means we reject these sequences:
GAD GTG MAM TTK SWG SWG TCT GG      d: 384      %rep: 12.5

Total number primers: 45
Total number of unique primers: 43
Total representation = 87.5
```

Primer sequences to make:

```
CAG ATG CAG CTT CAG GAG TCA GG
GAG GTC CAG CTG CAG CAG TTT GG
GAG GTG AAG CTG ATG GAG TCT GA
CAG GTC CAG CTG CAG CAG TTT GG
GAC GTG ATG CTG GTG GAG TCT GG
CAG GTT CAC CTA CAA CAG TCT GG
GAG GTG AAG CTG GTG GAA TCT GG
GAT GTG CAG CTT CAG CAG TCG GG
GAG GTG ATG CTG GTG GAG TCT GA
GAA GTG ATG CTG GTG GAG TCT GA
CAG GTT ACT CTG AAA GAG TCT GG
GAT GTG CAG CTT CAG CAG TCA GG
GAC GTG AAG CTG GTG GAG TCT GA
CAG GTC CAG CTG CAG CAG CCT GG
GAG GTG CAG CTT GTT GAG ACT GG
CAG GTA CAG CTG AAG CAG TCA GG
GAA GTG CAG CTG TTG GAG ACT GG
GAC GTG AAG CTG GTG GAG TCT GG
GAG GTG CAG CTT GTT GAG TCT GG
CAG GTA CAG CTG AAG GAG TCA GG
GAG GTG AAG CTG GTG GAG TCT GG
GAG GTG AAG CTG ATG GAA TCT GA
GAG GTC CAG CTG CAG CAG CTT GG
GAG GTC CAG CTG CAG CAG TCT GG
GAA GTG ATG CTG GTG GAG TCT GG
GAG GTG AAG CTG ATG GAG TCT GG
GAG GTG ATG CTG GTG GAG TCT GG
GAG GTT CAG CTG CAG CAG TCT GG
GAG GTG CAG CTT CAG CAG TCA GG
GAA GTG AAG CTG GTG GAG TCT GA
CAG GTC CAG CTG CAG CAG CTT GG
GAA GTG AAG CTT GAG GAG TCT GG
GAG GTT CAG CTG CAG CAG TCT GT
CAG GTG CAG CTG AAG GAG TCA GG
GAA GTG AAG CTG GTG GAG TCT GG
GAG GTG AAG CTG GTG GAA TCT GA
GAG GTG AAG CTG ATG GAA TCT GG
GAG GTC CAG CTG CAG CAG CCT GG
GAC GTG ATG CTG GTG GAG TCT GA
CAG GTC CAG CTG CAG CAG TCT GG
CAG GTG CAG CTG AAG CAG TCA GG
GAG GTG AAG CTG GTG GAG TCT GA
GAG GTG CAG CTT CAG CAG TCG GG
```

```
VL sequences ordered by representation:
                                                 d    %rep
Template: GAT GYT KTK VTG ACC CAA ACT CC        24    21.3
Template: SAA AWT BTK CTC WCC CAG TCT CC        48    17.9
Template: RAM ATT DTG WTG WCA CAG TCT MY       192    12.2
Template: GAT ATC CAG ATG ACA CAG ACT AC         1    10.8
Template: RRC ATT GTG ATG ACC CAG WCT CM        16    10.5
Template: GAM ATC MWG ATG ACC CAR TCT CC        16     7.7
Template: RAC ATT GTG CTG ACM CAA TCT CC         4     6.7
Template: GAT ATT GTG ATR ACB CAG GYT GM        24     2.4
Template: GAC ATC CAG ATG ACT CAG TCT CC         1     2.4
Template: GAY ATY BTG MTR ACY CAG TCT CC        96     2.4
Template: GAT GTC CAG ATA ACC CAG TCT CC         1     1.3
Template: GAC ATC CAG ATG AAC CAG TCT CC         1     1
Template: GAC ATC CAG ATG ACA CAG TCT CC         1     1
Template: GAA ACA ACT GTG ACC CAG TCT CC         1     0.7
Template: GAC ATC CAG ATG ACA CAA TCT TC         1     0.7
Template: GAC ATT GTG ATG ACT CAG TCT CC         1     0.7
Template: CAG ATC CAG ATG AYY CAG TCT CC         4     0.3

Code for wobbles:  R=A/G, Y=C/T, M=A/C, K=G/T, S=C/G, W=A/T).

If all templates are used:
Total rep: 100
total d: 432

The following restrictions were then put on the data:  Reject the
template sequence if:
(d>100) AND (rep<20) or
(d>20) AND (rep<5) or
(d>1) AND (rep<1)

This means we reject these sequences:
                                          d    %rep
RAM ATT DTG WTG WCA CAG TCT MY          192    12.2
GAY ATY BTG MTR ACY CAG TCT CC           96     2.4
GAT ATT GTG ATR ACB CAG GYT GM           24     2.4
CAG ATC CAG ATG AYY CAG TCT CC            4     0.3
Total number primers: 116
Total number of unique primers: 116
Total representation = 82.7

Primer sequences to make:
GAA AAT GTT CTC ACC CAG TCT CC
GAT GCT TTT ATG ACC CAA ACT CC
GGC ATT GTG ATG ACC CAG TCT CC
GAA AAT GTT CTC TCC CAG TCT CC
GAA ATT CTG CTC TCC CAG TCT CC
GAC ATC ATG ATG ACC CAA TCT CC
CAA AAT CTG CTC TCC CAG TCT CC
CAA AAT GTT CTC TCC CAG TCT CC
GAC ATC CAG ATG ACA CAA TCT TC
GAT GCT TTT GTG ACC CAA ACT CC
GAT GTT TTT CTG ACC CAA ACT CC
GAA ATT GTT CTC ACC CAG TCT CC
CAA AAT TTG CTC ACC CAG TCT CC
AAC ATT GTG ATG ACC CAG ACT CA
```

```
GAC ATC CTG ATG ACC CAA TCT CC
CAA ATT GTG CTC TCC CAG TCT CC
GAA AAT CTG CTC TCC CAG TCT CC
GAT GCT TTG GTG ACC CAA ACT CC
GAA AAT GTG CTC TCC CAG TCT CC
GAA ATC CAG ATG ACC CAA TCT CC
GAC ATT GTG ATG ACC CAG ACT CC
GAA ATC AAG ATG ACC CAG TCT CC
GAT GCT GTG GTG ACC CAA ACT CC
GAT GTT TTG CTG ACC CAA ACT CC
CAA AAT TTT CTC TCC CAG TCT CC
GAC ATC CAG ATG ACC CAG TCT CC
GGC ATT GTG ATG ACC CAG TCT CA
GGC ATT GTG ATG ACC CAG ACT CA
GAA ATT GTG CTC TCC CAG TCT CC
GAA ATT CTT CTC ACC CAG TCT CC
GAT GTT TTG ATG ACC CAA ACT CC
GAT GTT GTG ATG ACC CAA ACT CC
GAA ATC AAG ATG ACC CAA TCT CC
GAT GTT TTG GTG ACC CAA ACT CC
CAA ATT GTG CTC ACC CAG TCT CC
GAC ATC CAG ATG ACT CAG TCT CC
AAC ATT GTG ATG ACC CAG ACT CC
GAT GCT GTG CTG ACC CAA ACT CC
GAT GCT GTT CTG ACC CAA ACT CC
CAA ATT TTT CTC TCC CAG TCT CC
AGC ATT GTG ATG ACC CAG TCT CC
GAC ATT GTG CTG ACA CAA TCT CC
GAT GTT GTG CTG ACC CAA ACT CC
GAA ATC CAG ATG ACC CAG TCT CC
CAA ATT TTG CTC TCC CAG TCT CC
CAA AAT TTG CTC TCC CAG TCT CC
GAA ATC CTG ATG ACC CAA TCT CC
GAA ATT TTG CTC TCC CAG TCT CC
GAA AAT CTT CTC TCC CAG TCT CC
GAT GCT TTG CTG ACC CAA ACT CC
CAA AAT CTG CTC ACC CAG TCT CC
GAT GTT GTT GTG ACC CAA ACT CC
GAC ATC ATG ATG ACC CAG TCT CC
AGC ATT GTG ATG ACC CAG ACT CC
AAC ATT GTG ATG ACC CAG TCT CA
GAC ATC CAG ATG ACC CAA TCT CC
CAA ATT CTG CTC TCC CAG TCT CC
CAA AAT GTT CTC ACC CAG TCT CC
GAA AAT CTT CTC ACC CAG TCT CC
AGC ATT GTG ATG ACC CAG ACT CA
AAC ATT GTG CTG ACA CAA TCT CC
GAA ATC ATG ATG ACC CAG TCT CC
GAA ATC CTG ATG ACC CAG TCT CC
CAA ATT TTT CTC ACC CAG TCT CC
GAA AAT CTG CTC ACC CAG TCT CC
GAA ACA ACT GTG ACC CAG TCT CC
GAA AAT TTG CTC ACC CAG TCT CC
GAT GCT GTT ATG ACC CAA ACT CC
GAT GTT TTT ATG ACC CAA ACT CC
GGC ATT GTG ATG ACC CAG ACT CC
GAT GCT TTG ATG ACC CAA ACT CC
CAA AAT CTT CTC TCC CAG TCT CC
AAC ATT GTG ATG ACC CAG TCT CC
GAA ATT CTG CTC ACC CAG TCT CC
GAT GTT GTT ATG ACC CAA ACT CC
```

GAC ATC AAG ATG ACC CAG TCT CC
CAA AAT CTT CTC ACC CAG TCT CC
GAT ATC CAG ATG ACA CAG ACT AC
CAA AAT GTG CTC ACC CAG TCT CC
GAT GCT GTT GTG ACC CAA ACT CC
GAA ATT TTT CTC ACC CAG TCT CC
GAC ATC CAG ATG ACA CAG TCT CC
AGC ATT GTG ATG ACC CAG TCT CA
GAC ATT GTG CTG ACC CAA TCT CC
GAT GCT GTG ATG ACC CAA ACT CC
CAA AAT GTG CTC TCC CAG TCT CC
CAA ATT CTT CTC TCC CAG TCT CC
CAA AAT TTT CTC ACC CAG TCT CC
CAA ATT GTT CTC TCC CAG TCT CC
CAA ATT TTG CTC ACC CAG TCT CC
GAC ATT GTG ATG ACC CAG ACT CA
GAT GTT TTT GTG ACC CAA ACT CC
CAA ATT CTT CTC ACC CAG TCT CC
GAA ATT GTT CTC TCC CAG TCT CC
GAA AAT TTG CTC TCC CAG TCT CC
GAT GTT GTT CTG ACC CAA ACT CC
AAC ATT GTG CTG ACC CAA TCT CC
GAT GCT TTT CTG ACC CAA ACT CC
GAC ATT GTG ATG ACC CAG TCT CC
GAA AAT GTG CTC ACC CAG TCT CC
GAA ATT GTG CTC ACC CAG TCT CC
CAA ATT CTG CTC ACC CAG TCT CC
GAC ATC CAG ATG AAC CAG TCT CC
GAC ATT GTG ATG ACC CAG TCT CA
GAA ATT CTT CTC TCC CAG TCT CC
CAA ATT GTT CTC ACC CAG TCT CC
GAT GTC CAG ATA ACC CAG TCT CC
GAA ATC ATG ATG ACC CAA TCT CC
GAT GTT GTG GTG ACC CAA ACT CC
GAA AAT TTT CTC TCC CAG TCT CC
GAA ATT TTT CTC TCC CAG TCT CC
GAC ATT GTG ATG ACT CAG TCT CC
GAC ATC AAG ATG ACC CAA TCT CC
GAA AAT TTT CTC ACC CAG TCT CC
GAC ATC CTG ATG ACC CAG TCT CC
GAA ATT TTG CTC ACC CAG TCT CC

References

1. Kovari, L.C., Momany, C. and Rossmann, M.G. (1995) The use of antibody fragments for crystallization and structure determinations. *Structure*, **3**, 1291–1293.
2. Hunte, C. and Münke, C. (2005) In Dingermann, T., Steinhilber, D. and Folkers, G. (eds.), *Molecular biology in medicinal chemistry*, pp. 300–322.
3. Buchegger, F., Haskell, C.M., Schreyer, M., Scazziga, B.R., Randin, S., Carrel, S. and Mach, J.P. (1983) Radiolabeled fragments of monoclonal antibodies against carcinoembryonic antigen for localization of human colon carcinoma grafted into nude mice. *J Exp Med*, **158**, 413–427.
4. King, D.J., Turner, A., Farnsworth, A.P., Adair, J.R., Owens, R.J., Pedley, R.B., Baldock, D., Proudfoot, K.A., Lawson, A.D., Beeley, N.R. *et al.* (1994) Improved tumor targeting with chemically cross-linked recombinant antibody fragments. *Cancer Res*, **54**, 6176–6185.
5. Essono, S., Frobert, Y., Grassi, J., Creminon, C. and Boquet, D. (2003) A general method allowing the design of oligonucleotide primers to amplify the variable regions from immunoglobulin cDNA. *J Immunol Methods*, **279**, 251–266.
6. Krebber, A., Bornhauser, S., Burmester, J., Honegger, A., Willuda, J., Bosshard, H.R. and Pluckthun, A. (1997) Reliable cloning of functional antibody variable domains from hybridomas and spleen cell repertoires employing a reengineered phage display system. *J Immunol Methods*, **201**, 35–55.
7. Wang, Z., Raifu, M., Howard, M., Smith, L., Hansen, D., Goldsby, R. and Ratner, D. (2000) Universal PCR amplification of mouse immunoglobulin gene variable regions: the design of degenerate primers and an assessment of the effect of DNA polymerase 3′ to 5′ exonuclease activity. *J Immunol Methods*, **233**, 167–177.
8. Chardes, T., Villard, S., Ferrieres, G., Piechaczyk, M., Cerutti, M., Devauchelle, G. and Pau, B. (1999) Efficient amplification and direct sequencing of mouse variable regions from any immunoglobulin gene family. *FEBS Lett*, **452**, 386–394.
9. Kettleborough, C.A., Saldanha, J., Ansell, K.H. and Bendig, M.M. (1993) Optimization of primers for cloning libraries of mouse immunoglobulin genes using the polymerase chain reaction. *Eur J Immunol*, **23**, 206–211.
10. Ewert, S., Huber, T., Honegger, A. and Pluckthun, A. (2003) Biophysical properties of human antibody variable domains. *J Mol Biol*, **325**, 531–553.
11. Plückthun, A., Krebber, A., Horn, U., Knüpfer, U., Wenderoth, R., Nieba, L., Proba, K. and Riesenberg, D. (1996) In McCafferty, J., Hoogenboom, H. R. and Chiswell, D. J. (eds.), *Antibody engineering, a practical approach*. Oxford University Press, Oxford, pp. 203–252.
12. Persson, M.A.A., Samuelsson, A., Yari, F. and Hinkula, J. (1996) Comparisons of expression in procaryotic and eucaryotic hosts of human recombinant Fab molecules. *Immunotechnology*, **2**, 289.
13. Samuelsson, A., Yari, F., Hinkula, J., Ersoy, O., Norrby, E. and Persson, M.A.A. (1996) Human antibodies from phage libraries: neutralizing activity against human immunodeficiency virus type 1 equally improved after expression as Fab and IgG in mammalian cells. *Eur J Immunol*, **26**, 3029–3034.
14. Sambrook, J. and Russell, D.W. (2001), *Molecular cloning: a laboratory manual*. 3rd ed. Cold Spring Harbor Laboratory Press, New York, Vol. 3.
15. Iba, Y., Kaneko, T., Ekida, T., Miyata, K., Inouye, K., Kurosawa, Y. and Yasukawa, K. (1995) A new system for the expression of recombinant antibody in mammalian cells. *Biotechnol Lett*, **17**, 135–138.
16. Nettleship, J.E., Ren, J., Rahman, N., Berrow, N.S., Hatherley, D., Barclay, A.N. and Owens, R.J. (2008) A pipeline for the production of antibody fragments for structural studies using transient expression in HEK 293T cells. *Protein Expr Purif*, **62**, 83–89.
17. Frohman, M.A. (1993) Rapid amplification of complementary DNA ends for generation of full-length complementary DNAs: thermal RACE. *Methods Enzymol*, **218**, 340–356.
18. Aricescu, A.R., Lu, W. and Jones, E.Y. (2006) A time- and cost-efficient system for high-level protein production in mammalian cells. *Acta Crystallographica, Section D*, **62**, 1243–1250.
19. Burrows, P., LeJeune, M. and Kearney, J.F. (1979) Evidence that murine pre-B cells synthesise μ heavy chains but no light chains. *Nature*, **280**, 838–840.
20. Morrison, S.L. and Scharff, M.D. (1975) Heavy chain-producing variants of a mouse myeloma cell line. *J Immunol*, **114**, 655–659.
21. Nettleship, J.E., Rahman-Huq, N. and Owens, R.J. (2009) The production of glycoproteins

by transient expression in Mammalian cells. *Methods Mol Biol*, **498**, 245–263.

22. Strohal, R., Kroemer, G., Wick, G. and Kofler, R. (1987) Complete variable region sequence of a nonfunctionally rearranged kappa light chain transcribed in the nonsecretor P3-X63-Ag8.653 myeloma cell line. *Nucleic Acids Res*, **15**, 2771.
23. Carroll, W.L., Mendel, E. and Levy, S. (1988) Hybridoma fusion cell lines contain an aberrant kappa transcript. *Mol Immunol*, **25**, 991–995.
24. Cochet, O., Martin, E., Fridman, W.H. and Teillaud, J.L. (1999) Selective PCR amplification of functional immunoglobulin light chain from hybridoma containing the aberrant MOPC 21-derived V kappa by PNA-mediated PCR clamping. *Biotechniques*, **26**, 818–820, 822.
25. Carroll, W.L., Mendel, E. and Levy, S. (1988) Hybridoma fusion cell lines contain an aberrant kappa transcript. *Mol Immunol*, **25**, 991–995.
26. Irani, Y., Tea, M., Tilton, R.G., Coster, D.J., Williams, K.A. and Brereton, H.M. (2008) PCR amplification of the functional immunoglobulin heavy chain variable gene from a hybridoma in the presence of two aberrant transcripts. *J Immunol Methods*, **336**, 246–250.
27. Kutemeier, G., Harloff, C. and Mocikat, R. (1992) Rapid isolation of immunoglobulin variable genes from cell lysates of rat hybridomas by polymerase chain reaction. *Hybridoma*, **11**, 23–32.

Chapter 11

Generation of High-Expressing Cells by Methotrexate Amplification of Destabilized Dihydrofolate Reductase Selection Marker

Say Kong Ng

Abstract

A method combining the use of a destabilized dihydrofolate reductase (DHFR) selection marker with methotrexate (MTX) amplification to generate high-expressing cells is described here. The selection marker expression is weakened with the use of the murine ornithine decarboxylase PEST region and AU-rich element to target the DHFR protein and mRNA, respectively, for degradation in the cell. Cells that produce higher levels of DHFR protein, and the adjoining recombinant protein gene, can compensate for the more rapid turnover of the DHFR protein and survive the selection process. This effect can complement MTX amplification to reduce the amount of MTX and shorten the time needed to generate a high-expressing clone. The gene of interest is first inserted into an expression vector that contains a destabilized DHFR selection marker. The resulting expression vector is then linearized and transfected into suspension CHO-DG44 cells. Selection is performed by culturing the cells in a selection medium lacking hypoxanthine and thymidine. Low concentrations of MTX are then used to amplify the transfected genes for increased protein expression. A single cell cloning protocol is also described. This can be used after each stage of MTX amplification to isolate high-expressing clones that are also consistent producers over longer culture periods.

Key words: CHO, Mammalian cell, Recombinant protein expression, Dihydrofolate reductase, Methotrexate amplification, MTX, Murine ornithine decarboxylase, PEST, AU-rich element, Stable transfection

1. Introduction

Recombinant protein-expressing cells can be typically generated with the transfection of an expression vector driven by a strong viral promoter, such as the cytomegalovirus (CMV) promoter (1), as well as optimizing the codon usage of the gene to be expressed (2, 3). One strategy to obtain higher protein expression is to

James L. Hartley (ed.), *Protein Expression in Mammalian Cells: Methods and Protocols*, Methods in Molecular Biology, vol. 801, DOI 10.1007/978-1-61779-352-3_11, © Springer Science+Business Media, LLC 2012

improve selection stringency. This selects for cells with higher productivities of the selection marker, and thus correspondingly higher expression of the protein of interest (4–9). This is possible with higher concentrations of selective drugs, and two drugs that have been practically employed this way are methotrexate (MTX) which inhibits dihydrofolate reductase (DHFR) selection marker (4) and methionine sulfoximine (MSX) which inhibits glutamine synthetase (GS) selection marker (5). These drugs select for genomic changes in transfected Chinese hamster ovary (CHO) or murine NS0 cell lines, resulting in increased copy numbers of the expression vectors in these cells and correspondingly higher protein expression (4, 5). This process is also known as gene amplification.

Another approach to improve selection stringency is to weaken the expression of the selection marker. This also requires cells to be more productive to survive the selection process, leading to higher expression of the recombinant protein. One such method is to reduce the affinity of the neomycin phosphotransferase II selection marker to the neomycin drug by mutating the selection marker gene (6, 7). This is shown to improve specific monoclonal antibody productivities by 6- to 14-fold (7). Another reported method uses a weak Herpes simplex virus thymidine kinase (HSV-tk) promoter to express the selection marker (8).

Here, we describe a method that combines the use of a weakened DHFR selection marker with MTX amplification to generate high-expressing cells (9). The DHFR selection marker expression is weakened using elements that destabilize the DHFR mRNA and protein. The mRNA-destabilizing sequence used is the AU-rich element (ARE) commonly found on 3′ untranslated regions (UTRs) on short-lived mRNAs. The basic ARE is UUAUUUAUU and repeats of this sequence are found to be more effective in destabilizing mRNA than a single unit in murine NIH 3T3 fibroblast cells (10). More information about the functions of the ARE is beyond this chapter and can be found in literature (11–14). The protein degradation signal used here is a 37-amino acid sequence rich in proline, aspartate, glutamate, serine, and threonine (PEST region) in the carboxy-terminus of the short-lived murine ornithine decarboxylase (MODC) protein (15–17). This element confers a 2- to 40-fold increase in DHFR protein degradation in rabbit reticulocyte lysate and *Xenopus* egg extract (17), destabilizes green fluorescent protein (GFP) in various cell lines (18–20), and further reduces reporter expression when coupled with AREs (20, 21).

The gene of interest can be inserted into the CIHDpa (9) or IDP (unpublished data) expression vectors (Fig. 1), which contain destabilized DHFR selection markers. CIHDpa vector has a weak HSV-tk promoter driving its DHFR marker that is destabilized with PEST and six repeats of ARE. IDP vector has the PEST-destabilized DHFR marker linked to our recombinant human interferon-gamma (rhIFNγ) model gene via a weak *Encephalomyocarditis virus* (EMCV)

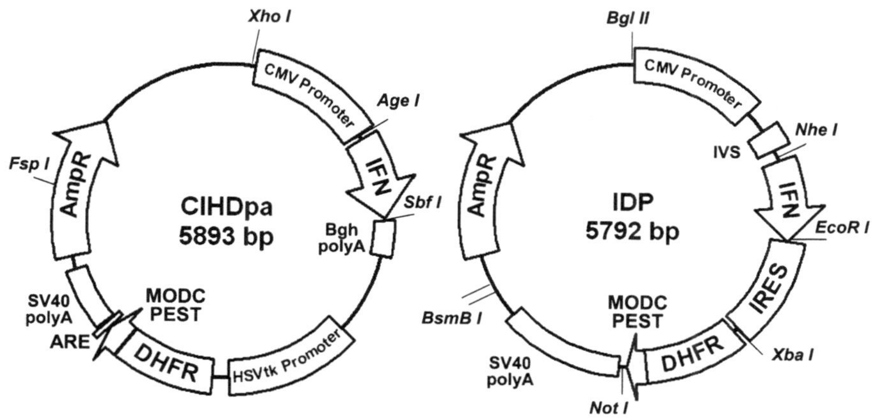

Fig. 1. Vector maps of ClHDpa and IDP expression vectors. The components of the vectors are indicated as follows: *CMV Promoter* cytomegalovirus immediate early promoter, *IFN* recombinant human interferon-gamma open reading frame, *BGH polyA* bovine growth hormone polyadenylation signal, *HSVtk Promoter* HERPES simplex virus thymidine kinase promoter, *DHFR* dihydrofolate reductase open reading frame, *MODC PEST* mouse ornithine decarboxylase PEST region, *ARE* AU-rich element, *SV40 polyA* Simian virus 40 polyadenylation signal, *AmpR* ampicillin resistance beta-lactamase gene, *IVS* intervening sequence, *IRES Encephalomyocarditis virus* internal ribosome entry site. Relevant restriction sites are also indicated on the vector maps in *italics*.

internal ribosome entry site (IRES) (Clontech). Both vectors currently express our model rhIFNγ protein driven by a strong CMV promoter. The expression vector is then linearized using a restriction enzyme and transfected into CHO-DG44 cells that are deficient in DHFR (22). The stable cell pool is obtained by selection using a medium lacking hypoxanthine and thymidine, followed by gene amplification with low concentrations of MTX. Single cell cloning can be performed after selection or after different stages of MTX amplification to isolate high-expressing clones that are consistent producers over long culture periods.

2. Materials

2.1. Preparations for Transfection

1. Standard molecular biology equipment, reagents, and materials.
2. ClHDpa or IDP vector.
3. Tris–HCl, pH 8.5, sterile filtered.
4. Suspension CHO-DG44 cells (Invitrogen, Gibco®, NY).
5. Maintenance medium: HyQ PF-CHO medium (Hyclone, Logan, UT) supplemented with 4 mM L-glutamine (Invitrogen), 0.1% Pluronic® F-68 (Invitrogen), and 0.1 mM sodium hypoxanthine and 16 nM thymidine (1× HT Supplement, Invitrogen).

6. Selection medium: HyQ PF-CHO medium (Hyclone) supplemented with 4 mM l-glutamine (Invitrogen) and 0. 1% Pluronic® F-68 (Invitrogen).
7. 125 ml Disposable Erlenmeyer flasks (Corning, Acton, MA; Catalog number 430421 or 431143).
8. Humidified CO_2 incubators, 37°C.
9. Shaker platform.
10. Hemocytometer.
11. Phosphate-buffered saline (PBS).
12. 0.4% (w/v) Trypan blue solution: 4.0 g of trypan blue (Sigma, St. Louis, MO) is dissolved in 1.0 L of 1× PBS. The solution is then filtered through 0.22-μm bottle-top filter and stored at room temperature (see Note 1).

2.2. Transfection

1. Amaxa® Nucleofector® device (Lonza, Germany).
2. Nucleofector® kit V (Lonza), comprising:
   (a) Cell Line Nucleofector® Solution V.
   (b) Nucleofector® Supplement.
   (c) Nucleofector®-certified cuvettes.
   (d) Plastic pipettes.
   (e) pmaxGFP® vector.
3. Non-treated suspension cell culture 6-well plates (BD Biosciences, San Jose, CA. BD Falcon™; Catalog number 351146).
4. Microcentrifuge tubes (Axygen, Union City, CA; Catalog numbers MCT-175 and MCT-200).

2.3. Selection and Amplification

1. Non-treated suspension cell culture well plates (BD Falcon™).
2. Conical centrifuge tubes (e.g., BD Falcon™; Catalog number 352096).
3. MTX (Sigma, St. Louis, MO): 5 mM MTX stocks are prepared by dissolving MTX in 1× PBS. This stock is then sterile filtered and stored at −20°C in the dark in aliquots of 1 ml (see Note 1).

2.4. Single Cell Cloning

1. Tissue-culture treated well plates (BD Falcon™).
2. DMEM/F12 medium (Invitrogen, Gibco®, NY).
3. Fetal bovine serum (FBS; Hyclone, Logan, UT).
4. Accutase (PAA Laboratories, Austria).

3. Methods

3.1. Preparations for Transfection

1. The gene encoding the recombinant protein can be inserted into the CIHDpa or the IDP vector by replacing the rhIFNγ gene using restriction sites, *Age*I and *Sbf*I or *Nhe*I and *Eco*RI for the respective vector. Alternatively, similar expression vectors can be constructed using standard molecular biology techniques.
2. The expression vector is linearized using an appropriate restriction enzyme (see Note 2). It is then purified by ethanol precipitation and dissolved in sterile-filtered Tris–HCl, pH 8.5, to obtain a DNA concentration of at least 0.5 μg/μl (see Note 3). The purity of the vector can be measured from the ratio of its absorbance at 260–280 nm using a UV spectrophotometer. This ratio is preferably greater than 1.8 for transfection.
3. Suspension CHO-DG44 cells are cultivated in maintenance medium by seeding at 3×10^5 cells/ml into 20-ml medium in 125-ml disposable Erlenmeyer flasks every 3 days or when cell densities are greater than 1.5×10^6 cells/ml. The cells are incubated on shaker platforms set at 110 rpm in a humidified 37°C/8% CO_2 incubator (see Note 4).
4. Since the cells are not attached to a surface, a hemocytometer is necessary to determine cell densities and viabilities using the Trypan blue exclusion method. Cell samples can be diluted with 1× PBS before gentle mixing with an equal volume of 0.4% Trypan blue solution in a microcentrifuge tube. The cells are left to stand for 5 min at room temperature, then gently mixed again, and pipetted into the hemocytometer. The viable unstained cells and the dead blue cells are then counted using a microscope to determine the viable cell density and viability (see Note 5).
5. Two days prior to transfection, seed the cells at 2.5–3.0×10^5 cells/ml in maintenance medium so that the cell density is 1.0–1.5×10^6 cells/ml on the day of transfection.

3.2. Transfection by Nucleofection

1. Set up the Nucleofector® device. With the Nucleofector® I device, the program U-023 works well for transfection into suspension CHO-DG44 cells. Other programs that are used for suspension CHO cells are U-024 and U-030, as described by the manufacturer.
2. Pipette 3 ml of maintenance medium into each well of the labeled nontreated suspension cell culture six-well plates for each sample to be transfected, and incubate these plates in a humidified 37°C/5% CO_2 incubator to warm up the medium

until they are needed. This incubation can last from 15 min to several hours.

3. Pipette 1–5 μg of expression vector or 2 μg pmaxGFP® control vector into a labeled microcentrifuge tube for each transfection (see Note 6).
4. Prepare the working Nucleofector® solution by mixing 20 μl of the provided Nucleofector® Supplement to 90 μl of Cell Line Nucleofector® Solution V (ratio 2:9) for each transfection to be done. Set aside at room temperature till needed (see Note 7).
5. Bring into the biological safety cabinet (BSC) the appropriate numbers of certified cuvettes and provided plastic pipettes for ease of use in later steps.
6. Pipette 1.5×10^6 cells into a sterile microcentrifuge tube for each transfection and centrifuge the cells at $300 \times g$ for 8 min. Pipette out the culture medium, taking care to minimize cell loss at this step. Note that the cells are relatively dry and deprived of nutrients from this step. Hence, the transfections should be completed fast: within 30 min from medium removal works in our hands (see Note 8).
7. Bring the prewarmed six-well plate (step 2) into the BSC.
8. Add 100 μl of the working Nucleofector® solution to an expression vector prepared previously (step 3), transfer the DNA/Nucleofector mixture into a tube containing a cell pellet, and resuspend the cells by pipetting the mixture very slowly 2–3 times with a 200-μl micropipette. Transfer all the cell suspension into a cuvette immediately and cap the cuvette.
9. Check that no air bubbles are trapped at the bottom of the cuvette and insert it into the Nucleofector® device.
10. Apply the transfection program by pressing the X-button.
11. Once the program is completed (in a few seconds), transfer the cuvette back into the BSC. Immediately add ~500 μl of the prewarmed medium from the corresponding well on the six-well plate into the cuvette. Gently transfer all the cell suspension from the cuvette into the same well using a plastic pipette from the Nucleofector® kit. Rinse the cuvette with medium from the same well to transfer all cells into the culture plate.
12. Steps 8–11 are repeated for each transfection.
13. Incubate the six-well plates in a humidified 37°C/5% CO_2 incubator.

3.3. Selection and Amplification

1. 48 h post-transfection, the cells should not be attached to the culture plate since suspension culture plates and serum-free medium are used. Cells transfected with pmaxGFP® vector can be observed under a fluorescence microscope to determine if

the transfection procedure has been successful. More than 60% of these cells should be fluorescent.

2. Harvest all the transfected cell pools by gentle mixing of the cell suspension using a pipette, transferring the cells into 15-ml conical centrifuge tubes, and centrifuging at 640 × *g* for 5 min. Gently resuspend each cell pellet with 5-ml selection medium, and transfer all the cell suspension into new labeled six-well plates (see Note 9).
3. Incubate the six-well plates in a humidified 37°C/5% CO_2 incubator.
4. Repeat the resuspension of cells into 3-ml fresh selection (i.e., without HT) medium into the same six-well plate every 5 days. Cell viabilities should reach a minimum (as low as 20%) after 10–15 days post-selection, after which viable cell counts and cell viabilities should start to increase (see Note 10). A 50–100 μl sample of cell culture can be sampled to determine the viable cell density and viability of the cells when passaging. Cells may be transferred to 12- or 24-well plates to attain at least 1×10^5 viable cells per ml (see Note 11).
5. Once the cells can be seeded at $2–3 \times 10^5$ viable cells per ml in 3- to 5-ml culture medium, the cells are seeded into six-well plates and incubated on a shaker platform at 110 rpm in a humidified 37°C/8% CO_2 incubator. Passaging is performed every 3–4 days subsequently.
6. The cells are scaled up to 20-ml selection medium in 125-ml disposable Erlenmeyer flasks. These flasks are similarly incubated on shaker platforms set at 110 rpm in a humidified 37°C/8% CO_2 incubator.
7. Once cell viability reaches 95% or more for two consecutive passages, MTX amplification can begin by passaging a flask of cells at $3–5 \times 10^5$ cells/ml into selection medium supplemented with 10 nM MTX. These cells are passaged in this supplemented medium every 2–5 days, depending on viable cell density. MTX concentration in the medium is increased stepwise to 50, 250, and 500 nM and 1 μM using the same criteria (i.e., 95% viability for two consecutive passages) and procedures (see Note 12).

3.4. Single Cell Cloning

Single cell cloning may be necessary to identify clones that can amplify and are producing the recombinant protein efficiently. This may be performed at different stages of the selection and MTX amplification processes when cell viabilities are good (i.e., greater than 90%). The single cell cloning protocol of these suspension cells using a serum medium is described here. Cloning with conditioned serum-free medium is also possible, although the efficiency is much lower (see Note 13).

1. The viable cell density of the cell to be cloned is first determined. 1-ml culture is then sampled as a working stock. A diluted cell sample is made by diluting the working stock to 5,000 cells/ml in another microcentrifuge tube using suspension culture medium (see Note 14).
2. Using a micropipette, pipette 0.5 μl of the diluted cell sample into the centers of 8 wells on a tissue culture-treated 96-well plate.
3. Count the number of cells in each droplet in each well under a microscope to obtain the average number of cells in each well (see Note 15).
4. An appropriate volume of fresh culture medium is then added to the diluted cell sample to obtain 1.5 cells per well. For example, a 500 μl diluted cell sample which gave an average of three cells per well should be diluted with another 500 μl of fresh culture medium.
5. Pipette 0.5 μl of the diluted cell sample into empty wells on half a tissue culture-treated 96-well plates, taking care not to touch the walls of the wells as that may leave a drop of cell sample which may contain cells (see Note 16).
6. Screen the droplets under a microscope to mark wells with single cell.
7. Top up the wells containing single cell with 100 μl of DMEM/F12 + 10% FBS with the appropriate concentration of MTX (see Note 17).
8. Repeat steps 5–7 till the desired number of wells with single cells is obtained.
9. Incubate the well plates in a humidified 37°C/5% CO_2 incubator.
10. Top up the medium in the wells every 5 days by adding 50- to 100-μl fresh serum medium per well.
11. To identify clones that are producing well, recombinant protein titers can be determined when the wells are 50–80% confluent. This is performed by replacing all medium with 150-μl fresh serum medium per well. This medium is harvested 24 h later to determine the recombinant protein concentration using ELISA or Western blot. The high-expressing clones thus identified can then be scaled up to six-well plates and adapted to suspension culture for further characterization.

4. Notes

1. Trypan blue and MTX are toxic and should be handled only with gloves and in a BSC or fume hood.
2. Vector linearization generates double-stranded breaks at a predetermined location to facilitate the vector's integration into

the cell genome. This lowers the probability of random breaks that may destroy the recombinant gene and generate antibiotic-resistant clones that do not express the gene. For CIHDpa, either *Fsp*I or *Xho*I is appropriate. For IDP, *Bgl*II or *Bsm*BI can be used. The restriction enzyme used to linearize the transfected DNA should cut in the vector backbone (e.g., ampicillin resistance, origin of replication), but not in the expression or selection cassettes.

3. A DNA concentration of at least 0.5 μg/μl provides flexibility during transfection since we are limited to adding a maximum of 10 μl DNA, and the maximum mass of DNA to be transfected in this protocol is 5 μg. Filtered micropipette tips can be used to manipulate the solution in a BSC to keep the solution sterile.
4. Regular cell passaging can be performed by seeding at 2.0, 3.0, or 5.0×10^5 cells/ml every 4, 3, or 2 days, respectively, to suit experimental needs. CHO-DG44 cells are more robust in a media mix comprising maintenance medium and CD-DG44 (Invitrogen) in a 1:1 ratio. For selected cells, the media mix is selection medium and CD-CHO (Invitrogen) also in a 1:1 ratio.
5. A typical dilution for cell counting consists of 50 μl cell sample, 50 μl 1× PBS, and 100 μl 0.4% Trypan blue solution. Using this 4× diluted sample, four sections of a typical hemocytometer (total volume of 0.4 mm^3) give the cell number corresponding to $10^{-4}\times$ the actual cell density in cells per ml.
6. Due to the inherent variability of the vector integration site and the heterogeneous cell population, expression levels of the transfected cell population may be varied. Multiple transfections (e.g., triplicates) can be set up for each recombinant protein to be expressed to obtain a high-expressing cell pool. Alternatively, single cell cloning (*see* Subheading 3.4) can be performed after selection to ensure that a high-expressing cell population is obtained.
7. A 10% excess of the working Nucleofector® solution is prepared to allow for easier pipetting. Leftover working solution can be stored at 4°C for future transfections as it is stable for 3 months, according to the manufacturer.
8. For pelleting the cells, 2-ml microcentrifuge tubes give tighter cell pellets which can minimize cell loss at this stage.
9. Cell loss during the cell transfers can be minimized by using some medium to rinse the cell culture plate and centrifuge tube after the initial transfer.
10. During selection, cells without integrated DHFR selection marker gene (including the pmaxGFP® transfected cells) and cells expressing low levels of DHFR die, resulting in the observed low viability. As the cells are in suspension, it is difficult to separate the dead cells from the surviving cells. Hence, recovery to

high cell viabilities partly depends on diluting out the dead cells over the next four to eight passages.

11. An important parameter to track during selection is the percentage viability of the cell culture. When the viability starts to increase, the culture should be monitored and passaged more frequently to prevent cells from dying due to nutrient deficiency.
12. After selection and at each stage of MTX amplification, a flask of cells can be seeded to determine its specific recombinant protein productivity. Cells can also be cryopreserved at each stage.
13. Serum contains proteins and macromolecules that enhance cell survivability and cell growth. While the absence of serum is mitigated by a moderate cell seeding density (e.g., greater than 2×10^5 cells/ml in this case) in serum-free cultures during normal passaging, single cell cloning subjects the cells to a low seeding density. Hence, many cells are not able to survive this process in serum-free medium, resulting in a low cloning efficiency. Less than 10% of the single cell clones survived in serum-free medium in our hands, whereas almost all clones survived in serum medium. However, the cells were attached to the plate at the end of the cloning process and it was necessary to detach the cells from the wells using a protease, such as trypsin or accutase, for clones to be expanded into larger culture vessels.
14. The aim of diluting the cell culture is to increase the probability of obtaining single cells in each well. By starting with a slightly higher cell density, it is easier to do the second dilution with cell culture medium to obtain 1.5 cells per well. The converse, starting with a lower cell density and trying to increase the cell density by adding cells from the stock, is much less accurate.
15. 50 μl of the original culture (containing many cells) can be pipetted into a well as a reference point to find the focal plane under the microscope.
16. The suggested 0.5 μl droplet facilitates the screening for single cells under the microscope because the droplet spreads over a small area. However, the small droplet also dries faster as compared to a larger volume (up to 2.0 μl) used. Hence, the single cell clones are prepared in batches of half a 96-well plate to prevent the droplets from drying up.
17. Culturing of single cell clones using DMEM/F12 + 1% FBS for nonamplified cells gave good clone yields in our hands. This reduced serum medium shortens the cell adaptation process to serum-free medium later.

Acknowledgments

The author would like to thank Professor Daniel I.C. Wang and Professor Miranda M.G.S. Yap for their guidance and advice, S.F. Lim for providing the rhIFNγ construct, and support from Bioprocessing Technology Institute (Biomedical Sciences Institutes, A*STAR) and Singapore-MIT Alliance.

References

1. Thomsen DR, Stenberg RM, Goins WF, Stinski MF. (1984) Promoter-regulatory region of the major immediate early gene of human cytomegalovirus. *Proc Natl Acad Sci U S A*. **81**(3):659–63.
2. Kim CH, Oh Y, Lee TH. (1997) Codon optimization for high-level expression of human erythropoietin (EPO) in mammalian cells. Gene. **199**(1–2):293–301.
3. Kalwy S, Rance J, Young R. (2006) Toward more efficient protein expression: keep the message simple. Mol Biotechnol. **34**(2):151–6.
4. Kaufman, R.J., Sharp, P.A. (1982) Amplification and expression of sequences cotransfected with a modular dihydrofolate reductase cDNA gene. J. Mol Biol. **159**:601–621.
5. Cockett MI, Bebbington CR, Yarranton GT. (1990) High level expression of tissue inhibitor of metalloproteinases in Chinese hamster ovary cells using glutamine synthetase gene amplification. Biotechnology (N Y). **8**(7): 662–7.
6. Chen, L., Xie, Z., Teng, Y., Wang, M., Shi, M., Qian, L., Hu, M., Feng, J., Yang, X., Shen, B., Guo, N. (2004) Highly efficient selection of the stable clones expressing antibody-IL-2 fusion protein by a dicistronic expression vector containing a mutant neo gene. J. Immunol. Methods. **295**:49–56.
7. Sautter, K., Enenkel, B. (2005) Selection of high-producing CHO cells using NPT selection marker with reduced enzyme activity. Biotechnol. Bioeng. **89**(5):530–538.
8. Niwa, H., Yamamura, K., Miyazaki, J. (1991) Efficient selection for high-expression transfectants with a novel eukaryotic vector. Gene. **108**:193–200.
9. Ng SK, Wang DI, Yap MG. (2007) Application of destabilizing sequences on selection marker for improved recombinant protein productivity in CHO-DG44. Metab Eng. **9**(3):304–16.
10. Zubiaga, A.M., Belasco, J.G., Greenberg, M.E. (1995) The nonamer UUAUUUAUU is the key AU-rich sequence motif that mediates mRNA degradation. Mol. Cell. Biol. **15**(4):2219–2230.
11. Ross, J. (1995) mRNA stability in mammalian cells. Microbiol. Rev. **59**(3):423–450.
12. Guhaniyogi, J., Brewer, G. (2001) Regulation of mRNA stability in mammalian cells. Gene **265**(1–2):11–23.
13. Bakheet, T., Frevel, M., Williams, B.R.G., Greer, W., Khabar, K.S.A. (2001) ARED: human AU-rich element-containing mRNA database reveals an unexpectedly diverse functional repertoire of encoded proteins. Nucleic Acids Res. **29**(1):246–254.
14. Bakheet, T., Williams, B.R.G., Khabar, K.S.A. (2003) ARED 2.0: an update of AU-rich element mRNA database. Nucleic Acids Res. **31**(1):421–423.
15. Ghoda, L., van Daalen Wetters, T., Macrae, M., Ascherman, D., Coffino, P. (1989) Prevention of rapid intracellular degradation of ODC by a carboxyl-terminal truncation. Science **243**:1493–1495.
16. Ghoda, L., Sidney, D., Macrae, M., Coffino, P. (1992) Structural elements of ornithine decarboxylase required for intracellular degradation and polyamine-dependent regulation. Mol. Cell. Biol. **12**(5):2178–2185.
17. Loetscher, P., Pratt, G., Rechsteiner, M. (1991) The C terminus of mouse ornithine decarboxylase confers rapid degradation on dihydrofolate reductase. J. Biol. Chem. **266**(15):11213–11220.
18. Li, X., Zhao, X., Fang, Y., Jiang, X., Duong, T., Fan, C., Huang, C.C., Kain, S.R. (1998) Generation of destabilized green fluorescent protein as a transcription reporter. J. Biol. Chem. **273**(52):34970–34975.
19. Corish, P., Tyler-Smith, C. (1999) Attenuation of green fluorescent protein half-life in mammalian cells. Protein Eng. **12**(12):1035–1040.

20. Zdanovsky, A., Zdanovskaia, M., Ma, D., Wood, K.V., Almond, B., Wood, M.G. (2004) Rapidly degraded reporter fusion proteins. United States Patent US2004/0146987 A1.
21. Voon, D.C., Subrata, L.S., Baltic, S., Leu, M.P., Whiteway, J.M., Wong, A., Knight, S.A., Christiansen, F.T., Daly, J.M. (2005) Use of mRNA- and protein-destabilizing elements to develop a highly responsive reporter system. Nucleic Acids Res. **33**(3):e27.
22. Urlaub, G., Kas, E., Carothers, A.M., Chasin, L.A. (1983) Deletion of the diploid dihydrofolate reductase locus from cultured mammalian cells. Cell. **33**:405–412.

Chapter 12

Tools for Coproducing Multiple Proteins in Mammalian Cells

Zahra Assur, Wayne A. Hendrickson, and Filippo Mancia

Abstract

Structural and functional studies of many mammalian systems are critically dependent on abundant supplies of recombinant multiprotein complexes. Mammalian cells are often the most ideal, if not the only suitable host for such experiments. This is due to their intrinsic capability to generate functional mammalian proteins. This advantage is frequently countered by problems with yields in expression, time required to generate overexpressing lines, and elevated costs. Coexpression of multiple proteins adds another level of complexity to these experiments, as cells need to be screened and selected for expression of suitable levels of each component. Here, we present an efficient fluorescence marking procedure for establishing stable cell lines that overexpress two proteins in coordination, and we validate the method in the production of recombinant monoclonal antibody Fab fragments. This procedure may readily be expanded to systems of greater complexity, comprising more than two components.

Key words: Coexpression, Monoclonal antibody, Fab fragments, Protein complexes, Protein–protein interactions, Mammalian cells

1. Introduction

The importance of macromolecular assemblages over individual proteins in determining eukaryotic cell function is becoming ever clearer, as evidenced by recent proteomic studies performed in yeast (1, 2). Justifiably as a result, macromolecular complexes are receiving an increasing amount of attention. Biophysical characterization and structure determination of ensembles of two or more protein components requires the capability to successfully generate stable complexes (3). These studies, more often than not, rest their chances of success on the ability to coexpress every element of the complex in the same cell of an appropriate host. While reconstitution of the individually expressed components remains a viable option,

James L. Hartley (ed.), *Protein Expression in Mammalian Cells: Methods and Protocols*, Methods in Molecular Biology, vol. 801, DOI 10.1007/978-1-61779-352-3_12,

frequently observed instability of the single, isolated proteins can only be overcome with a successful coexpression strategy.

Recombinant protein expression in bacteria, typically *Escherichia coli*, has been by far the most successful strategy for generating material to support biophysical and structural studies, and there are established methods for coexpression of genes in this organism (3, 4). However, bacteria are often not suitable hosts for expression of functional eukaryotic proteins (5). Difficulties may arise due to toxicity of the foreign protein to the host, due to differences in the protein folding machineries, or due to the need for posttranslational modifications, absent in bacteria.

Expression systems based on yeasts or on baculovirus-infected insect cells are viable, often successful choices for producing recombinant proteins of mammalian origin (6, 7). Undoubtedly, however, mammalian cells provide the closest match to a native environment for ectopically expressed mammalian genes. Expression of recombinant proteins in mammalian cells can be achieved through transient transfection, viral infection, or stable integration of expression constructs into the host's genome. Transient cotransfection of multiple, separate plasmids carrying individual genes represents the most frequently and successfully employed technique for functional studies. This approach is unfortunately unsuitable for most structural studies, and other large-scale efforts, when abundant amounts of material have to be generated on a routine basis. These applications require a system based on viral infection (8) or on the generation of cell lines in which stable integration of the transfected DNA in the host's genome has occurred. Stable integrants can be identified readily by antibiotic selection. However, the integration of the transfected constructs into the genome of the host cell inevitably leads to a wide spectrum in protein synthesis levels within the same batch of cells, depending primarily on the number of integrants and their sites of integration.

The most critical and often time-consuming step in generating a high-expressing cell line comes down to selecting those cells capable of producing the most protein. The task is made exponentially more complex when one needs to identify cells that can generate high levels of each of two or more protein components Coupling the expression of each polypeptide chain with that of different antibiotic-resistant markers, targeting a specific genomic locus for integration, or incorporating all the components in a single plasmid are valid strategies utilized to increase the frequency of integrants able to synthesize all the desired components. These methods, however, offer little relief in the time consuming step of identifying cells with the required expression characteristics (9).

Discussed here in detail is a fluorescence-based method for the rapid selection of mammalian cells that coexpress two associated proteins (10). The expression of each protein chain is coupled to a separate fluorescent marker via an internal ribosome entry site element (11) present in the vector. One chain is linked to green fluorescent protein (GFP; (12)) and the other to red fluorescent protein (RFP; (13)). Gene expression is driven by a constitutively active strong promoter, derived from cytomegalovirus (CMV; (14)). Figure 1a shows a schematic of the expression constructs, adapted for the production of recombinant antibody Fab fragments, the example discussed here. Expression levels of the two chains are directly proportional to GFP and RFP fluorescence levels (Fig. 1b). The task of identifying cells expressing the highest levels of both components is thus reduced to simple detection and isolation of the most double-fluorescent cells. These can be picked manually when colonies grow out of individual cells after antibiotic selection. Alternatively, the process can be automated with a fluorescence activated cell sorter (FACS), where individual cells, bright for both GFP and RFP can be identified and deposited in single wells of 96-well plates (Fig. 1c). The selected, double-fluorescent colony or cell can then be expanded into a production cell line.

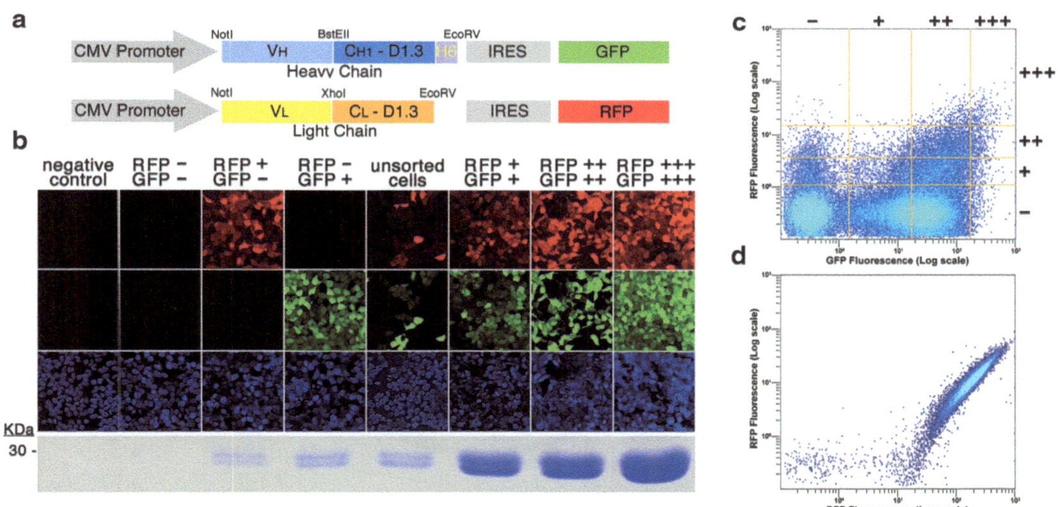

Fig. 1. Coexpression of two polypeptide chains with fluorescent markers. (**a**) Schematic of the expression constructs. Expression is driven by the cytomegalovirus (CMV) promoter. The gene for the heavy chain of the Fab is in blue colors (VH and CH1 D1.3), followed by a hexahistidine tag (H6) fused to its C-terminus. The light chain, in a separate vector is shown in *yellow* (VL) and orange (CL D1.3). Expression of the heavy and light chains is coupled to that of *green* (GFP) and *red* (RFP) protein, respectively, by an internal ribosome entry site (IRES). (**b**) Correlation of Fab 2A11 expression levels with fluorescence. Cells were sorted according to their fluorescence profile and the pools were photographed by confocal microscopy in the emission channels corresponding (from top to bottom) to RFP, GFP, and the nuclear stain TOTO-3. The resulting Fab is shown under the respective columns, run out, after purification, on a Coomassie blue-stained SDS-PAGE gel. (**c**) Fluorescence profile of sorted cells. Fluorescence data from 100,000 viable HEK293-T cells expressing 2A11 after antibiotic selection are shown. Fluorescence is plotted in logarithmic scale for RFP (*vertical axis*) and GFP (*horizontal axis*). The *orange lines* are representative of the 4 × 4 grid used to select negative, +, ++, and +++ cells, for RFP fluorescence (*horizontal lines* from *bottom* to *top*, in increasing order) and for GFP fluorescence (*vertical lines left* to *right*, in increasing order). (**d**) Fluorescence profile of a production cell line. A clonal cell line expressing 2A11 Fab was analyzed by FACS as described for (**c**). Reprinted from Assur et al. [10] with permission from Elsevier.

The fluorescence profile obtained through the FACS, of a production cell line secreting high levels (>20 mg/L) of an Fab fragment, is shown in Fig. 1d. For the example discussed here, a hexahistidine tag genetically fused to the C-terminus of the H chain (Fig. 1a), allows for rapid purification of the Fab complex. The entire procedure is schematized in Fig. 2. By working with additional fluorophores, we expect this method to be readily expandable to cells simultaneously expressing more than two polypeptide chains.

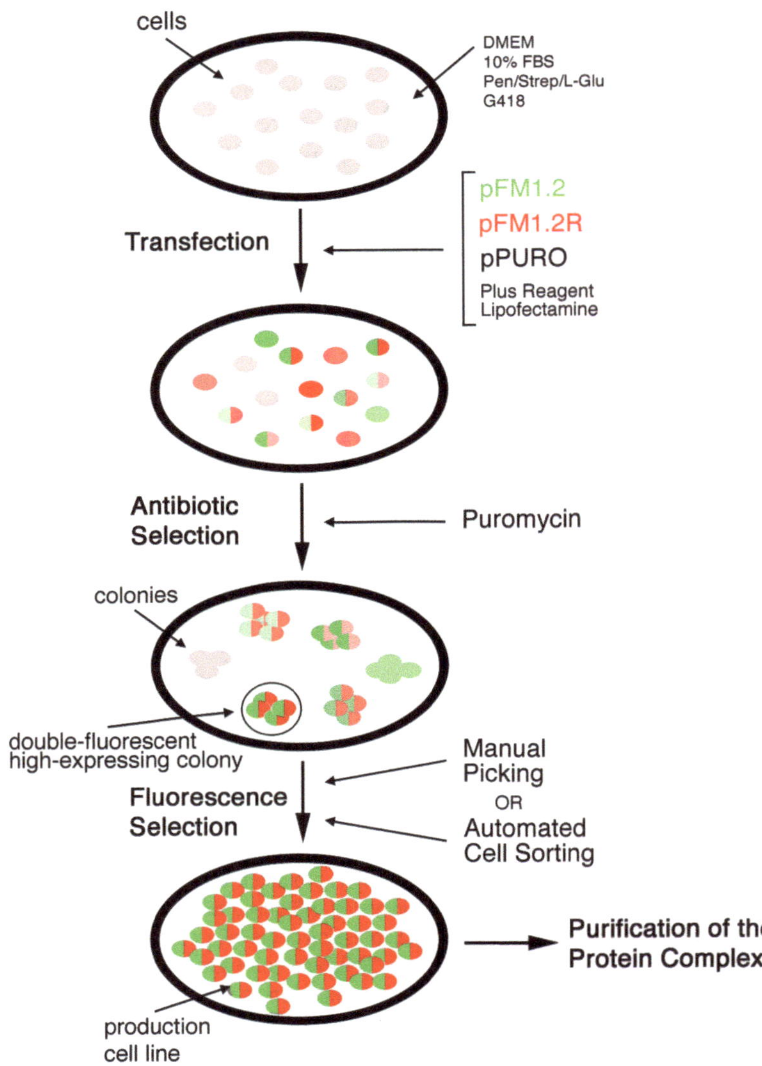

Fig. 2. Flowchart of the two-color selection system. Required procedures from transfection to antibiotic selection to fluorescence selection, are schematized. The intensity of the colors is intended to represent the intensity of the emitted fluorescence for GFP (in *green*) and RFP (in *red*). Double-fluorescent cells are drawn with two colored halves. Cells expressing only one fluorescent marker are represented with the matching color. Untransfected cells and cells expressing neither GFP nor RFP are drawn in *light gray*.

2. Materials

2.1. Construction of Expression Vectors (Add Sub-numbers)

1. pFM1.2 (15) (see Note 1).
2. pIRES2-DsRed-Express (Clontech, cat. no. 632463) (see Note 2).
3. pASK84 (16), pBluescript (Stratagene, cat. no. 212205-8), and Fv fragments cloned in pGEM-T easy (Promega, cat. no. A1360) (see Note 3).
4. pPURO (see Note 4).
5. Standard, commercially available molecular biology reagents (Stratagene, Promega, New England Biolabs, Roche, Qiagen) (see Note 5).
6. DH5α Max Efficiency Competent Cells (Invitrogen, cat. no. 18258-012) (see Note 6).
7. Ampicillin-resistant LB-Agar plates.
8. LB liquid medium with 100 μg/mL of Ampicillin.

2.2. Preparation of DNA for Transfection

1. Maxi prep kit (Qiagen, cat. no. 12263).
2. S-200 DNA Microspin Columns (GE Healthcare, cat. no. 27-5120-01).

2.3. Cell Culture and Generation of Stable Lines

1. T-antigen transformed Human Embryonic Kidney 293 cells (HEK293-T; ATCC, cat. no. CRL-11268) (see Note 7).
2. Dulbecco's modified Eagle's medium (DMEM; Chemicon, cat. no. SLM-020-B) (see Note 8).
3. Fetal bovine serum (FBS; Hyclone, cat. no. SH30071.03).
4. Penicillin, streptomycin, and L-glutamine (Pen/Strep/L-Glu; Sigma, cat. no. G1146).
5. Puromycin (Puro, prepared as a 0.5 mg/mL 100× stock with ultrapure H_2O and filtered; Sigma, cat. no. P8833).
6. Geneticin (G418, prepared as a 50 mg/mL 100× stock with ultrapure H_2O and filtered; Gibco, cat. no. 11811-031).
7. Plus® Reagent (Invitrogen, cat. no. 11514-015) (see Note 9).
8. Lipofectamine (Invitrogen, cat. no 18324-012).
9. Phosphate buffer saline, without Ca and Mg (PBS; Chemicon, cat. no. BSS-1006-B).
10. Ultrapure H_2O (Chemicon, cat. no. TMS-006-B).
11. 100-mm diameter Tissue culture-treated circular dish (Corning, cat. no. 430167).
12. 150-cm^2 Tissue culture-treated flasks (Corning, cat. no. 430825).

2.4. Selecting High-Expressing Cells

2.4.1. Manual Picking of Double-Fluorescent Colonies

1. Trypsin–EDTA (Chemicon, cat. no. SM-2005-C).
2. 0.1–10 μL Sterile filter tips (USA Scientific, cat. no. 1121-4810).
3. 0.1–10 μL Pipette.
4. 96-Well round-bottom tissue culture-treated plates (Corning, cat. no. 3799).
5. 48-Well flat-bottom tissue culture-treated plates (Corning, cat. no. 3548).
6. Inverted fluorescence microscope (see Note 10).

2.4.2. FACS Sorting

1. Beckman Coulter Altra flow cytometer equipped with Autoclone functionality (see Note 11).
2. Leibovitz's L-15 medium (L15; Gibco, cat. no. 11415-064).
3. 5-mL Snap-cap polypropylene round bottom sterile tubes (BD Falcon, cat. no. 352063).
4. 96-Well tissue culture-treated flat bottom plates (Corning, cat. no. 3595).
5. Cell strainer 40 μM nylon (BD Falcon, cat. no. 352340).

2.5. Large-Scale Protein Expression

CellBind surface HYPERflask (Corning, cat. no. 10010) (see Note 12).

2.6. His-Tag Based Fab Purification

1. Ni-NTA agarose resin (Qiagen, cat. no. 30210).
2. Disposable columns (Bio-Rad, cat. no. 731-1553) fitted with 22G1/2 gauge needles (Becton Dickinson, cat. no. 305156).
3. Na/HEPES 1 M, pH 7.5, NaCl 5 M and imidazole/HCl 2 M, pH 7.5 stock solutions (see Note 13).
4. Slide-a-lyzer dialysis cassettes 10,000 kDa cut-off, 3–12 mL capacity (Pierce, cat. no. 66810) (see Note 14).
5. 10 mL Luer-Lok™ tip syringe (Becton Dickinson, cat. no. 309604).
6. Bio-Rad Protein Assay (Bradford Dye; Bio-Rad, cat. no. 500-0006).

3. Methods

3.1. Cloning and Construction of Expression Vectors

To generate the RFP-expressing pFM1.2R from the GFP-expressing pFM1.2, the region coding for the IRES-RFP elements was excised from pIRES2-DsRed-Express to replace IRES-GFP in pFM1.2. This was achieved in five steps, and ultimately an *Xho*I-*Spe*I fragment (IRES-RFP, after the introduction of a 3′ *Spe*I site by site-directed mutagenesis, and a 5′ *Xho*I site via a short oligonucleotide linker) was cloned *Xho*I-*Xba*I into pFM1.2 (see Note 15).

Fv_H and Fv_L regions of anti-serotonin receptor subtype 2c (17) monoclonal antibodies (18) were cloned into pBluescript containing the C_{H1} and C_L regions of anti-lysozyme D1.3 (19) monoclonal antibody respectively. The genes were fused using *Bst*EII (H chain) and *Xho*I (L chain) as reported for the expression of Fab fragments in *E. coli* (16). The two chains were then cloned as *Not*I/*Eco*RV fragments into A1.2 (H chain) and A1.2R (L chain) prepared by digestion with the same restriction enzymes (Fig. 1a).

3.2. Preparation of DNA for Transfection

1. Retransform miniprep quality DNA into *E. coli* and plate on LB/Amp plates.
2. Inoculate a 250 mL LB/Amp culture with a single colony.
3. Purify plasmid DNA using a maxi prep kit (see Note 16).
4. Spin 50 μL of plasmid DNA through an S-200 column, prepared following the manufacturers instructions, for 1 min at 800 × *g* in a benchtop centrifuge (see Note 17).

3.3. Cell Culture and Generation of Stable Lines

HEK293-T cells are grown in DMEM supplemented with 10% FBS, 1:100 Pen/Strep/L-Glu and 500 μg/mL G418 are maintained at all times in temperature controlled incubators at 37°C, in a humidified environment enriched with 5% CO_2.

For transfection of these cells to generate stable lines:

1. Plate HEK293-T cells on a 100-mm diameter tissue culture dish to 30–40% confluency the evening before the transfection (see Note 18).
2. The next morning, to a 15-mL sterile tube add 1 μg of pPURO and 5 μg of the two expression plasmids containing the genes to be coexpressed (see Note 19). The total volume of the three DNAs should ideally be less than 10 μL.
3. Add 750 μL of serum-free DMEM (without any supplements) and 20 μL of Plus® Reagent to the DNA. Mix or vortex gently and incubate at room temperature for a minimum of 15–20 min.
4. Add 750 μL of serum-free DMEM (without any supplements) and 30 μL of Lipofectamine to the same tube. Mix or vortex gently and incubate at RT for at least 15–20 min.
5. Replace media from the cells with 5 mL of serum-free DMEM (without any supplements). Perform this step immediately before or after step 4.
6. Add 5 mL of serum-free DMEM (without any supplements) to the transfection mixture, mix well, and add to the dish containing the cells after having removed their previous media.
7. Allow the cells to incubate with the transfection mixture for a minimum of 5 h at 37°C in the humidified incubator, enriched with 5% CO_2.

8. Add 10 mL of fully supplemented media and leave overnight.
9. The next morning, replace with 10 mL of fresh media (see Note 20).
10. The following day, supplement the media by adding 5 μg/mL of puromycin to the growth medium (see Note 21).
11. Replace media every 3–4 days (see Note 22).
12. Monitor for the formation of puromycin-resistant, double-fluorescent colonies by visual inspection of the cells under the fluorescence microscope (see Note 23).

3.4. Selecting High-Expressing Cells

After antibiotic selection, the only cells to survive are those in which stable integration of the transfected DNA in the host's genome has occurred. Cells that failed to transfect and cells in which the uptake of plasmid has only been transient will not survive the antibiotic selection step (see Note 24). Stable integrants will lead to colony formation. Each colony will typically grow from a single cell, and hence, colonies are considered of clonal purity. Protein expression levels vary dramatically from colony to colony, as a function of the site(s) of integration and the copy number, just to mention two most important parameters. The investigator must, therefore, identify colonies for which expression of the two recombinant proteins is maximal. In this system, expression levels correlated with fluorescence levels (Fig. 1b). The task is, therefore, to select colonies that exhibit high levels of both GFP and RFP derived fluorescence. This can be achieved by visual inspection and manual isolation of colonies (Subheading 3.4.1) or, in an automated way, by FACS and cloning of individual cells (Subheading 3.4.2) (see Note 25).

3.4.1. Manual Picking of Double-Fluorescent Colonies

1. Carefully scan the plate by visual inspection under the fluorescence microscope for bright, double-fluorescent colonies (see Note 26).
2. Pick individual colonies by gentle aspiration with a 10-μL tip/pipette (see Note 27).
3. Transfer each aspirated colony to a well of a 96-well tissue culture plate containing 15 μL of Trypsin–EDTA (see Note 28).
4. Transfer the contents of each 96-well to a 24-well tissue culture plate containing 500 μL of fully supplemented DMEM media (with puromycin).
5. Leave in the incubator to grow to approximately 80% confluency.
6. Expand to 6-well format, allow growth to confluency, and test for expression of the proteins of interest (see Note 29).

3.4.2. FACS Sorting

We present here a protocol describing how to prepare the cells for sorting, and the steps that should be followed for this experiment.

However, operation of a cell sorter requires appropriately trained personnel. Details of how a cell sorter is operated depend on the make and model of the instrument and are beyond the scope of this work.

1. Wait for the colony-containing plate to reach 50–80% confluency before scheduling the experiment.
2. The day before the experiment split the cells 1:3 into fresh medium (see Note 30).
3. Dislodge the cells by gentle trypsinization.
4. Resuspend the cells with 5–10 mL of fully supplemented DMEM, transfer to a 15-mL tube and pellet by centrifugation for 5 min at 800 × *g*.
5. Remove the medium by aspiration and resuspend the pellet by pipetting up and down with 4 mL of serum-free, L15 medium (see Note 31).
6. Filter the cells through a nylon cell strainer (40–70 μM mesh size) (see Note 32).
7. Transfer cells to a 5-mL snap-cap tube.
8. In parallel, perform steps 2–6 for an untransfected control and run these cells first through the cell sorter (see Note 33).
9. Determine the viable cell population using their forward and side scatter characteristics. Set the gates accordingly.
10. The fluorescent cells are excited with the 488 nm line of a krypton–argon laser. RFP emission is detected using a 590/20 nm band pass filter and GFP emission detected with a 525/30 nm band pass filter (see Note 34).
11. Select the top fluorescent cells of the double positive population (corresponding to 0.1% of the viable cell population) for cloning to single cell purity (Fig. 1c).
12. Using the cell sorter's Autoclone mode, deposit one selected cell in each well of a 96-well plate containing 100 μL of fully supplemented DMEM media in each well (see Note 35).
13. Incubate the plate in a temperature controlled incubator at 37°C, in a humidified environment enriched with 5% CO_2 for approximately 1 week.
14. Check the plate for bright, double-fluorescent colonies under a fluorescence microscope. Expand a limited number of colonies and check for expression of proteins (see Note 36).

3.5. Large-Scale Protein Expression

This protocol refers to how we scale-up expression of recombinant Fabs.

1. Expand cells from 96- to 24- to 6-well plates.
2. Expand 6-well plates to 100-mm diameter dishes to 150-cm^2 flasks.

3. Use 2–4 confluent 150-cm^2 flasks of cells to set up each HYPERflask.
4. Fill the HYPERflask to the brim (approximately 500 mL of medium).
5. Leave for 7–14 days (until the medium turns orange/yellow).
6. Harvest the medium and proceed to purification of the recombinant protein(s).

3.6. His-Tag Based Protein Purification

The recombinant Fab fragments carry a genetically engineered hexahistidine tag fused to the C-terminus of the heavy chain. We use this simple purification procedure both to screen colonies for the highest expressing ones and for milligram-scale protein purification (see Note 37).

1. Add 20 mM Na HEPES (pH 7.5), 400 mM NaCl, and 10 mM imidazole (pH 7.5) to the harvested medium.
2. Equilibrate Ni-NTA resin on a column, with five column volumes of buffer containing 20 mM Na HEPES (pH 7.5), 400 mM NaCl, and 10 mM imidazole (pH 7.5). Fit a 22- or 23-gauge needle to the column's outlet (see Note 38).
3. Add recombinant-protein containing medium to the resin (see Note 39).
4. Wash the resin with 10 column volumes of 20 mM Na HEPES (pH 7.5), 400 mM NaCl, 25 mM imidazole.
5. Elute the purified protein with 20 mM Na HEPES (pH 7.5), 400 mM NaCl, 200 mM imidazole. Collect 0.5 column volume fractions and monitor for the presence of protein (see Note 40).
6. Pool the protein-containing fractions and dialyze overnight at 4°C against a buffer of choice (see Note 41).

4. Notes

1. pFM1.2 (15) is an in-house generated vector to express proteins under the control of a strong constitutively active promoter (CMV), linking the expression of the gene of interest to that of GFP via an IRES element. Similar vectors are available commercially.
2. We took the RFP gene from this vector and cloned it into pFM1.2 to replace GFP and generate pFM1.2R (10). This parental vector could most likely serve the same purpose (untested in our hands) as pFM1.2R.
3. These plasmids were required to generate the heavy and light chains of the Fabs that were then cloned into pFM1.2 and

pFM1.2R respectively. Fv regions were cloned from cDNA generated from RNA extracted from hybridoma cells using kits from Roche. The CH1 and CL regions of the Fab were taken from the anti-lysozyme D1.3 antibody (in pASK84) and fused to the respective variable regions as reported by Skerra (16).

4. pPURO is an in-house generated selection vector that confers puromycin resistance to eukaryotic cells. The expression of the puromycin-resistance gene is driven by the RSV promoter. Equivalent plasmids are readily available from various commercial sources (for example, pPUR, Clontech, cat. no. 631601).
5. In our laboratory, we typically use the QuickChange site-directed mutagenesis kit (Stratagene, cat. no. 200515) to introduce mutations, miniprep kit from Promega (cat. no. A1460), restriction enzymes from New England Biolabs, DNA ligation kit from Roche (cat. no. 11 635 379 001), QIAquick gel extraction kit from Qiagen (cat. no. 28704).
6. Any other bacterial competent cell that is endA would work. DNA from non-endA strains is quickly degraded.
7. HEK293-T cells transfect well, are robust, and relatively easy to handle. SV40 large T-antigen transformation (20) is supposed to boost expression levels of ectopically expressed proteins.
8. Whatever source for DMEM you choose to utilize, make sure that it is supplied with high glucose. We also recommend keeping cells in culture with DMEM from the same supplier.
9. We have also tested other transfection kits and obtained comparable results.
10. Our laboratory is equipped with a Nikon Diaphot 300 inverted fluorescence microscope.
11. This cell sorter has excitation capabilities at 488 nm, and emission detection is achieved using 590/20 nm (for RFP) and 525/30 nm (for GFP) band-pass filters.
12. As an alternative to HYPERflask, we also recommend two chamber CellStack flasks (Corning, cat. no. 3269). HYPERflasks are excellent for secreted proteins.
13. Ingredients to make these solutions are from molecular biology grade chemicals from Sigma-Aldrich.
14. The approximate molecular weight of an Fab fragment is 50,000 Da. These dialyzing units with a 10,000 Da molecular weight cut-off safe to use.
15. Sequencing and PCR amplification across IRES elements most often does not work due to their high GC contents. In this

instance, the IRES element of pIRES2-DsRed-Express was removed by digestion with *Pst*I and reinserted after the PCR to introduce the 3′ *Spe*I site.

16. We recommend washing the DNA twice with 70% EtOH after precipitation with isopropanol. This step improves the quality of the purified DNA. We typically resuspend the DNA pellet in 300 μL of ultrapure H_2O to yield concentrations between 0.5 and 3 μg/μL as measured by absorbance at 280 nm.
17. This step increases the purity of the DNA to be used for transfection. Purity of the DNA is a very important parameter to maximize the efficiency of transfection. Also, it serves to minimize the risk of any residual bacteria contaminating the mammalian cells following transfection.
18. The cells should look healthy, not clumpy, perfectly distributed and at approximately 50–60% confluency the following morning.
19. This and all other procedures to be carried out under a tissue culture hood following standard sterile techniques.
20. Alternatively, one can transfect later in the day, leave the cells incubated with only the transfection mix overnight, and replace this with fully supplemented DMEM the next morning. Results are comparable.
21. Puromycin is added to select for stable integrants. At this point, RFP and GFP should be clearly visible under the fluorescent microscope. Fluorescence of the cells monitored at this stage is a useful indicator of the efficiency of transfection. Puromycin should be kept continuously present from this point on.
22. Untransfected cells, and cells where integration has failed to occurred, should start to die a couple of days after the addition of puromycin. Occasionally, the dead or dying cells do not lift off and stay clumped on the dish. Should this occur, wash cells with PBS every other day.
23. Colonies should start to appear approximately a week after transfection. The number of colonies and their fluorescence profile varies according to several parameters including the efficiency of transfection and the toxicity of the expressed proteins to the cells. Do not expect all colonies to be double fluorescent.
24. Plasmids that are not integrated in the genome are rapidly eliminated from the cell within a few days following transfection. Recombinant protein expression typically peaks 48–72 h after transfection and declines thereafter.
25. The two methods have advantages and disadvantages. Manual picking saves time, as colonies are selected as opposed to single cells that then need to grow into colonies. However, timing is critical as colonies need to be picked when large enough to

be aspirated, but small enough not to have grown into neighboring colonies. Also, only a limited number of colonies can be picked manually, and one always runs the risk of failing to identify the brightest, best expressing cells. FACS, on the contrary, is quicker to perform, and provides a more accurate readout of the fluorescence profile of the cells. Ultimately, however, the procedure from transfection to a scaled-up cell line is lengthier as individual cells are selected and it takes time to expand these into a cell-line of clonal purity. FACS is the only feasible option when either colonies overgrow or dead cells remain attached to the dish.

26. Care should be taken to ensure that selected colonies are not only bright for both GFP and RFP fluorescence but also homogeneous. Colonies that show nonfluorescent or partially fluorescent patches should be discarded.
27. Ideally, the fluorescence microscope should be placed under a sterile hood. This is most likely not feasible. Care should be taken to minimize the chances of contamination. The time the lid is kept off the cell-containing dish should be kept to a minimum, and the microscope should not be located around potential sources of contamination such as A/C ventilation outlets.
28. This step is performed to disperse the colony into individual cells. Care should be taken not to leave the cells in Trypsin–EDTA more than a minute.
29. We recommend testing more than one double-fluorescent, bright colony, for protein expression. We typically screen 3–6. The degree of expansion of the cells required to test for expression depends on the protein levels and on the assay.
30. Make sure that the cells have been well trypsinized and that they are evenly dispersed.
31. L15, unlike DMEM, exhibits no [CO_2]-dependent pH variability.
32. This step is necessary to remove clumps from the cells. Clumps often clog the cell sorter and/or negatively impact its accuracy.
33. Untransfected cells are required to determine correct settings for the cell sorter. Cells exhibit background fluorescence. This can be set to zero by running cells expressing neither GFP nor RFP.
34. Settings depend on the emission spectra of the fluorescent proteins, and on the lasers and filters available with the instrument. If excitation is performed with one laser, the wavelength must be chosen carefully to compromise adequately between the two fluorophores. In our experiments, we were obliged to choose 488 nm for excitation. At this wavelength, GFP excitation is close to maximum, while RFP excitation is

at approximately 40%. We could also have performed the experiment at 514 nm where RFP excitation would have been optimal, but GFP would have been below 10%. The reader is encouraged to consult http://www.metroflow.org under tools & tutorials for a better understanding on how to set up a two-color experiment.

35. To increase viability of the cloned cells, we recommend adding 30–50% of conditioned medium to the wells. To generate conditioned medium, split cells 1:5 into fresh medium, remove and sterile filter this medium after overnight incubation. Alternatively, consider using a higher percentage of serum (15–20%).
36. Do not expect to find a bright, double-fluorescent colony in each well. Success rates depend primarily on cell viability (empty wells), and the accuracy of the instrument (empty wells and multiple colonies per well).
37. For small-scale (colony screening) purification we typically use 2 mL of medium from cells grown in 6-well format. For milligram-scale expression we purify medium from 1 HYPERflask (approximately 500 mL).
38. We use approximately 200 μL and 5 mL of Ni-NTA resin for screening and production purposes respectively. Column sizes are chosen accordingly.
39. For screening purposes, we fit the column with a 23-gauge needle and allow the medium to slowly pass through the resin by gravity flow. For large-scale purification, we load the column overnight with a peristaltic pump.
40. Protein can be monitored by Bradford assay or by absorbance at 280 nm. Purified Fab typically elutes in fractions 2–4.
41. If the sample needs to be greater than 95% pure, we recommend concentrating the protein and loading it onto a size-exclusion chromatography column (for example a Superdex 200, GE Healthcare). This second purification step is very efficient at removing residual contaminants.

Acknowledgments

We are grateful to Richard Axel for discussions and to Ira Schieren for sharing with us his invaluable expertise on cell sorting. This work was supported in part by NIH grants GM68671 and GM75026.

References

1. Gavin, A. C., Aloy, P., Grandi, P., Krause, R., Boesche, M., Marzioch, M., Rau, C., Jensen, L. J., Bastuck, S., Dumpelfeld, B., Edelmann, A., Heurtier, M. A., Hoffman, V., Hoefert, C., Klein, K., Hudak, M., Michon, A. M., Schelder, M., Schirle, M., Remor, M., Rudi, T., Hooper, S., Bauer, A., Bouwmeester, T., Casari, G., Drewes, G., Neubauer, G., Rick, J. M., Kuster, B., Bork, P., Russell, R. B., and Superti-Furga, G. (2006) Proteome survey reveals modularity of the yeast cell machinery, *Nature* ***440***, 631–636.
2. Krogan, N. J., Cagney, G., Yu, H., Zhong, G., Guo, X., Ignatchenko, A., Li, J., Pu, S., Datta, N., Tikuisis, A. P., Punna, T., Peregrin-Alvarez, J. M., Shales, M., Zhang, X., Davey, M., Robinson, M. D., Paccanaro, A., Bray, J. E., Sheung, A., Beattie, B., Richards, D. P., Canadien, V., Lalev, A., Mena, F., Wong, P., Starostine, A., Canete, M. M., Vlasblom, J., Wu, S., Orsi, C., Collins, S. R., Chandran, S., Haw, R., Rilstone, J. J., Gandi, K., Thompson, N. J., Musso, G., St Onge, P., Ghanny, S., Lam, M. H., Butland, G., Altaf-Ul, A. M., Kanaya, S., Shilatifard, A., O'Shea, E., Weissman, J. S., Ingles, C. J., Hughes, T. R., Parkinson, J., Gerstein, M., Wodak, S. J., Emili, A., and Greenblatt, J. F. (2006) Global landscape of protein complexes in the yeast Saccharomyces cerevisiae, *Nature* ***440***, 637–643.
3. Romier, C., Ben Jelloul, M., Albeck, S., Buchwald, G., Busso, D., Celie, P. H., Christodoulou, E., De Marco, V., van Gerwen, S., Knipscheer, P., Lebbink, J. H., Notenboom, V., Poterszman, A., Rochel, N., Cohen, S. X., Unger, T., Sussman, J. L., Moras, D., Sixma, T. K., and Perrakis, A. (2006) Co-expression of protein complexes in prokaryotic and eukaryotic hosts: experimental procedures, database tracking and case studies, *Acta Crystallogr D Biol Crystallogr* ***62***, 1232–1242.
4. Tan, S. (2001) A modular polycistronic expression system for overexpressing protein complexes in Escherichia coli, *Protein Expr Purif* ***21***, 224–234.
5. Yin, J., Li, G., Ren, X., and Herrler, G. (2007) Select what you need: a comparative evaluation of the advantages and limitations of frequently used expression systems for foreign genes, *J Biotechnol* ***127***, 335–347.
6. Geisse, S., Gram, H., Kleuser, B., and Kocher, H. P. (1996) Eukaryotic expression systems: a comparison, *Protein Expr Purif* ***8***, 271–282.
7. Eifler, N., Duckely, M., Sumanovski, L. T., Egan, T. M., Oksche, A., Konopka, J. B., Luthi, A., Engel, A., and Werten, P. J. (2007) Functional expression of mammalian receptors and membrane channels in different cells, *J Struct Biol* ***159***, 179–193.
8. Lundstrom, K. (2003) Semliki Forest virus vectors for rapid and high-level expression of integral membrane proteins, *Biochim Biophys Acta* ***1610***, 90–96.
9. Birch, J. R., and Racher, A. J. (2006) Antibody production, *Adv Drug Deliv Rev* ***58***, 671–685.
10. Assur, Z., Schieren, I., Hendrickson, W. A., and Mancia, F. (2007) Two-color selection for amplified co-production of proteins in mammalian cells, *Protein Expr Purif* ***55***, 319–324.
11. Vagner, S., Galy, B., and Pyronnet, S. (2001) Irresistible IRES. Attracting the translation machinery to internal ribosome entry sites, *EMBO Rep* ***2***, 893–898.
12. Tsien, R. Y. (1998) The green fluorescent protein, *Annu Rev Biochem* ***67***, 509–544.
13. Knop, M., Barr, F., Riedel, C. G., Heckel, T., and Reichel, C. (2002) Improved version of the red fluorescent protein (drFP583/DsRed/RFP), *Biotechniques* ***33***, 592–602.
14. Thomsen, D. R., Stenberg, R. M., Goins, W. F., and Stinski, M. F. (1984) Promoter-regulatory region of the major immediate early gene of human cytomegalovirus, *Proc Natl Acad Sci U S A* ***81***, 659–663.
15. Mancia, F., Patel, S. D., Rajala, M. W., Scherer, P. E., Nemes, A., Schieren, I., Hendrickson, W. A., and Shapiro, L. (2004) Optimization of protein production in mammalian cells with a coexpressed fluorescent marker, *Structure (Camb)* ***12***, 1355–1360.
16. Skerra, A. (1994) A general vector, pASK84, for cloning, bacterial production, and single-step purification of antibody Fab fragments, *Gene* ***141***, 79–84.
17. Julius, D., MacDermott, A. B., Axel, R., and Jessell, T. M. (1988) Molecular characterization of a functional cDNA encoding the serotonin 1c receptor, *Science* ***241***, 558–564.
18. Mancia, F., Brenner-Morton, S., Siegel, R., Assur, Z., Sun, Y., Schieren, I., Mendelsohn, M., Axel, R., and Hendrickson, W. A. (2007) Production and characterization of monoclonal antibodies sensitive to conformation in the 5HT2c serotonin receptor, *Proc Natl Acad Sci U S A* ***104***, 4303–4308.
19. Amit, A. G., Mariuzza, R. A., Phillips, S. E., and Poljak, R. J. (1986) Three-dimensional structure of an antigen-antibody complex at 2.8 A resolution, *Science* ***233***, 747–753.
20. Ali, S. H., and DeCaprio, J. A. (2001) Cellular transformation by SV40 large T antigen: interaction with host proteins, *Semin Cancer Biol* ***11***, 15–23.

Chapter 13

Identification and Characterization of Protein Glycosylation Using Specific Endo- and Exoglycosidases

Paula Magnelli, Alicia Bielik, and Ellen Guthrie

Abstract

Enzymatic deglycosylation followed by SDS-PAGE is a valuable method to detect glycan modifications on protein samples. Specific glycosidases were used to remove sugars from glycoproteins in a controlled fashion leaving the protein core intact; the resulting change in molecular weight could be detected as shifts in gel mobility. Alternatively, glycan-sensitive reagents were used to visualize the intensity of glycoprotein bands before and after enzyme treatment. The ease of use of these techniques, which require only basic laboratory instrumentation and reagents, makes them the methodology of choice for initial glycobiology studies. These protocols are also well suited to screen for optimal expression conditions, since multiple glycoprotein samples can be processed at once.

Key words: Glycoprotein, *N*-Glycan, *O*-Glycan, *O*-GlcNAc, PNGase F, *O*-Glycosidase, Deglycosylation

1. Introduction

Glycosylation, the addition of covalently linked sugars, is a major posttranslational modification of eukaryotic proteins and can significantly affect in vivo activities of biopharmaceuticals. These sugar moieties (glycans) influence binding, internalization, and serum clearance of therapeutic agents such as immunoglobulins, erythropoietin, and replacement enzymes. Improper or variable glycosylation on drug formulations is undesirable for regulatory and intellectual property reasons (1, 2).

Although glycosylation is traditionally considered in the context of secreted proteins (Fig. 1, top panel), glycosylation of proteins resident in the nucleus and cytoplasm (i.e., modification of proteins not destined for secretion or the cell surface) is now acknowledged as a signaling code fundamental to many biochemical processes (3). Nuclear and cytoplasmic proteins may be dynamically

James L. Hartley (ed.), *Protein Expression in Mammalian Cells: Methods and Protocols*, Methods in Molecular Biology, vol. 801,
DOI 10.1007/978-1-61779-352-3_13,

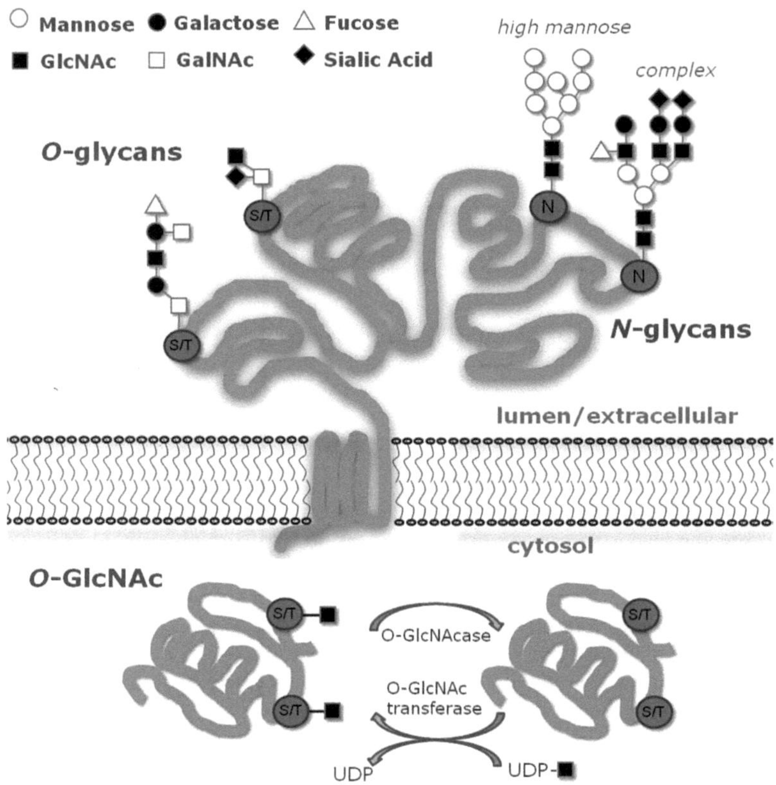

Fig. 1. Typical glycosylation patterns for secreted or cell-surface proteins (*top panel*) or cytoplasmic or nuclear proteins (*bottom panel*).

substituted at serine or threonine residues with a single *N*-acetylglucosamine (GlcNAc) (Fig. 1, bottom panel), often exchanging with *O*-phosphorylation at the same amino acids (3, 4). Thus, protein *O*-GlcNAcylation has been found to be a key component of cellular pathways such as Tau phosphorylation, transcription factor regulation, and proteosomal degradation (5). *O*-GlcNAcylation fluctuates in response to nutritional and other stresses, and is considered a promising marker for the diagnosis of neurodegeneration and diabetes (6, 7).

Proteins that are glycosylated in the secretary pathway are modified at consensus asparagine residues (*N*-linked) or at serine or threonine (*O*-linked) (Fig. 1, top panel). The initiation of *N*-glycan synthesis is highly conserved in eukaryotes, while the end products can vary greatly among different species, tissues, or proteins. Some glycans remain unmodified ("high mannose *N*-glycans") or are further processed in the Golgi ("complex *N*-glycans"). A greater diversity is found for *O*-glycans, which start with a common *N*-acetylgalactosamine (GalNAc) residue in animal cells but are very different in lower organisms.

In summary, the simple *O*-GlcNAc modification of nuclear and cytoplasmic proteins is a signaling mechanism, while larger *O*- and *N*-glycans are fundamental to maintaining structure and function of secreted and membrane-associated proteins. Other important glycosylated molecules, such as glycolipids and glycosaminoglycans, have been previously treated in this series (8, 9). The reader seeking further reference on the subject area would find exhaustive information elsewhere (10) (see Note 1).

The detailed analysis of the glycosylation of proteins is a field unto itself and requires extensive resources to execute properly. However, the availability of a spectrum of enzymes that remove sugars (glycosidases) allows researchers to get a general idea of the glycosylation status of their particular protein of interest. Here, we illustrate the use of these glycosidases for the analysis of model glycoproteins: (1) *N*- and *O*-glycosylation in recombinant human chorionic gonadotropin beta (hCGβ), (2) *O*-GlcNAcylation in a total human cell extract (HeLa), and (3) *O*-GlcNAc in α-Crystallin, a protein modified with a single *O*-GlcNAc. The techniques require only standard laboratory instrumentation and consumables, allowing parallel analysis of multiple glycoprotein samples.

2. Materials

2.1. Removal of N- and O-Glycans from a Recombinant Glycoprotein Hormone, Expressed in a Mouse Cell Line

1. Human chorionic gonadotropin β (hCGβ), vial of 150 μg (Sigma Aldrich, St. Louis, MO, #C6572).
2. PNGase F (#P0704, New England Biolabs, Inc., Ipswich, MA).
3. Protein Deglycosylation Mix (#P6039, New England Biolabs, Inc., Ipswich, MA) supplied with 10× G7 reaction buffer (0.5 M Sodium Phosphate, pH 7.5), 10× Glycoprotein Denaturing Buffer (5% SDS, 400 mM DTT), and 10% NP-40.
4. α-*N*-Acetylgalactosaminidase (#P0734, New England Biolabs, Inc., Ipswich, MA). α1-2 Fucosidase (#P0724 New England Biolabs, Inc., Ipswich, MA); β1-3 Galactosidase (#P0726, New England Biolabs, Inc., Ipswich, MA); α1-3,6 Galactosidase (#P0731), supplied with 100X Bovine Serum Albumin (BSA) 10 mg/ml.
5. SDS loading buffer (187.5 mM Tris–HCl pH 6.8, 6% SDS, 30% glycerol and 0.03% bromophenol blue), (#B7703, New England Biolabs, Inc., Ipswich, MA, supplied with 1.25 M DTT).
6. ColorPlus Prestained Protein Marker (#P7709, New England Biolabs, Inc., Ipswich, MA).
7. 10–20% Tris–Glycine Multigel (Cosmo Bio Co., Tokyo, Japan, #DCB-414893).

8. Tris–Glycine SDS running buffer (25 mM Tris, 192 mM Glycine, 0.1% SDS, pH 8.3).
9. Mini-gel cell and power source for protein electrophoresis.
10. Pro-Q® Emerald 300 Glycoprotein Stain kit (Invitrogen, Carlsbad, CA #P-21857).
11. Brilliant Blue R solution: 1 g of brilliant blue (Sigma Aldrich, St. Louis, MO, #B0149), 100 ml glacial acetic acid, 500 ml methanol, 400 ml dH_2O.
12. Destain solution: 100 ml glacial acetic acid, 300 ml methanol, 600 ml dH_2O.
13. PCR tubes (VWR®, West Chester, PA, #20170-004).
14. PCR Thermocycler.
15. Methanol.
16. Acetic acid.
17. Rocking platform.
18. UV transilluminator.

2.2. Detection of O-GlcNAcylated Proteins on a Total Cell Lysate Using a Monoclonal Anti-O-GlcNAc Antibody, and Removal of O-GlcNAc with β-N-Acetylglucosaminidase

1. HeLa (IP) Cell Lysate, 2 mg/800 μl RIPA lysis buffer (Santa Cruz Biotechnologies, Inc., Santa Cruz, CA #sc-24785) (see Note 2).
2. β-*N*-Acetylglucosaminidase (#P0732, New England Biolabs, Inc., Ipswich, MA), supplied with 10× G1 reaction buffer (50 mM sodium phosphate, pH 7.5) and 100× Bovine Serum Albumin (BSA) 10 mg/ml.
3. β-*N*-Acetylhexosaminidase (#P0721, New England Biolabs, Inc., Ipswich, MA).
4. 10% NP-40 (# B0701S, New England Biolabs, Inc., Ipswich, MA) (see Note 3).
5. Leupeptin, dissolved to 10 mg/ml in dH_2O (1,000×) (Sigma Aldrich, St. Louis, MO, #L2884).
6. Aprotinin, dissolved to 10 mg/ml in dH_2O (1,000×) (Sigma Aldrich, St. Louis, MO, #A6103).
7. SDS loading buffer (187.5 mM Tris–HCl pH 6.8, 6% SDS, 30% glycerol and 0.03% bromophenol blue) (New England Biolabs, Inc., Ipswich, MA, #B7703, supplied with 1.25 M DTT).
8. ColorPlus Prestained Protein Marker (New England Biolabs, Inc., Ipswich, MA, #P7709).
9. 10–20% Tris–Glycine Multigel (Cosmo Bio Co., Tokyo, Japan, #DCB-414893).
10. Tris–Glycine SDS running buffer (25 mM Tris, 192 mM Glycine, 0.1% SDS, pH 8.3).
11. Mini-gel cell and power source for protein electrophoresis.

12. Methanol.
13. PDVF membrane, Immobilion™ (Millipore, Billerica, MA, #IPVH304F0).
14. Blot Filter paper.
15. Tris–Glycine transfer buffer. Dilute 40 ml of 25× transfer buffer (18.2 g Tris base, 90 g glycine to 500 ml with dH_2O, pH is 8.3, do not adjust) with 200 ml methanol to 1,000 ml dH_2O. Final concentration: 12 mM Tris, 96 mM Glycine, 20%v/v methanol.
16. XCell *SureLock*™ Mini-Cell and XCell II™ blot module (Invitrogen, Carlsbad, CA, #EI0002).
17. Power source.
18. Phosphate-Buffered Saline 10× (PBS 10×): 80 g NaCl, 2.0 g KCl, 14.4 g Na_2HPO_4, 2.4 g KH_2PO_4, to 1,000 ml dH_2O.
19. PBS-T: add 0.5 ml Tween 20–100 ml PBS 10×, mix gently (see Note 4). Adjusted to 1,000 ml with dH_2O.
20. IgG-free BSA (Jackson ImmunoResearch Laboratories, Inc., West Grove, PA, #001-000-162).
21. Monoclonal Antibody against *O*-GlcNAc Clone CTD110.6 (IgM) (Covance, Princeton, NJ, #MMS-248R).
22. *N*-Acetylglucosamine (GlcNAc) (Sigma Aldrich, St. Louis, MO, #A-8625).
23. Goat Anti-Mouse IgM, HRP conjugated (Jackson Immuno Research Laboratories, Inc., West Grove, PA, #115-035-075).
24. Amersham™ ECL™ Western Blotting detection reagent (GE Healthcare, Buckinghamshire, UK, #RPN2209).
25. Amersham™ Hyperfilm™ ECL chemiluminescent film (GE Healthcare, Buckinghamshire, UK, #28906838).
26. Autoradiography cassette and film developer.
27. Restore™ PLUS Western Blot Stripping Buffer (Thermo, Rockford, IL, # 46430).
28. Brilliant Blue R solution: 1 g Brilliant Blue (Sigma Aldrich, St. Louis, MO, #B0149), 100 ml glacial acetic acid, 500 ml methanol, 400 ml dH_2O.
29. Destain solution: 100 ml glacial acetic acid, 300 ml methanol, 600 ml dH_2O.
30. PCR tubes (VWR®, West Chester, PA, #20170-004).
31. PCR Thermocycler.
32. Water bath or heating block.
33. Rocking platform.
34. Transilluminator and camera.

2.3. Detection of O-GlcNAcylation of α-Crystallin Using Click-iT®, and Removal of O-GlcNAc with β-N-Acetylglucosaminidase

1. α-Crystallin (Sigma Aldrich, St. Louis, MO, #C4163).
2. β-*N*-Acetylglucosaminidase (New England Biolabs, Inc., Ipswich, MA, #P0732), supplied with 10× G1 reaction buffer (50 mM sodium phosphate, pH 7.5) and 100× Bovine Serum Albumin (BSA) 10 mg/ml.
3. 10× Glycoprotein denaturing buffer (5% SDS, 400 mM DTT, New England Biolabs, Inc., Ipswich, MA #B0701S) (see Note 3).
4. 10% NP-40 (#B0701S, New England Biolabs, Inc., Ipswich, MA) (see Note 3).
5. SDS loading buffer (187.5 mM, Tris–HCl pH 6.8, 6% SDS, 30% glycerol and 0.03% bromophenol blue) (New England Biolabs, Inc., Ipswich, MA, #B7703, supplied with 1.25 M DTT).
6. ColorPlus Prestained Protein Marker (New England Biolabs, Inc., Ipswich, MA, # P7709).
7. 10–20% Tris–Glycine Multigel (Cosmo Bio Co., Tokyo, Japan, #DCB-414893).
8. Tris–Glycine SDS running buffer (25 mM Tris, 192 mM Glycine, 0.1% SDS, pH 8.3).
9. Mini-gel cell and power source for protein electrophoresis.
10. Click-It® Enzymatic Labeling System (Invitrogen, Carlsbad, CA, #C33368).
11. Click-It® TAMRA Detection Kit (Invitrogen, Carlsbad, CA, #C33370).
12. 20 mM HEPES buffer pH 7.9.
13. Millipore Type VS, 0.025-μM Filter Pack (Millipore, Billerica, MA, #VSWP02500).

3. Methods

3.1. Removal of N- and O-Glycans from a Recombinant Glycoprotein Hormone, Expressed in a Mouse Cell Line

Secreted proteins often contain large glycans that may be *O*- or *N*-linked. Depending on the size of the parental protein and/or the number of glycans attached to it, treatment of such proteins with appropriate glycosidases may significantly change the protein's electrophoretic mobility. Additionally, as sugars are removed the intensity of staining with sugar-specific reagents may be noticeably diminished. Here we illustrate both principles in the analysis of a model protein.

Human chorionic gonadotropin (hCG) is produced by placental cells and is readily isolated from the urine of pregnant women. This heterodimer consists of two subunits: hCGα and hCGβ. The beta subunit of hCG is *N*-glycosylated at N13 and N30, and

O-glycosylated at S121, S127, S132, and S138 (11). Glycosylation of hCG (from both natural and recombinant sources) has been extensively studied, making it an excellent model protein to illustrate glycosylation analysis. We demonstrate the use of enzymatic deglycosylation with PNGase F, or with a mixture of glycosidases (Protein Deglycosylation Mix) to detect *N*-and *O*-glycosylation in the hCGβ subunit.

PNGase F is able to remove all types of *N*-linked glycans (see Note 5). The Protein Deglycosylation Mix contains all of the enzymes needed to remove almost all *N*-linked and simple *O*-linked glycans, as well as some complex *O*-linked glycans (see Note 6). The deglycosylation mix was also supplemented with a mixture of additional exoglycosidases, which sometimes help remove otherwise resistant sugars. SDS-PAGE/Coomassie blue used to visualize differences in protein migration with and without glycosidase treatment. In addition, a sugar-specific staining method, ProQ® Emerald-300, shows diminished signal as glycans are successively removed.

This protocol is designed for the analysis of small amounts of recombinant glycoprotein (0.5–2 μg), although enzymatic deglycosylation can be scaled up to accommodate larger quantities of protein as needed (see Note 7). Similar protocols have been described elsewhere (12) using lectins for glycoprotein detection.

3.1.1. Enzymatic Deglycosylation

1. Label one set of PCR tubes 1–7 (see Note 8).
2. Thaw 10× G7 buffer, 10× glycoprotein denaturing buffer, and 10% NP-40, gently tapping to mix the contents. Keep at room temperature.
3. Keep enzyme-containing vials on ice at all times. Minimize thaw–freeze cycles (see Note 9).
4. Dissolve the contents of the hCGβ vial (150 μg) in 600 of dH_2O, keep on ice.
5. Prepare 1 ml of 1× G7 buffer.
6. Dilute 0.5 μl PNGase F in 25 μl 1× G7 buffer and keep on ice.
7. Prepare an exoglycosidase mix (EG mix) combining 2 μl each of α-*N*-Acetylgalactosaminidase, α1-2 Fucosidase, β1-3 Galactosidase, and α1-3, 6 Galactosidase (see Note 10).
8. Dilute 100× BSA to 2.5 mg/ml (5 μl BSA in 15 μl dH_2O) (see Note 11).
9. Set up PCR tubes as indicated (see Note 12):

Tube	1	2	3	4	5	6	7
hCGβ (0.25 mg/ml)	9 μl	9 μl	9 μl	9 μl	–	–	–
10× Glycoprotein denaturing buffer	1 μl	1 μl	1 μl	1 μl	1 μl	1 μl	1 μl
dH_2O	–	–	–	–	9 μl	9 μl	9 μl

10. Place in thermocycler, close, and denature the protein by incubating for 10 min at 94°C, followed by a 4°C hold (see Note 13).
11. Remove tubes from thermocycler. Open carefully (see Note 14).
12. Add the remainder of the reagents as indicated:

Sample#	1	2	3	4	5	6	7
10% NP-40 (see Note 15)	2.5 μl	2.5 μl	2.5 μl	2.5 μl	2.5 μl	2.5 μl	2.5 μl
10× G7 buffer	2.5 μl	2.5 μl	2.5 μl	2.5 μl	2.5 μl	2.5 μl	2.5 μl
PNGase F 1:50 dil.	–	2 μl	–	–	2 μl	–	–
Degl. Mix	–	–	2 μl	2 μl	–	2 μl	2 μl
EG mix	–	–	–	2 μl	–	–	2 μl
25× BSA	1 μl	1 μl	1 μl	1 μl	1 μl	1 μl	1 μl
dH_2O	9 μl	7 μl	7 μl	5 μl	7 μl	7 μl	5 μl
Total reaction vol.	*25 μl*	*25 μl*	*25 μl*	*25 μl*	*25 μl*	*25 μl*	*25 μl*

13. Close PCR tubes using new caps (see Note 16).
14. Mix by gently tapping four times and spin down.
15. Incubate in thermocycler at 37°C for 4 h. Place the samples at 4°C following incubation.

3.1.2. SDS-PAGE of Deglycosylated Samples

1. Prepare fresh 3× reducing SDS loading buffer (4 μl of 1.25 M DTT, 130 μl 3× SDS Loading Buffer) (see Note 17).
2. Add 17 μl of 3× reducing SDS loading buffer to each sample.
3. Close with new caps (see Note 16). Gently tap to mix.
4. Incubate in thermocycler at 94°C for 5 min, hold at 4°C.
5. Load 30 μl of each sample and protein marker on a 10–20% Tris–Glycine gel. Save remaining 10 μl for steps in Subheading 3.1.3. Run electrophoresis at 130 V.
6. When the gel has finished running, remove the gel from the cast and place it in a small plastic box. Add enough Coomassie blue solution to cover the gel, and incubate for 1 h with gentle agitation.
7. Wash three times for 30 min in 50 ml of destain solution.
8. To record images use a white light transilluminator or scanner. Alternatively, dry the gel between sheets of cellophane in a frame.

3.1.3. Pro-Q® Emerald 300 for Detection of Glycosylated Proteins in SDS-PAGE Gels

1. Load 10 μl from step 3.1.2.5, above, for SDS-PAGE.
2. When electrophoresis is complete, remove gel from cast and place it in a plastic box.
3. Following the product's manual and using the reagents provided with the kit, fix the gel, oxidize, and stain glycoproteins with Emerald Green reagent as indicated by the manufacturer (see Note 18).
4. Record images with an UV transilluminator at 300 nm.
5. Alternatively, after Emerald Green imaging, the gel can be stained with Coomassie blue as described above (see Note 19).

3.1.4. Results

The changes in protein migration after enzymatic deglycosylation are shown in Fig. 2. Compare the control sample (panel (a), lane 1) with the PNGase F treatment (removal of *N*-glycans, lane 2), and Deglycosylation Mix (PNGase F plus endo-and exoglycosidases to remove *O*-glycans, lane 3). No further reduction in size is seen after digesting with additional glycosidases (lane 4). Besides a change in mass, bands become sharper as glycans are removed (see Note 20). A band running under the 17-kDa marker (asterisk) likely represents the fully deglycosylated hCGβ polypeptide (see Note 21). Other bands might derive from incomplete deglycosylation or from unidentified proteins present in the hCGβ sample (see lane 1). Lanes 5–7 (controls) show the bands corresponding to the glycosidase/BSA reagents.

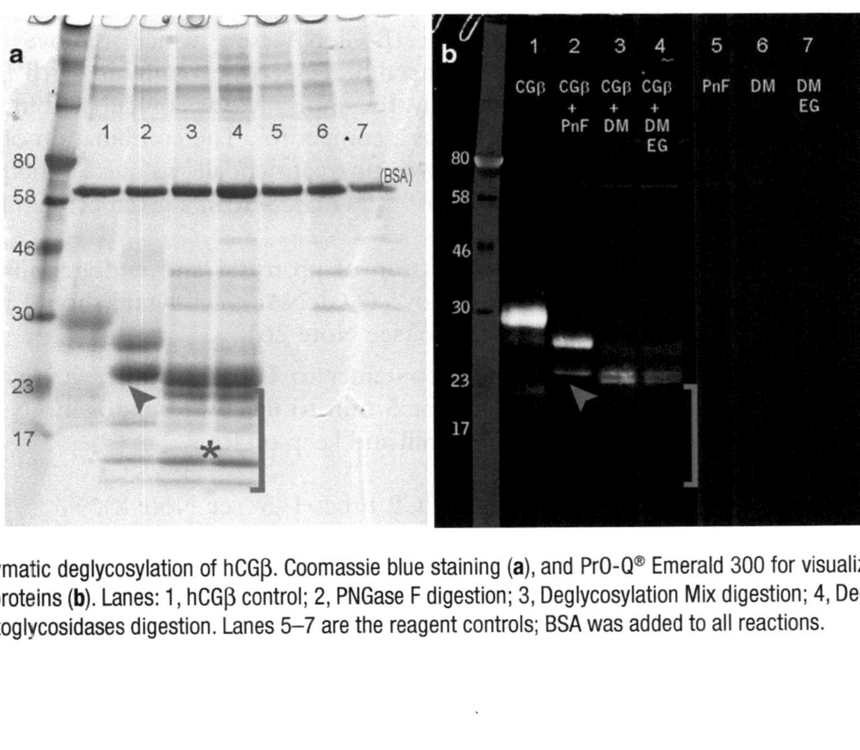

Fig. 2. Enzymatic deglycosylation of hCGβ. Coomassie blue staining (**a**), and PrO-Q® Emerald 300 for visualization of glycosylated proteins (**b**). Lanes: 1, hCGβ control; 2, PNGase F digestion; 3, Deglycosylation Mix digestion; 4, Deglycosylation Mix plus exoglycosidases digestion. Lanes 5–7 are the reagent controls; BSA was added to all reactions.

The glycoprotein staining by Emerald Green is shown in panel (b). This reagent oxidizes and stains all glycans present in a protein molecule. Therefore, the intensity of the signal decreases as hCGβ is enzymatically deglycosylated (lane 1 to lane 4). The residual signal in lanes 3 and 4 indicate the presence of glycan motifs, which are resistant to the enzymes used (see Note 22). The additional glycosidases used in lane 4 remove a few extra sugar residues: the protein migration is the same, but a slight reduction in intensity can be seen. Resistant sugar moieties were not present in all protein species: some bands are not detected by Emerald Green (absent in the UV image, in brackets), indicating they were extensively deglycosylated. Additional data support the conclusion that hCGβ is heterogeneously glycosylated. The lower band on lane 2 (*arrow*) is faint on the Emerald Green image, while the upper band on lane 2 is bright, indicating that many glycans groups are still present. These data support the conclusion that recombinant hCGβ expressed in mouse cells contains multiple glycoforms (13) (see Note 23).

3.2. Detection of O-GlcNAcylated Proteins on a Total Cell Lysate Using a Monoclonal Anti-O-GlcNAc Antibody, and Removal of O-GlcNAc with β-N-Acetylglucosaminidase

In contrast with the often large carbohydrates present in secreted glycoproteins, nuclear and cytoplasmic proteins are modified by a single GlcNAc residue. Even when several *O*-GlcNAcylation sites are present in a protein, the shift in molecular weight after enzymatic deglycosylation is minor. Therefore, detecting loss of *O*-GlcNAcylation requires GlcNAc-specific reagents, such as chemoenzymatic labels (see protocol 3), lectins (see Note 24), or antibodies.

In this section, we describe a protocol to detect *O*-GlcNAcylation in a human (HeLa) whole cell lysate using the highly specific *O*-GlcNAc monoclonal antibody CTD110 (14). To confirm the identity of the positive bands, samples were preincubated with β-*N*-Acetylglucosaminidase (β-GlcNAcase), which removes GlcNAc residues, or with a general β-*N*-Acetylhexosaminidase (β-Hex) (see Note 25). Alternatively, the antibody was coincubated in the presence of free GlcNAc as a competitor. The combination of the specific antibody and enzymatic treatment allows one to unequivocally confirm the presence of the *O*-GlcNAc in a given protein band.

3.2.1. Preparation of Cell Lysate for Digestion

1. To prevent the action of proteases and endogenous glycosidases (which remove *O*-GlcNAc), cell lysate should be stored at –80°C until use (see Note 26).
2. Preheat a water container to 100°C, transfer vial with HeLa lysate, and boil for 5 min to inactivate endogenous enzymes (see Note 27). Chill and keep on ice.

3.2.2. Removal of O-GlcNAc

1. Label one set of PCR tubes 1–5 (see Note 8).
2. Thaw vials tapping gently to mix the contents. Keep NP-40 and G1 and G2 buffers at room temperature, and keep BSA and protease inhibitors on ice.

3. Keep β-GlcNAcase (β-*N*-Acetylglucosaminidase) and β-Hex (β-*N*-Acetylhexosaminidase) vials on ice at all times. Minimize thaw–freeze cycles (see Note 9).
4. Dilute 100× BSA to 25× (5 μl BSA in 15 μl dH_2O).
5. Dilute 1,000× protease inhibitors to 25× (1 μl Aprotinin and 1 μl Leupeptin in 40 μl dH_2O).
6. Set up PCR tubes as indicated:

Sample#	1	2	3	4	5
HeLa lysate	20 μl	20 μl	20 μl	–	–
10% NP-40 (see Note 15)	2.5 μl	2.5 μl	2.5 μl	2.5 μl	2.5 μl
10× Buffer	2.5 μl G1	2.5 μl G2	2.5 μl G1	2.5 μl G1	2.5 μl G2
25× Protease Inh.	1 μl	1 μl	1 μl	1 μl	1 μl
25× BSA	1 μl	1 μl	1 μl	1 μl	1 μl
Glycosidase	β-GlcNAcase 10 μl	β-Hex 10 μl	–	β-GlcNAcase 10 μl	β-Hex 10 μl
dH_2O	3 μl	3 μl	13 μl	23 μl	23 μl
Total reaction vol.	*40 μl*	*40 μl*	*40 μl*	*40 μl*	*40 μl*

7. Close PCR tubes and mix by tapping.
8. Incubate in thermocycler at 37°C for 16 h; keep at 4°C following incubation.

3.2.3. SDS-PAGE and Transfer to PVDF Membrane

1. Prepare fresh 3× reducing SDS loading buffer (4 μl of 1.25 M DTT, 130 μl 3× SDS Loading Buffer) (see Note 17). Add 20 μl to each sample. Incubate in thermocycler at 94°C for 5 min and hold at 4°C.
2. Load 30 μl of each sample and protein marker on a 10–20% Tris–Glycine gel, run at 130 V. Use the remaining 30 μl to load a duplicate gel (for GlcNAc competition control).
3. When the gels have finished running, remove them from cast and wash each gel separately with 50 ml of transfer buffer for 5–10 min to rinse SDS.
4. Wet two PVDF membranes with methanol, rinse for 5 min in transfer buffer. Wet two sets of filter paper and transfer pads with transfer buffer.
5. Following manufacturer's instructions, assemble transfer block.
6. Transfer for 2 h at 26 V.
7. After transfer, keep membrane wet by placing on a tray with PBS-T.

3.2.4. Western Blot

1. Block the membranes for at least 1 h with 50 ml of PBS-T 5% IgG-free BSA (see Note 28).
2. Incubate one membrane with 30 ml of anti *O*-GlcNAc 1:2,000 in PBS-T 1% IgG-free BSA for 2 h at room temperature in a rocking platform (see Note 28). Incubate the other membrane under the same conditions but add GlcNAc (to a final concentration of 100 mM) for binding competition.
3. Wash three times for 20 min in 50 ml of PBS-T.
4. Incubate with 30 ml of anti IgM 1:2,000 in PBS-T, 1% IgG-free BSA for 1 h at room temperature in a rocking platform.
5. Wash three times for 20 min in 50 ml of PBS-T.
6. Remove the membranes and develop with 1–2 ml of detection reagent following manufacturer's instructions.
7. Keep the membranes between plastic sheets (sheet protectors for instance) and expose to film in an autoradiography cassette.

3.2.5. Protein Staining (See Note 29)

1. After autoradiography, rinse the PVDF membranes in dH_2O.
2. Strip the membrane (removal of antibodies and blocking agent) by incubating with 20 ml of Restore™ stripping buffer. Rock gently for 30 min.
3. Wash the PVDF membrane extensively with dH_2O.
4. Incubate the PVDF membrane in 20 ml Coomassie Blue staining solution, rocking for up to 5 min. Do not overstain.
5. Wash two times for 10 min in 50 ml of destain solution (30% v/v methanol, 10% v/v acetic acid in dH_2O).
6. Keep membrane between plastic sheets and record pictures with a transilluminator.

3.2.6. Results

After the enzymatic removal of *O*-GlcNAc with βGlcNAcase or β-Hex, several bands are no longer recognized by the anti-*O*-GlcNAc monoclonal antibody CTD110.6, indicating its specific affinity for the *O*-GlcNAc modification (Fig. 3 panel (b): compare lane 3, bands marked by *black arrows*, with lanes 1 and 2). For unknown reasons, β-Hexosaminidase digestion is more extensive than βGlcNAcase digestion, yet the enzymatic treatment never reaches completion. This shows that some *O*-GlcNAc sites are not accessible. One band in particular (*white arrow*) seems to remain almost unaffected. Competition of anti-*O*-GlcNAc antibody with free GlcNAc (Fig. 3, panel (a)) shows that the antibody preparation binds unspecifically to some bands (all in the low-molecular-weight range) that do not carry the *O*-GlcNAc modification (see Note 30). The protein stain control shows that the polypeptides remained unaffected by the enzymatic treatment (Fig. 3 panel (c)).

This technique allows one to readily identify many protein bands that carry *O*-GlcNAc, and it is well suited to compare the

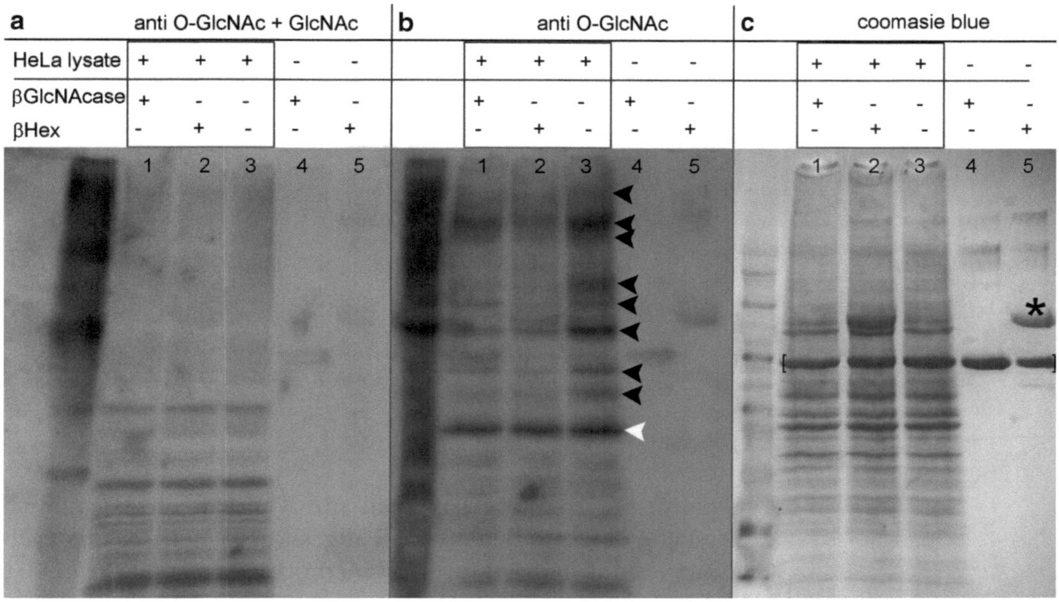

Fig. 3. Enzymatic removal of *O*-GlcNAc from HeLa lysates, using βGlcNAcase or β-Hex for 16 h incubation. Panel (**a**), western blot with anti-*O*-GlcNAc antibody coincubated with 100 mM free GlcNAc, showing unspecific bands, which are in turn insensitive to enzyme treatment. Panel (**b**), anti-*O*-GlcNAc antibody, recognition of specific bands which are sensitive to enzymes (*black arrows*). One band is mostly resistant to enzyme treatment (*white arrow*). Panel (**c**), protein staining, BSA was added to all reactions (*brackets*), βHex band (*asterisk*). βGlcNAcase band is not detectable.

overall levels of *O*-GlcNAc between samples. The use of specific enzymes confirms the identity of the bands, a suitable control since the antibody preparation binds to some unknown epitopes. The protocol, however, does not provide a quantitation of the *O*-GlcNAc modification.

3.3. Detection of O-GlcNAcylation of α-Crystallin Using Click-iT®, and Removal of O-GlcNAc with β-N-Acetyl glucosaminidase

The chemoenzymatic labeling of GlcNAc is a very sensitive assay to detect GlcNAcylated proteins. This technique uses a mutant galactosyltransferase (GalT) to transfer an azido-galactose (GalNAz) to GlcNAc, followed by the reaction (labeling) between the azide group and an alkyne (15). Finally, the labeled glycoproteins are analyzed (gels, western blots, or mass spectroscopy). This technique can be applied to any protein with at least one *O*-GlcNAc modification site.

The following protocol uses the Invitrogen's Click-It *O*-GlcNAc Enzymatic Labeling System and TAMRA Detection Kit, to monitor the release of *O*-GlcNAc from pure α-Crystallin treated with β-*N*-Acetylglucosaminidase and separated by SDS-PAGE. The protein α-Crystallin consists of four subunits in which a single *O*-GlcNAc modification is present on the alpha subunit at S164 (16). The α-Crystallin protein is reported to be 2–10% *O*-GlcNAc modified (16). Therefore, according to Invitrogen's Click-It protocol, 1 μg α-Crystallin is equal to approximately 50 pmol of

Table 1
Native conditions

Sample # (conc. α-Crystallin)	1 (10 μg)	2 (5 μg)	3 (2.5 μg)	4 (1.25 μg)	5 (0.625 μg)
α-Crystallin (1.5 mg/ml)	6.7 μl	3.3 μl	1.7 μl	0.9 μl	0.4 μl
BSA (1 mg/ml)	5 μl	5 μl	5 μl	5 μl	5 μl
10× G1 Buffer	2 μl	2 μl	2 μl	2 μl	2 μl
dH_2O	6.3 μl	9.7 μl	11.3 μl	12.1 μl	12.6 μl
Total reaction vol.	*20 μl*	*20 μl*	*20 μl*	*20 μl*	*20 μl*

O-GlcNAc sugar. There has been no *N*-glycan modification in α-Crystallin detected by PNGase F digestion (17); therefore, every GlcNAc residue labeled by this method corresponds to *O*-Glc NAcylated sites (see Note 31).

3.3.1. Release of O-GlcNAc from α-Crystallin Under Native Conditions

1. Label ten microcentrifuge tubes as sample #1–5 (+) and #1–5 (−). Samples labeled (+) are incubated with β-*N*-Acetylglucosaminidase. Samples labeled (−) are not incubated with β-*N*-Acetylglucosaminidase.
2. Dilute 100× BSA to 1 mg/ml (1 μl BSA in 10 μl dH_2O).
3. Combine the appropriate amounts of α-Crystallin (1.5 mg/ml in water), BSA, 10× G1 Reaction Buffer, and dH_2O to every microcentrifuge tube, as listed in Table 1.
4. Add 2 μL of β-*N*-Acetylglucosaminidase (4,000 U/ml) to the samples labeled #1–5 (+). Do not add enzyme to samples labeled (−).
5. Incubate all ten samples (+) and (−) for 4 h at 37°C.

3.3.2. Release of O-GlcNAc from α-Crystallin Under Denaturing Conditions

1. Label ten microcentrifuge tubes as sample #1–5 (+) and #1–5 (−). Samples labeled (+) are incubated with β-*N*-Acetylglucosaminidase. Samples labeled (−) are not incubated with β-*N*-Acetylglucosaminidase.
2. Dilute 100× BSA to 1 mg/ml (1 μl BSA in 10 μl dH_2O).
3. Combine the appropriate amounts of α-Crystallin, 10× Glycoprotein Denaturing Buffer, and dH_2O to every microcentrifuge tube, as listed in Table 2.
4. Incubate the tubes for 5 min at 95°C. Chill on ice 3 min.
5. Add 10× G1 Reaction Buffer, 10% NP-40, and BSA to every microcentrifuge tube.

Table 2
Denatured conditions

Sample # (conc α-crystallin)	1 (10 μg)	2 (5 μg)	3 (2.5 μg)	4 (1.25 μg)	5 (0.625 μg)
α-Crystallin (1.5 mg/ml)	6.7 μl	3.3 μl	1.7 μl	0.9 μl	0.4 μl
BSA (1 mg/ml)	5 μl	5 μl	5 μl	5 μl	5 μl
10× G1 Buffer	2 μl	2 μl	2 μl	2 μl	2 μl
10% NP-40	2 μl	2 μl	2 μl	2 μl	2 μl
10× Glycoprotein denaturing buffer	1 μl	1 μl	1 μl	1 μl	1 μl
dH_2O	3.3 μl	6.7 μl	8.3 μl	9.1 μl	9.6 μl
Total reaction volume	*20 μl*	*20 μl*	*20 μl*	*20 μl*	*20 μl*

6. Add 2 μL of β-*N*-Acetylglucosaminidase (4,000 U/ml) to the samples labeled #1–5 (+). (Do not add enzyme to samples labeled (−)).
7. Incubate all ten samples (+) and (−) for 4 h at 37°C.
8. Following incubation, place all samples on ice.

3.3.3. Drop Dialysis into Click-It Reaction Buffer (see Note 32)

1. Label one Petri dish per sample with the specific sample name.
2. Pour 50 ml of 20 mM HEPES, pH 7.9 into each Petri dish.
3. Place one Millipore Filter (Type VS, 0.025 μM, Millipore #VSWP02500) per Petri dish gently on the surface of the buffer (shiny side up). Pipette each sample onto its corresponding filter very gently (do not touch the filter with the pipette tip as this can cause the filter to submerge into the buffer).
4. Cover the Petri dish and let sample sit on filter for 1 h at room temperature.
5. Using a pipette, very gently pull each sample off of its filter and put it back into its original labeled microcentrifuge tube (see Note 33).

3.3.4. Click-It Enzymatic Labeling

1. All β-*N*-Acetylglucosaminidase treated (+) and untreated (−) α-Crystallin samples are labeled using Invitrogen's Click-It Enzymatic labeling system (#C33368). A modified version of Invitrogen's Click-It Protocol is used to accommodate the smaller sample size (20 μl reactions) used here.
2. Subject 15 samples to enzymatic labeling: the ten samples treated in Subheading 3.3.1 or Subheading 3.3.2 (under native

or denatured conditions), plus five α-Crystallin samples (10 μg, 5 μg, 2.5 μg, 1.25 μg, 0.625 μg) to create a standard curve.

3. Make a 0.5 mM solution of UDP-GalNAz by adding 144 μl of 10 mM HEPES, pH 7.9 to component A. Mix well to ensure complete reconstitution.
4. To every sample add 20 μl labeling buffer (component C) and 3.75 μl of 100 mM $MnCl_2$ (component D).
5. Vortex samples briefly to mix and then briefly centrifuge.
6. Add 2.5 μl of UDP GalNAz (made in step 3) to every sample. Pipette up and down to mix.
7. Add 2.5 μl of Gal-T1 (Y289L) enzyme (component B) to every sample. Pipette up and down to mix.
8. Incubate all reactions at 4°C overnight (18–24 h).
9. Store samples at −20°C until ready to analyze using the Click-It TAMRA Detection Kit.

3.3.5. Click-It TAMRA Detection

1. Following *O*-GlcNAc labeling, all samples are detected using a Click-It TAMRA Detection Kit (#C33370). One kit is sufficient material for the detection of 15 samples (see Note 34).
2. Add 60 μl of the alkyne solution (component A) to the Click-It Reaction Buffer (component B).
3. Add 500 μl dH_2O to the Click-It Reaction Buffer Additive 2 (component E).
4. Add 100 μl dH_2O to one vial of Click-It Reaction Buffer Additive 1 (component D). This solution must be made fresh the day of use.
5. To all 15 samples, add 82 μl of Click-It Reaction Buffer containing the alkyne detection reagent prepared in step 2.
6. Add 20 μl of 1% SDS in 50 mM Tris–HCl, pH 8 to all samples.
7. Cap the tubes and vortex briefly.
8. Add 10 μl of $CuSO_4$ (component C) to every sample and vortex briefly.
9. Add 10 μl of Click-It Reaction Buffer Additive 1 (component D), made in step 4, to every sample and vortex briefly.
10. Wait for 2–3 min before proceeding to next step.
11. Add 20 μl of Click-It Reaction Buffer Additive 2 (component D), made in step 3, to every sample and vortex briefly. The solution turns bright orange.
12. Cover all tubes with foil to minimize light exposure and rotate end-over-end for 20 min using a rotator at room temperature.

3.3.6. Sample Preparation and SDS-PAGE

1. Add 600 μl of methanol to every reaction and vortex briefly.
2. Add 150 μl of chloroform to every reaction and vortex briefly.
3. Add 400 μl of dH_2O to every reaction and vortex briefly.
4. Centrifuge for 5 min at 14,000×*g*. Then, carefully remove and discard as much of the upper aqueous phase as possible while leaving the interface layer with the protein precipitate intact. Note: the upper phase is orange and the lower phase is pink.
5. Add 450 μl of methanol to every reaction and vortex briefly.
6. Centrifuge for 5 min at 14,000×*g* to pellet the protein, then remove and discard the supernatant.
7. Let samples dry with lids open for 10 min.
8. Prepare fresh 3× reducing SDS loading buffer (4 μl of 1.25 M DTT, 130 μl 3× SDS Loading Buffer).
9. Resolubilize every sample in 20 μl of loading buffer.
10. Vortex samples and incubate for 5 min at 95°C.
11. Centrifuge samples for 1 min at 14,000×*g*.
12. Load samples on a 10–20% gradient Tris–glycine gel and run electrophoresis at 160 V.
13. When the gel has finished running, place in a plastic tray with dH_2O.

3.3.7. Analysis

1. Analyze all 15 samples on a Typhoon 9400 Variable Mode Imager. Use an excitation wavelength of 555 nm and an emission wavelength of 580 nm (see Note 35).
2. Perform quantitation of the amount of *O*-GlcNAc sugar released from the treated α-Crystallin samples using Image Quant TL Software (GE Healthcare).
3. Calculate quantitation by comparing nonincubated standards to incubated samples.

3.3.8. Results

The β-*N*-Acetylglucosaminidase catalyzes the removal of the *O*-GlcNAc modification directly from α-Crystallin (Fig. 4). There appears to be no major advantage over native versus denatured conditions. However, native conditions may require more units of β-*N*-Acetylglucosaminidase for larger amounts of protein. Only partial GlcNAc removal was obtained under the conditions of this assay; longer incubation times (as described in protocol 2) might be necessary for a complete deglycosylation.

3.4. Interpretation of Results

Enzymatic deglycosylation and SDS-PAGE can provide valuable information about the glycosylation state of a protein of interest, while glycan-specific reagents facilitate the interpretation of the data. The protocols described in this chapter are intended for the

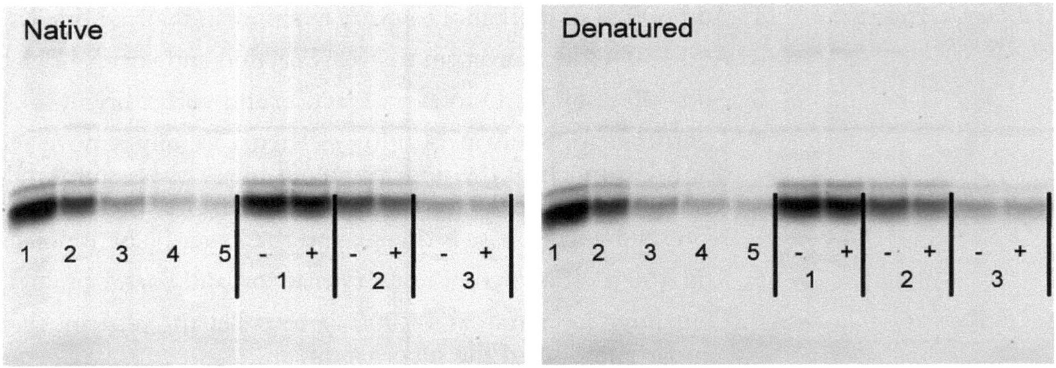

Standards	α-Crystallin Quantity (μg)	*O*-GlcNAc Amount (pmol)
1	10	500
2	5	250
3	2.5	125
4	1.25	62.5
5	0.625	31.25

Samples	α-Crystallin Quantity	Amount *O*-GlcNAc (-) lane	Amount *O*-GlcNAc (+) lane	Amount *O*-GlcNAc Released
Denatured				
1	10.0 μg	421.0 pmol	370.0 pmol	51.0 pmol
2	5.0 μg	240.5 pmol	172.0 pmol	68.5 pmol
3	2.5 μg	88.5 pmol	82.5 pmol	

Fig. 4. The standard curve samples are labeled numbers 1–5 with 1 being 10 μg and 5 being 0.625 μg for both native and denatured samples. The α-Crystallin samples are labeled (–) or (+), in which the (–) refers to samples not treated with β-*N*-Acetylglucosaminidase and (+) refers to samples treated with β-*N*-Acetylglucosaminidase.

initial studies of protein glycosylation. To elucidate the rate of occupancy (which amino acids are glycosylated), extent of glycosylation, or to determine the fine structure of glycans, more sophisticated techniques (mass spectrometry, liquid chromatography, NMR) are required.

Because of their simplicity, the protocols can be adjusted and/or combined to accommodate various experimental needs. However, to obtain reliable results it is important to understand their strengths and limitations.

First, the specificity and purity of the glycosidases is crucial. Well-characterized enzymes tested to be free of proteases and other contaminating activities should be used. Second, a careful choice of detection is necessary: (a) Protein staining reagents are useful only if deglycosylation results in a significant shift on molecular mass (although it can be always used as a control). (b) Sugar detection with antibodies presents unique challenges limiting their applicability (see Note 36); one exception is the widely used anti O-GlcNAc monoclonal antibody CTD110.6 (see Note 37) (14). (c) Lectins (proteins with intrinsic sugar affinity) are well suited for sugar detection, plus they give insight into glycan structure (see Note 38). Lectin or antibody-binding studies are performed along with controls coincubated with a free sugar (substrate competition), and/or controls previously treated with glycosidases. (d) Chemical labeling kits (based on periodate oxidation of sugars) are the method of choice to stain all glycoproteins. (e) The chemoenzymatic labeling of terminal GlcNAc allows the direct identification of *O*-GlcNAc and it is exceptionally useful for high-throughput mass spectroscopy, providing a "handle" to isolate and identify *O*-GlcNAcylated peptides. Yet, it is also very effective in common methods such as SDS-PAGE (18). Controls are critical to avoid false positives: because terminal GlcNAc residues (from *N*- or *O*-glycans) are labeled as well, it is essential to fractionate or purify protein extracts, and/or to release *N*-glycans (see Note 31).

4. Notes

1. The second edition of the Essentials of Glycobiology is now available at the NCBI Bookshelf. A valuable resource for all glycobiologists combines concise yet up-to-date information with a well-organized introduction to each area of research. (http://www.ncbi.nlm.nih.gov/bookshelf/br.fcgi?book=glyco2).
2. RIPA buffer formulation: 50 mM Tris–HCl pH 8, 150 mM NaCl, 1% NP40, 0.1% SDS, 1% deoxycholic acid, 2 mM PMSF, 2 mM Na orthovanadate, and protease inhibitors (Pepstatin A, Leupeptin, Aprotinin, 10 μg/ml each).
3. Supplied with PNGase F or Protein Deglycosylation Mix from New England Biolabs.
4. Tween20 easily dissolves in 10×PBS.
5. Some α1,3 fucose *N*-glycan cores found in plant and insect cells are resistant to PNGase F and require PNGase A treatment.
6. This mix contains: PNGase F (*N*-glycosidase); endo-α-*N*-Acetylgalactosaminidase (*O*-Glycosidase, to remove core 1 and core 3 *O*-glycans), neuraminidase (sialidase), β1-4 Galactosidase, and β-*N*-Acetylglucosaminidase (β-GlcNAcase). A detailed characterization of this mix, as well as detailed characterization of the mode of action of these enzymes, is found at http://www.neb.com (Glycobiology and protein tools, endoglycosidases/exoglycosidases).
7. The standard reaction for the Deglycosylation Mix takes 5 μl of enzyme cocktail to digest 100 μg of glycoprotein in a total volume of 50 μl.
8. PCR tubes minimize pipetting errors and evaporation.
9. Glycosidases are generally stable if stored according to the manufacturer's instructions. However, temperature fluctuations might affect their activity. Enzyme aliquots can be prepared in

advance (using an appropriate dilutant). Never freeze enzymes indicated for storage at 4°C.

10. Residues such as α-GalNAc, fucose, and α-galactose are known to be present in some mammalian *O*-glycans; therefore, additional exoglycosidases were added to the mix.
11. BSA is required to improve stability of some of the enzymes used in this protocol; however, it is not needed when only PNGase F or Deglycosylation Mix digestions are performed. In those cases, replace the BSA volume by an equivalent volume of dH_2O.
12. Close tubes tightly and mix gently and spin down.
13. Heat is required for optimal denaturation in the presence of SDS. A thermocycler provides a good temperature control and evaporation prevention (heated lid).
14. Spin down if condensation is present.
15. SDS (in denaturing buffer or RIPA buffer) inhibits the action of PNGase F, and also seems to reduce β-*N*-Acetylglucosaminidase activity. It is essential to have NP-40 present in the reaction mixture. Why this nonionic detergent counteracts the SDS inhibition is unknown at present.
16. After the high temperature incubation, the caps might not fit well.
17. Any SDS gel loading buffer can replace the 3× loading dye reagent.
18. The complete procedure takes several hours, which makes difficult to finish the experiment in a workday. However, we use a modified procedure: leaving the gel in fresh fixing solution overnight instead of fixing it twice for 45 min (as indicated by the manual) gives identical results and allows restarting on the next day.
19. The Emerald Green works best when the total protein amount remains low. Thus, some bands are barely visible by Coomassie. If desired, use a high-sensitivity protein stain instead.
20. The heterogeneity of glycans typically makes glycoprotein bands fuzzy and difficult to quantify. Protein bands become sharper after deglycosylation.
21. Such a clear result is not always obtained. For other proteins, we have seen abnormal migration after deglycosylation (far from the predicted molecular weight, or even *slower migration*). This phenomenon is not well understood, but it can be said that *any change* in migration is evidence that the protein has been deglycosylated.
22. Although it is convenient to use several enzymes at once, formulating a mix where all components are equally active in a

same buffer is challenging. Sequential deglycosylation (along with buffer adjustments) can be an alternative solution. Molecular constrains also influence activity: for instance, the fucosidase used in this protocol will not remove residues from a branched oligosaccharide.

23. Glycosylation is heterogeneous: some polypeptides do not receive a glycan in each consensus site, and/or some glycans are extended while others (even on the same protein) are not.
24. Succinylated Wheat germ agglutinin (sWGA), a lectin that recognizes GlcNAc, has been used for *O*-GlcNAc detection (19). However, it does not discriminate *O*-GlcNAc from terminal GlcNAc in *N*- or *O*-glycans.
25. A thorough characterization of these enzymes can be found at http://www.neb.com (Glycobiology and protein tools, endoglycosidases/exoglycosidases).
26. To minimize endogenous deglycosylation, *O*-GlcNAcase inhibitors can be used during preparation of cell lysates; however, they have to be removed before incubation with exogenous hexosaminidase or β-GlcNAcase. The authors prefer a simpler protocol without inhibitors and used a heat inactivation step instead.
27. Some proteins might precipitate, so disperse by vortexing.
28. Milk glycans interfere with *O*-GlcNAc binding; use only BSA as blocking agent. Keep the antibody concentration low to minimize background. The first antibody incubation can be performed overnight at 4°C.
29. To prove that signal loss is due to GlcNAc removal and not to protein degradation, it is recommended to perform a total protein staining. Reusing the western blot membrane saves time and provides a perfect side-by-side comparison to identify protein bands.
30. The available CTD110.6 preparations are crude ascites fluid. The antibody can be purified prior to use (20).
31. When this reagent is used with crude extracts, the samples ought to be pretreated with PNGase F, to remove *N*-glycans which might contain terminal GlcNAc. *Although GlcNAc could also be present in O-Glycans, using a deglycosylation mix (which contains β-GlcNAcase) should be avoided.* Another way to prevent false positives is by fractionating cells into cytosolic and nuclear extracts, which are virtually free of secretory and membrane glycoproteins.
32. The drop dialysis step is necessary as it removes the citrate buffer and leaves the samples in the *O*-GlcNAc labeling buffer. If one does not remove the citrate buffer, *O*-GlcNAc labeling will not be successful.

33. Following drop dialysis, a small increase in sample volume may occur, this is to be expected.
34. The TAMRA alkyne detection kit is not the only detection kit that can be used in conjunction with this protocol. The Biotin alkyne detection kit (Invitrogen #C33372) can also be used. If the biotin alkyne detection kit is used, perform a western transfer procedure, rather than imaging the gel.
35. A Typhoon Imager is not the only mode for analysis. When using the TAMRA alkyne reagent, the Click-It labeled protein must be immediately imaged on any imager that can read an excitation wavelength of 555 nm and an emission wavelength of 580 nm.
36. It has been very difficult to generate general antiglycan antibodies; they are usually raised against a complex glycan target, which restricts their use. Moreover, several monoclonal antiglycan antibodies display undesired cross reactivity (21).
37. This antibody has been raised against a synthetic glycopeptide carrying a single *O*-GlcNAc residue, to overcome two limitations: (a) GlcNAc affinity reagents (lectins/chemoenzymatic tagging) do not discriminate *O*-GlcNAc from other terminal GlcNAc (*N*- or *O*-glycans) and (b) anti-*O*-GlcNAc antibodies require additional peptide epitopes for recognition (and thus are not universal). However, we have observed that some *O*-GlcNAcylated proteins, α-Crystallin for instance, are weakly bound by CTD110.6.
38. Lectins usually bind small sugar motifs. However, not all of them have a narrow specificity, and many are only partially characterized (meaning that they could have unknown affinities). As a consequence, positive lectin provides indication, not proof, of the presence of a given sugar.

References

1. Rich JR, Withers SG. Emerging methods for the production of homogeneous human glycoproteins. *Nat Chem Biol.* 2009; 5(4):206–15.
2. Hossler P, Khattak SF, Li ZJ. Optimal and consistent protein glycosylation in mammalian cell culture. *Glycobiology.* 2009; 19(9):936–49.
3. Wells L, Vosseller K, Hart GW. Glycosylation of nucleocytoplasmic proteins: signal transduction and O-GlcNAc. *Science.* 2001, 291(5512):2376–8.
4. Butkinaree C, Park K, Hart GW. O-linked β-N-acetylglucosamine (O-GlcNAc): Extensive crosstalk with phosphorylation to regulate signaling and transcription in response to nutrients and stress. *Biochim Biophys Acta.* 2010; 1800(2):96–106.
5. Love DC, Hanover JA. The hexosamine signaling pathway: deciphering the "O-GlcNAc code". *Sci STKE.* 2005; 2005(312): re13.
6. Wang Z, Udeshi ND, Slawson C, Compton PD, Sakabe K, Cheung WD, Shabanowitz J, Hunt DF, Hart GW. Extensive crosstalk between O-GlcNAcylation and phosphorylation regulates cytokinesis. *Sci Signal.* 2010, 3(104): ra2.
7. Lazarus BD, Love DC, Hanover JA. O-GlcNAc cycling: implications for neurodegenerative disorders. *Int J Biochem Cell Biol.* 2009, 41(11):2134–46
8. Bielik AM, Zaia J. Extraction of chondroitin/dermatan sulfate glycosaminoglycans from connective tissue for mass spectrometric

analysis. *Methods Mol Biol.* 2010, 600: 215–25.

9. Azzouz N, Gerold P, Schwarz RT. Metabolic labeling and structural analysis of glycosylphosphatidylinositols from parasitic protozoa. *Methods Mol Biol.* 2008, 446:183–98.
10. Varki A, Cummings RD, Esko JD, Freeze HH, Stanley P, Bertozzi CR, Hart GW, Etzler ME. (Eds). *Essentials of Glycobiology*, 2nd ed. 2008. Cold Spring Harbor Laboratory Press. Plainview (NY).
11. Carlsen RB, Bahl OP, Swaminathan N. Human chorionic gonadotropin. Linear amino acid sequence of the beta subunit. *J Biol Chem.* 1973, 248(19):6810–2.
12. Lectin analysis of proteins blotted onto filters. Freeze HH. *Curr Protoc Mol Biol.* 2001; Chapter 17:Unit17.7.
13. Thakur D, Rejtar T, Karger BL, Washburn NJ, Bosques CJ, Gunay NS, Shriver Z, Venkataraman G. Profiling the glycoforms of the intact alpha subunit of recombinant human chorionic gonadotropin by high-resolution capillary electrophoresis-mass spectrometry. *Anal Chem.* 2009, 81(21):8900–7.
14. Comer FI, Vosseller K, Wells L, Accavitti MA, Hart GW. Characterization of a mouse monoclonal antibody specific for O-linked N-acetylglucosamine. *Anal Biochem.* 2001; 293(2):169–77.
15. Boeggeman E, Ramakrishnan B, Kilgore C, Khidekel N, Hsieh-Wilson LC, Simpson JT, Qasba PK. Direct identification of nonreducing GlcNAc residues on N-glycans of glycoproteins using a novel chemoenzymatic method. *Bioconjug Chem.* 2007; 18(3):806–14.
16. Roquemore EP, Dell A, Morris HR, Panico M, Reason AJ, Savoy LA, Wistow GJ, Zigler JS Jr, Earles BJ, Hart GW. Vertebrate lens alpha-crystallins are modified by O-linked N-acetylglucosamine. *Journal of Biological Chem.* 1992, 267:555–63.
17. Bielik, A.M. New England Biolabs, Inc., Unpublished Results.
18. Clark PM, Dweck JF, Mason DE, Hart CR, Buck SB, Peters EC, Agnew BJ, Hsieh-Wilson LC. Direct in-gel fluorescence detection and cellular imaging of O-GlcNAc-modified proteins. *J Am Chem Soc.* 2008; 130(35):11576–7.
19. Kelly WG, Hart GW. Glycosylation of chromosomal proteins: localization of O-linked N-acetylglucosamine in Drosophila chromatin. *Cell.* 1989; 57(2):243–51.
20. Zachara NE. Detection and analysis of O-linked β-N-acetylglucosamine-modified proteins. *Methods Mol Biol.* 2009; 464:227–54.
21. Park S, Lee MR, Shin I Carbohydrate microarrays as powerful tools in studies of carbohydrate-mediated biological processes. *Chem Commun* (Cambridge). 2008;(37):4389–99.

Chapter 14

Strategies for Efficient Transfection of CHO-Cells with Plasmid DNA

Renate Kunert and Karola Vorauer-Uhl

Abstract

Stable cell lines of Chinese hamster ovary (CHO) cells are the predominant source of commercial biopharmaceutical proteins. Because making suitable CHO cell lines is time-consuming and costly, preliminary experiments with transient expression are usually performed to optimize as many protein production parameters as possible. Here, we describe protocols for optimizing expression in transient expression experiments and isolating stable CHO cell lines using two types of self-made reagents, namely, lipoplexes and polyplexes.

Key words: Transfection, Chinese hamster ovary, Transient, Stable, Lipoplex, Polyplex

1. Introduction

Chinese hamster ovary (CHO) cells were isolated in 1957, and for more than two decades this epithelial-like cell line has been used as a major cellular host for the expression of biopharmaceutical proteins, because of their high productivity even in chemically defined cultivation media. In addition to their manufacturing advantages, their mammalian origin results in proteins that are highly native in structure and have posttranslational modifications that are closely related to human proteins, which is not always the case with bacterial hosts.

A prerequisite for recombinant protein expression is the transfer of the transgene to the cellular nucleus for transcription into mRNA, followed by its export to the cytoplasm and subsequent translation into protein. The steps following initial transfection depend on the ultimate protein production goal. Large-scale manufacturing requires the isolation of a stable cell line with a favorable balance of

James L. Hartley (ed.), *Protein Expression in Mammalian Cells: Methods and Protocols*, Methods in Molecular Biology, vol. 801,
DOI 10.1007/978-1-61779-352-3_14, © Springer Science+Business Media, LLC 2012

productivity (per cell), health (high density growth with rapid cell division), and high clone stability through many generations. Unfortunately, generating and qualifying such a clonal cell line is so costly that typically only high-value targets make it worth attempting. An alternative approach is transient protein production, with the aim to accumulate enough recombinant protein to characterize a newly developed molecule. Transient expression allows multiple DNA constructs to be tested quickly and at small scale. Transfection for both stable and transient expression can be accomplished with commercially available transfection kits and protocols but, nevertheless, need to be evaluated and optimized carefully. The limiting factor especially in large-scale transient expression experiments is usually the cost of commercial available transfection reagents, as well as the cost of preparing large amounts of DNA. Here, we describe the comparison of transfection methods with self-made lipoplexes or polyplexes. We illustrate how such transfection methods can be evaluated for transient expression and stable clone selection using enhanced green fluorescent protein (eGFP) and a heterodimeric immunoglobulin molecule as models.

Different methods have been developed to transfer extracellular transgenes to the cellular nucleus. The crossing of the cellular membrane can be accomplished by viral, physical, or chemical approaches, each of them displaying different characteristics, advantages, and shortcomings. Viral vectors are the most efficient vehicles for DNA transfer, but they are not permitted for the development of recombinant cell lines expressing proteins for clinical applications due to potential risks such as immunogenicity, mutagenesis, and upregulation or downregulation of tumor-inducing or tumor-suppressing genes (1–4). Electroporation permeabilizes cell membranes through the application of electric field pulses. Electric forces generate pores in the membrane through which DNA passes either uncoupled or after association to sphingosine (5, 6). For this approach, the operator needs an electroporation apparatus. Capacity and voltage of the electric pulse, which have to be chosen carefully for different cells lines, and optimal salt concentration are critical issues, since cell viability and efficacy of DNA transfer are diverging trends.

Mammalian cells take up particles they encounter in their environment through the endocytosis pathway. The earliest method for introducing DNA into eukaryotic cells involved making a coprecipitate of calcium phosphate (CaP) and DNA, and then simply applying the coprecipitate to cells (7). Owing to its low immunogenicity and low toxicity, CaP is also suitable for in vivo applications (8). Despite the fact that key factors underlying the CaP mediated gene transfer are poorly understood, it was determined that a minimum ratio of CaP nanoparticles to plasmid DNA (pDNA) is required to completely bind pDNA. Another critical issue is the presence of calcium chloride in the transfection cocktail combined with a distinct size of nanoparticles, leading to optimal transfection efficiencies probably due to the surface charge of the complex (9). Another class of delivery vehicles are spherical colloidal particles called liposomes. The surrounding lipid bilayer consists of hydrophobic lipid chains with variable hydrophilic polar head groups, which are orientated toward the aqueous media. Liposomes are primarily uni- or multilamellar, positively or negatively charged and in a size range of 50–150 nm, corresponding to approximately 1% of the diameter of a typical mammalian cell. The gene delivery approach is predominantly based on the formation of complexes generated between the negatively charged DNA and positively charged transport vehicles. Such vehicles enable cellular uptake by different internalization pathways and carry the DNA into the cytosol of the cells.

Lipoplexes are complexes of liposomes and DNA with distinct properties according to the lipid composition and the method of their assembly. Conventional liposomes have no specific reactivity. By contrast, reactive liposomes carrying the ability to change their structure upon interaction are known as polymorphic liposomes (10). The best known are pH-sensitive liposomes, which can undergo lamellar to hexagonal or micellar phase transition. Further examples are cationic liposomes, which disintegrate and restructure during complexation with nucleic acids (11). Since most laboratories do not have appropriate equipment to produce such defined liposomes, commercially available cationic lipid-based transfection reagents such as Lipofectamine 2000, DMRIE-C, and Lipofectin have been used efficiently for transfection.

Another class of positively charged transfer vehicles are dendrimers that are characterized by a variety of functional groups, enabling high solubility, reactivity, and DNA uptake into empty cavities (12, 13).

Polyplexes are efficient gene delivery vehicles of condensed DNA attached to cationic polymers such as polyethyleneimine (PEI). PEI consists of ethylene groups and amino groups, which are partly protonated at physiological pH providing a high positive charge density for electrostatic interaction with the negatively charged DNA. The condensed DNA anneals to the cationic polymers, thereby forming the polyplex (PEI/DNA) with an overall positive charge, which supports entry via the negatively charged cell membrane. Endocytosis triggers transport into the cellular endolysosomal compartment. There, the main function of PEI is to protect the DNA from degradation, which is avoided by buffering the acid in the lysosomes (14–16).

Despite the fact that various transfection methods have been optimized for different plasmids expressing reporters as well as proteins of distinct interest, it is difficult to directly compare transfection methods for production of a specific protein. Among the reasons for this difficulty are the following: (1) the random location of integration into the genome and the associated uncertainty about the epigenetic marks of any particular locus, (2) the heterogeneity of different isolates of the same cell line after years of propagation in different laboratories, and (3) the stress of protein overexpression on the metabolic and secretory processes of the host cell, which leads to selection of lower-producing clones.

In general, we can say that all the transfection methods described in literature have their advantages and drawbacks in regard to costs, effort in time, and accessory equipment needed. Therefore, a plausible plan of action is to decide on one method and adapt critical parameters as results indicate. Here, we illustrate transient and stable transfection of CHO-cells with liposomes and polyplexes and compare the costs and outcomes with commercial transfection kits (see Note 1).

2. Materials

2.1. Plasmids

Plasmid DNA is purified using a Qiagen Maxi-prep kit and stored in water with an A_{260}–A_{280} ratio typically between 1.7 and 1.8. All plasmids are adjusted to a concentration of 100 ng/μl.

1. Plasmid peGFP-N3 (Clontech, Palo Alto, CA, USA).
2. Plasmid p2F5_LC: mAb 2F5 light chain (LC) cDNA in a eukaryotic expression vector controlled by a viral promoter (17).
3. Plasmid p2F5_HC: mAb 2F5 heavy-chain (HC) cDNA in a eukaryotic expression vector controlled by a viral promoter (17).
4. Plasmid p2_dhfr: expressing mouse dihydrofolate-reductase (dhfr) under the control of a SV40 promoter (18), equivalent to pSV2-dhfr (ATCC 37146).

2.2. Cell Line and Media

1. Chinese hamster ovary cells, CHO-DUKX-B11 (ATCC CRL-9096).
2. Cultivation medium: DMEM (Dulbecco's modified Eagle's medium, protein free) (Biochrom KG, Berlin, Germany), supplemented with final concentrations of the following: 4 mM L-glutamine (Life Technologies, Grand Island, NY, USA); 0.25% Soya-peptone/UF (HY-SOY/UF Quest International GmbH, Erfstadt-Lechenich, Germany); 0.1% Pluronic-F68 (Sigma-Aldrich Handels GmbH, Vienna, Austria); PF-supplement (Ethanolamin, 2.5 mM; Ferric-Citrate, 25 mM; L-Ascorbic acid, 2.0 mM; Sodium Selenite, 5.0 μM; Polymun Scientific, Vienna, Austria); HT (hypoxanthine 8 mM and thymidine 0.8 mM: Sigma-Aldrich).
3. Selection medium: cultivation medium without HT.

4. Amplification medium: selection medium with increasing amounts of methotrexate (MTX) as described below (Subheading 3.3.2).

2.3. Other Materials

1. DOTAP: 1,2-Dioleoyloxy-3-trimethylammonium-propane chloride: Merck Eprova AG (Darmstadt, Germany) (see Note 2).
2. DOPE: 1,2-dioleoyl-sn-glycero-3-phosphoethanolamine: Lipoid (Ludwigshafen, Germany).
3. HBS-buffer: 10 mM HEPES and 150 mM NaCl, pH 7.5.
4. Hamilton Syringe 750RN, 500 μl (Sigma Aldrich, Cat.no.: 24539).
5. Extruder: LIPEX™ 10 ml Thermobarrel Extruder; (Northern Lipids Inc.).
6. Whatman membranes: 400 nm: Nuclepore Track-Etch Membrane PC MB 25 mm; 0.4 μm (Whatman, Cat.no.: 110 607); 200 nm: Nuclepore Track-Etch Membrane PC MB 25 mm; 0.2 μm (Whatman, Cat.no.: 110 606); 100 nm: Nuclepore Track-Etch Membrane PC MB 25 mm; 0.1 μm (Whatman, Cat.no.: 110 605).
7. 20-ml polystyrene tube: Nunc, Cat. no.: 364238.
8. 24-well plates: Greiner Bio-One, Cat. no.: 662160.
9. PEI: Polyethylenimine, linear, 25 kDa (PolySciences Inc., Warrington, Pennsylvania, USA).
10. Coating buffer: 0.1 N Na_2CO_3/$NaHCO_3$, pH 9.5–9.8 (8.4 g $NaHCO_3$ and 4.2 g Na_2CO_3 in 1,000 ml H_2O).
11. Washing buffer: phosphate-buffered saline (PBS: 1.15 g $Na_2HCO_3 \times 2H_2O$, 0.2 g KH_2PO_4, 0.2 g KCl, 8 g NaCl in 1,000 ml H_2O) containing 0.1% Tween 20.
12. Dilution buffer: 1% bovine serum albumin (BSA, Sigma, 98% purity) in washing buffer.
13. pNPP staining solution: p-nitrophenylphosphate (1 mg/ml, Sigma-Aldrich) in coating buffer.
14. Mouse anti-human κ-chain antibody conjugated with Quantum Red (Sigma-Aldrich).
15. Fluorescein isothiocyanate (FITC)-labeled goat anti-human γ-chain antiserum (Sigma-Aldrich).
16. Tris-buffer: 100 mM Tris–HCl, 2 mM $MgCl_2$, pH 7.4.
17. Tris-FCS buffer: 10% FCS (heat-inactivated fetal calf serum, cell culture grade) added to Tris-buffer (2.16).
18. Commercial transfection reagents: Lipofectin, Lipofectamine 2000, and DMRIE-C, all purchased from Invitrogen.

3. Methods

3.1. Expression Plasmids and Host Cell Lines

The plasmid peGFP-N3 carrying the reporter gene eGFP is used in transient transfection experiments. For stable cell line development p2F5_LC, p2F5_HC, and p2_dhfr are cotransfected (see Note 3) and amplified via increasing concentrations of methotrexate, as follows (19).

1. Dihydrofolate-reductase (dhfr) deficient CHO-cells DUKX-B11 (20) are propagated in static suspension culture (T-flasks, Nunc) in DMEM (i.e., protein-free) cultivation medium and split twice a week 1:5 to 1:10. Seeding density of cells is typically $1–2 \times 10^5$ cells per ml to maintain cells exponentially growing by splitting twice a week.
2. Unless otherwise noted, all cultivation and incubation steps are performed at 37°C and 7% CO_2 in a humidified incubator.

3.2. Preparation of Liposomes Consisting of DOTAP/DOPE (21, 22)

Lipid vesicles are formed spontaneously by injection of dissolved lipids into aqueous phases. During this process, the ethanolic lipid solution and the buffer must be adjusted to 37°C ± 2°C.

1. Prepare lipid solution: 26.2 mg DOTAP and 28 mg DOPE are dissolved in 0.5 ml 96% (v/v) ethanol in a glass vessel and incubated for 2 h at 37°C ± 2°C without agitation.
2. 5 ml HBS buffer adjusted to 37°C ± 2°C are stirred with a magnetic bar in a 20 ml polystyrene tube while the lipid solution is continuously injected with a tight Hamilton needle into the solution and kept for 1 h at 37°C ± 2°C in an incubator without stirring. This initial liposome preparation contains 11 mg/ml lipids as heterogeneous large unilamellar vesicles (LUVs) with a molar ratio of 4:6 of DOTAP and DOPE.
3. Small unilamellar vesicles (SUVs) are prepared from the LUVs by extrusion through polycarbonate membranes (Whatman) in an extruder (Lipex). Extrusion cycles are repeated three times for 400 and 200 nm membranes and seven times for 100 nm membranes at a working pressure of 2–50 bar (23). Repeated extrusion of liposome preparations through those filters generates finally SUVs with mean diameters representing that of the filter pores of (80–110 nm).
4. For the verification of the extrusion process, the mean size and homogeneity of SUVs are controlled using DLS (dynamic light scattering; Zetasizer Nano-ZS, Malvern Instruments, Malvern, United Kingdom) (24).
5. Hydrated unilamellar lipid vesicles are stored at 4°C in a tightly closed container.

3.3. Preparation of Lipoplexes and Transfection of CHO-Cells

3.3.1. Transient Transfection (25) (see Note 4)

1. Exponentially growing CHO-cells are passaged 1:2 the day before transfection to assure that they are actively dividing.
2. Preparation of the transfection cocktail: 11 μg DOTAP/DOPE (DD) SUVs with a mean diameter of 80–110 nm (1 μl of the SUVs prepared above) and 1 μg pDNA are adjusted to a final volume of 100 μl with HBS buffer in Sarstedt tubes and incubated at 22°C for 30 min. This results in the formation of lipoplexes with a molar charge ratio of 2.5:1 (cationic lipid to DNA).
3. Before transfection, the cells are washed twice by centrifugation at $180 \times g$ for 10 min and subsequent suspending the cell pellet in 10 ml PBS. After the second washing step, the cells are adjusted to a concentration of 5×10^5 cells per 1 ml of fresh cultivation medium in 24-well-plates.
4. The lipoplexes are added to 1 ml CHO cell suspension by up and down pipetting and incubated for 4 h.
5. Medium is exchanged to cultivation medium to get rid of the transfection cocktail (see Note 5).
6. Transfectants are cultivated for 72 h in 24-well plates before flow cytometry analysis for eGFP expression. Transfection rates are found to be 15–20% for eGFP with this method (see Note 6).

3.3.2. Stable Transfection and Selection of Recombinant Cell Lines (see Note 4)

During stable transfection experiments, 5×10^6 cells in 10 ml medium are transfected with a tenfold upscale of DNA (including a DHFR gene either in *cis* or in *trans* for selection and amplification) expressing the protein of interest and DD, respectively.

1. Passage CHO-cells 1:2 the day before transfection.
2. Prepare lipoplexes with 110 μg liposomes and 10 μg pDNA in a total volume of 1 ml HBS in Sarstedt tubes.
3. Transfect 10 ml washed CHO-cells (5×10^6 cells) with 1 ml lipoplex by incubating at 37°C for 4 h in a Nunc T25-flask.
4. Medium is exchanged to cultivation medium to get rid of the transfection cocktail (see Note 5).
5. After removing the transfection cocktail, incubate for 24 h in 10 ml cultivation medium.
6. Start clone selection by exchanging the cultivation medium for 50 ml selection medium (i.e., cultivation medium without HT) and transfer the suspended cells to five 96-well plates (100 μl per well) resulting in a seeding density of 10,000 cells per well. This should lead to one to five growing clones per well. Since cells do not attach to the vessels under protein free growth conditions, changes of the cultivation medium should be minimized to avoid loss of cells by aspiration.
7. After 10–14 days, growing clones are adapted to 0.1 μM MTX by feeding with 100 μl 0.2 μM MTX containing amplification

medium. After viability has reached approximately 80% – typically after 7–10 days – the 0.1 μM MTX medium is aspirated off and replaced with 0.2 μM MTX medium. When some clones (not all will survive 0.2 μM MTX) cover half of the bottom of the wells with cells, the selection medium should be changed to 0.3 μM MTX.

8. Test wells with good growth at 0.3 μM MTX for the presence of the protein of interest.
9. The best wells regarding volumetric titers are propagated in T-flasks with amplification medium containing 0.4 μM MTX. In the case of antibody expression, typical growth rates are in the range of 0.3–0.4 doublings per day and volumetric titers are approximately 10 μg/ml.
10. In a second subcloning round, the best performing clone is applied to limiting dilution cloning with 3–30 cells per well in 96-well-plates. The six best performing clones resulting after ELISA product screening are cultivated in 125 ml spinner vessels (Techne) by passaging twice a week with an initial cell concentration of 2×10^5 cells per ml. The final clone is selected by evaluation of specific growth rate, volumetric titer and specific productivity.
11. This clone is called 2F5/DD.

3.4. Preparation of Polyplexes and Transfection of CHO-Cells

3.4.1. Transient Transfection (see Note 4)

1. Prepare PEI stock solution by dissolving PEI powder in water generating a stock solution with a concentration of 1 mg/ml and store at 22°C in plastic tubes (see Note 7).
2. Exponentially growing CHO-cells are passaged 1:2 the day before transfection.
3. Preparation of polyplexes: 25 μl plasmid DNA (2.5 μg pDNA) are added to preheated (60°C) PEI solution containing 90 μg PEI (90 μl), mixed gently and adjusted to 200 μl with HBS buffer of room temperature (see Note 8).
4. The prepared polyplexes are added to 1 ml CHO cell suspension (5×10^5 cells/ml in a 24-well-plate) by gentle pipetting and the plate is incubated at 37°C for 4 h.
5. Medium is exchanged to cultivation medium to get rid of the transfection cocktail (see Note 5).
6. The transfectants are cultivated for 72 h before analysis of eGFP expression by flow cytometry (see Note 6).

3.4.2. Stable Transfection and Selection of Recombinant Cell Lines (See Note 4)

During the stable transfection experiments, 5×10^6 cells in 10 ml are transfected with a tenfold upscale of DNA and PEI, respectively.

1. Passage CHO-cells 1:2 the day before transfection.
2. Prepare polyplexes with 900 μg PEI and 25 μg of total pDNA in a total volume of 2 ml.

3. Transfect 10 ml CHO-cells (5×10^6 cells) with 2 ml polyplexes in cultivation medium and incubate for 4 h at 37°C before changing the cultivation medium.
4. Medium is exchanged to cultivation medium to get rid of the transfection cocktail (see Note 5).
5. After removing the transfection cocktail, incubate the initial transfectants for 24 h in a T-flask.
6. Clone selection, adaptation to amplification medium and screening for best performing clones is described in Subheading 3.3.2.
7. The final clone is called 2F5/PEI.

3.5. Flow Cytometry Analysis for Antibody Heavy (γ) and light (κ) Chains (See Note 9)

1. 5–10×10^5 cells are washed once in PBS and suspended in 70% ethanol for membrane permeabilization. Such fixed cells can be stored at 4°C for at least 1 month. Owing to this procedure, intracellular targets are detectable and homogeneity of a cell population can be evaluated.
2. Fixed cells are washed twice in Tris–FCS buffer.
3. Cells are suspended in 100 μl Tris–FCS buffer and incubated for 30 min at 37°C.
4. Add 100 μl Tris–FCS buffer containing mouse anti-human κ-chain antibody conjugated with Quantum Red (diluted 1:30 in Tris–FCS) and goat anti-human γ-chain antiserum conjugated with fluorescein isothiocyanate (1:30) and incubate for 1 h at 37°C.
5. After washing with HBS, the cells are analyzed on a flow cytometer such as FACS-Calibur (Becton Dickinson) equipped with a 5-W argon laser tuned to 488 and 350 nm by excitation with 100 mW laser power at both wave lengths, and the fluorescence emissions are measured with a 530/30 and 660/20 filter, respectively.
6. Nonproducing (i.e., nontransfected) CHO-cells serve as negative controls and compensation of spectral overlap between Fl1 and Fl3 is adjusted using cells stained only with anti γ-chain FITC antiserum.

3.6. ELISA Quantification of IgG

1. Anti human γ-chain specific antiserum is diluted 1:1,000 in coating buffer and used as catcher antibody on microtiter plates (ELISA grade I; Nunc, Denmark).
2. After each incubation step, the plates are washed three times with washing buffer.
3. Standard human IgG is serially diluted, starting with 200 ng/ml dilution buffer.
4. Standard and serially diluted samples (50 μl/well) are allowed to bind to the precoated well for 1 h at room temperature, and

after washing the plates, 50 μl of alkaline phosphatase-labeled goat anti-human γ antiserum (1:1,000 in dilution buffer) is used as detection antibody.

5. After 1 h incubation, plates are stained with 100 μl pNPP staining solution and read at 405 nm.

4. Notes

1. Several companies provide predominantly lipoplex based kits for efficient mammalian cell transfection. Three commercial systems were selected and compared with the DOTAP/DOPE system described here using each manufacturer's recommended conditions and eGFP as a readout plasmid. As shown in Fig. 1, the self-made DD reagent and Lipofectin gave comparable results with this CHO cell line. DMRIE-C yielded fewer transfectants and Lipofectamine 2000 was quite inefficient.
2. Liposomes by definition and positively charged liposomes in particular are able to modulate the cell membrane and thus enable lipoplex uptake by different and not completely elucidated mechanisms. The charge of different medium components contributes to the optimal condition and the concentration of

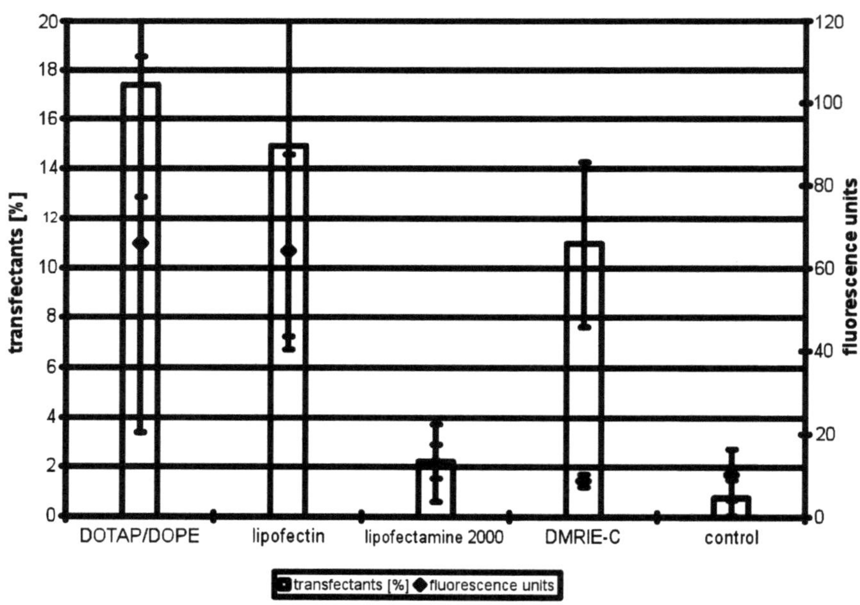

Fig. 1. Comparison of four methods regarding transient transfection efficiencies. 5×10^5 CHO-cells were transfected with commercial lipids according to manufacturer's instructions and cultivated for 72 h. Results are collected from three independent experiments and averaged.

lipoplexes influences the toxicity of the transfection cocktail. Generally, the composition of a transfection cocktail is defined by the amount of DNA necessary for gene delivery to a distinct number of target cells, the composition of the liposomes, and the molar charge ratio of positively charged liposome and negatively charged DNA. In case of DOTAP/DOPE, the optimal liposome–DNA ratio is 2.5:1 for transfection of 500,000 cells with 1 μg DNA (24) in the described protein free supplemented DMEM medium. If different media and supplements are used, the amount of liposomes and DNA has to be evaluated.

3. Cotransfection of two or three plasmids is commonly applied in cell culture transfection. The ratio of HC and LC plasmid needs to be determined to maximize the expression of each antibody. Therefore, only the total amounts of plasmid DNA are described in transfection methods.

4. The aim of transient gene expression is to generate a maximal amount of recombinant protein in the shortest possible time. Therefore a maximal number of cells must be transfected (often more than 10^9 cells are used in one experiment). This approach is often limited by the amount of DNA available and therefore protocols need to be optimized. The necessity for maximum transfection efficacy in stable cell line development is the variability of the initial integration locus of the transgene. Owing to this variability and other events occurring during the transfection process, the number of selectable clones must be maximized. The easiest way is to use the limiting dilution method to isolate a maximal number of initial clones and to screen these growing clones for protein secretion. Afterward, gene amplification, most typically with methotrexate, is applied to select high copy number clones. This subcloning and amplification procedure can be repeated several times to generate a monoclonal and homogenous population of a recombinant expression cell line. Even if transient protein expression of a distinct protein is rather low it is possible to generate adequate stable clones with the same transfection procedure as for transient experiments. The development from early transfectants to stably expressing cell clones is accomplished by MTX selection, subcloning, and screening.

5. To get rid of the transfection cocktail, the cell suspension is centrifuged for 10 min at $180 \times g$ and cells are suspended in fresh cultivation medium. Alternatively, the cultivation medium is exchanged by careful aspiration and medium is replenished accordingly.

6. A direct comparison of the two described transfection methods is illustrated in Fig. 2 indicating the transfection rate of each method after 3 days. Both strategies are suitable for optimization of transient transfection protocols despite in this example

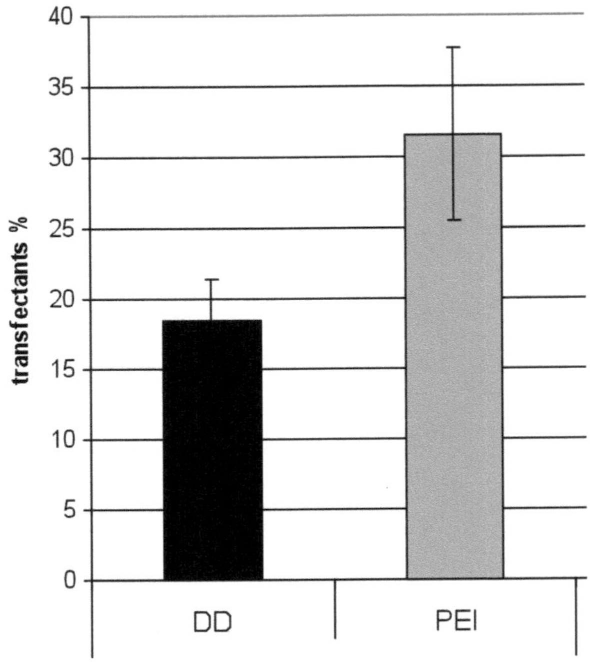

Fig. 2. Comparison of the number of transfectants 72 h after DD (*filled square*) and PEI (*filled square*) transfection using eGFP (peGFP) to determine the percentage of transfectants of viable cells via flow cytometry.

higher transfection rates were detected for the PEI method. *The choice of the transfection method is primarily dependent on the technical equipment available rather than on results obtained during preliminary experiments with reporter proteins since additional factors such as gene translocation or complexity of proteins influence the protein production and secretion rate significantly.*

7. Polyplexes are used for transfection immediately after preparation and are not stored.
8. In case of PEI transfections, the balance of PEI and DNA must be evaluated for each particular medium to avoid cell death caused by polyethylenenimine (26).
9. The homogeneity of the cell population is most often evaluated by flow cytometry analysis. Figure 3 shows intracellular content of heavy and light chain of the independently generated recombinant cell lines, 2F5/DD and 2F5/PEI. More than 95% of cells stained positively for both, heavy and light chain, indicating that both transfection systems are suitable for the development of stable cell lines.

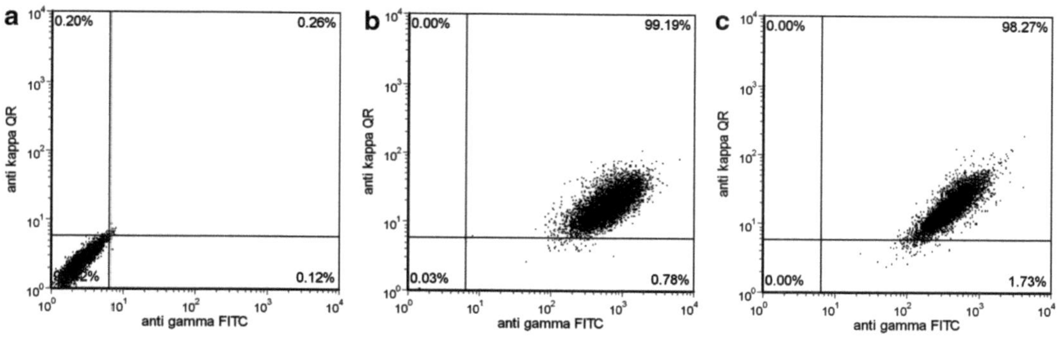

Fig. 3. Intracellular light and heavy chain content of cell lines 2F5/DD and 2F5/PEI cultivated in spinner vessels. The intracellular light chain was labeled with an anti-κ Quantum red-labeled antibody and the heavy chain was labeled with an anti-γ FITC-labeled antibody. Negative control (**a**), 2F5/DD (**b**), and 2F5/PEI (**c**). Host cells and single stained cell lines served as controls to set the gate and adjust the compensation.

Acknowledgments

We thank Hannes Reisinger for technical help and Marion Tschernutter for her help with manuscript preparation.

References

1. Crystal, R. (1995) Transfer of genes to humans: early lessons and obstacles to success. *Science* **270**, 404–410.
2. Tripathy, S., Black, H., Goldwasser, E., and Leiden, J. (1996). Immune responses to transgene-encoded proteins limit the stability of gene expression after injection of replication-defective adenovirus vectors. *Nat. Med.* **2**, 545–550.
3. Sadelain, M. (2004) Insertional oncogenesis in gene therapy: how much of a risk? *Gene Ther.* **11**, 569–573.
4. Marshall E. (2002) What to do when clear success comes with an unclear risk? *Science* **298**, 510–511.
5. Golzio, M., Teissié, J., and Rols, M. (2001) Control by membrane order of voltage-induced permeabilization, loading and gene transfer in mammalian cells. *Bioelectrochemistry* **53**, 25–34.
6. Golzio, M., Teissie, J., and Rols, M. (2002) Direct visualization at the single-cell level of electrically mediated gene delivery. *Proc. Natl. Acad. Sci. USA* **99**, 1292–1297.
7. Graham, F. L., and Van der Eb A. J. (1973) A new technique for the assay of infectivity of human adeno-virus 5 DNA. *Virology* **52**, 456–467
8. Kim, Y., Park, J., Lee, M., Kim, Y., Park, T., and Kim S. (2005) Polyethylenimine with acid-labile linkages as a biodegradable gene carrier. *Journal of Controlled Release* **103**, 209–219.
9. Pedraza, C., Bassett, D., McKee, M., Nelea, V., Gbureck, U., and Barralet, J. (2008) The importance of particle size and DNA condensation salt for calcium phosphate nanoparticle transfection. *Biomaterials* **29**, 3384–3392.
10. Lasic D. D., Templeton N. S. (1996) Liposomes in gene therapy. *Advanced Drug Delivery Reviews*, **20**, 2-3, 221-266
11. Lasic, D. (1997) Recent developments in medical applications of liposomes: sterically stabilized liposomes in cancer therapy and gene delivery in vivo. *Journal of Controlled Release* **48**, 203–222.
12. Dufès, C., Uchegbu, I., and Schätzlein, A. (2005) Dendrimers in gene delivery. *Adv. Drug. Deliv. Rev.* **57**, 2177–2202.
13. Shcharbin, D., Klajnert, B., and Bryszewska, M. (2009) Dendrimers in gene transfection. *Biochemistry (Mosc)* **74**, 1070–1079.

14. Boussif, O., Lezoualc'h, F. Zanta, M., Mergny, M., Scherman, D., Demeneix, B., and Behr, J. (1995) A versatile vector for gene and oligonucleotide transfer into cells in culture and in vivo: polyethylenimine. *Proc. Natl. Acad. Sci. USA* **92**, 7297–7301.
15. Lungwitz, U., Breunig, M., Blunk, T., and Göpferich, A. (2005) Polyethylenimine-based non-viral gene delivery systems. *Eur. J. Pharm. Biopharm.* **60**, 247–266.
16. Breunig, M., Lungwitz, U., Liebl, R., Klar, J. Obermayer, B., Blunk, T., and Goepferich, A. (2007) Mechanistic insights into linear polyethylenimine-mediated gene transfer. *Biochim. Biophys. Acta* **1770**, 196–205.
17. Kunert, R., Steinfellner, W., Purtscher, M., Assadian A., Katinger, H., (2000) Stable Recombinant Expression of the Anti HIV-1 Monoclonal Antibody 2F5 After IgG3/IgG1 Subclass Switch in CHO Cells. *Biotechnology and Bioengineering* **67** (1): 97–103
18. Gach, JS., Quendler, H., Weik, R., Katinger, H., Kunert, R., (2007) Partial humanisation of an anti-idiotypic antibody against monoclonal antibody 2F5, a potential HIV-1 vaccine? *Aids Research and Human Retroviruses*; **23** (11), 1405–15
19. Alt, F. W., Kellems, R. E., Bertino, J. R., and Schimke, R. T. (1978) Selective multiplication of dihydrofolate reductase genes in methotrexate-resistant variants of cultured murine cells. *J. Biol. Chem.* **253**, 1357–1370.
20. Urlaub, G., and Chasin, L. A. (1980) Isolation of Chinese hamster cell mutants deficient in dihydrofolate reductase activity. *Proc. Natl. Acad. Sci. USA* 77, 4216–4220.
21. Zuidam, N., and Barenholz, Y. (1997) Electrostatic parameters of cationic liposomes commonly used for gene delivery as determined by 4-heptadecyl-7-hydroxycoumarin. *Biochim. Biophys. Acta* **1329**, 211–222.
22. Simberg, D., Danino, D., Talmon, Y., Minsky, A., Ferrari, M., Wheeler, C., and Barenholz, Y. (2001) Phase behavior, DNA ordering, and size instability of cationic lipoplexes. Relevance to optimal transfection activity. *J. Biol. Chem.* **276**, 47453–47459.
23. D.G. Hunter, D.G. and Frisken, B.J. (1998) Effect of Extrusion Pressure and Lipid Properties on the Size and Polydispersity of Lipid Vesicles. *Biophysical Journal*, **74**, 2996–3002.
24. Reisinger, H., Sevcsik, E., Vorauer-Uhl, K., Lohner, K., Katinger, H., and Kunert, R. (2007) Serum-free transfection of CHO-cells with tailor-made unilamellar vesicles. *Cytotechnology* **54**, 157–168.
25. Regelin, A. E., Fankhaenel, S., Gurtesch, L., Prinz, C., von Kiedrowski, G., and Massing, U. (2000) Biophysical and lipofection studies of DOTAP analogs. *Biochim. Biophys. Acta* **1464**, 151–164.
26. Reisinger, H., Steinfellner, W., Katinger, H., and Kunert, R. (2009) Serum-free transfection of CHO cells with chemically defined transfection systems and investigation of their potential for transient and stable transfection. *Cytotechnology* **60**, 115–123.

Chapter 15

Methods for Constructing Clones for Protein Expression in Mammalian Cells

Takefumi Sone and Fumio Imamoto

Abstract

Multisite Gateway technology is a DNA cloning method based on in vitro site-specific recombination that is becoming increasingly popular because it allows quick and highly efficient assembly of multiple DNA fragments into a vector backbone. In the conventional Gateway Multisite strategy, cloning of multiple DNA fragments requires recombination of multiple entry clones with a single destination vector. The limitation of this approach is that as the number of entry clones increases, the efficiency of the assembly reactions decreases due to difficulty in successfully recognizing multiple pairs of matched *att* signals simultaneously. To address this problem, we have devised methods to generate modular expression clones, modular entry clones, and modular destination vectors. These allow many DNA fragments to be assembled stepwise into complex expression clones. We describe here how to construct these intermediate clones and vectors, and how to use these modules to construct expression clones comprising ten or more DNA segments. These principles can be applied to make multicomponent DNAs for many applications.

Key words: Multisite Gateway, Modular expression clones, Modular entry clones, Modular destination vectors, Multi-DNA fragments expression clone, Stepwise construction

1. Introduction

MultiSite Gateway is a recombinational cloning method that uses bacteriophage lambda integrase to assemble multiple DNA fragments into a single vector backbone in a predefined order and orientation (1–3). We have developed a method for tandem assembly of 5–15 DNA fragments into a single vector to construct multi-cDNA expression clones containing two or three tandem arrayed cDNA expression elements with multicolored fluorescent protein tags (4, 5). These multielement constructs can then be controllably inserted into the cellular genome to provide stable high content cell-based assay platform in tissue culture cells (6, 7).

James L. Hartley (ed.), *Protein Expression in Mammalian Cells: Methods and Protocols*, Methods in Molecular Biology, vol. 801, DOI 10.1007/978-1-61779-352-3_15, © Springer Science+Business Media, LLC 2012

In this chapter, we describe how six different *att* signals (*att*1, *att*2, *att*3, *att*4, *att*5, and *att*6) can be used to accomplish stepwise cloning of up to 11 DNA fragments into a standard Gateway Destination vector.

Kits that allow assembly of three and four fragments in one LR reaction are commercially available from Invitrogen (see Note 1). However, as the number of reacting molecules increases the proportion of desired product molecules decreases (4). This limitation can be resolved by stepwise LR reactions or by using modular destination (pDEST) vectors, both of which are described below. The essential idea behind these schemes is that one can use LR reactions to string together subsets of fragments, separated by *att*B sites. Then BP reactions with specialized vectors are performed to convert the *att*B sites within the string into *att*L or *att*R sites. Only these new *att*L and *att*R sites are reactive in the next round of LR reactions, which can be used to assemble strings of DNA segments.

We first describe how to construct the specialized Multisite vectors used in our method (Subheading 3.2). Then, the steps used to assemble strings of DNA segments into a pDEST vector are described in Subheading 3.3. The pDEST vectors harboring the other accessory elements, which are needed for integration of resultant pEXPR clones into chromosomal genome and successful expression of cDNA, are available from Invitrogen (see Note 2).

2. Materials

1. Source plasmids: pDONR™201, pDONR™ P4-P1R, and pDONR™ P2R-P3 (see Note 1), pUC18 (8), template DNA for amplification of *att*B-flanked PCR products, pDEST vectors for desired expression system (see Note 2).
2. Platinum® Taq DNA Polymerase High Fidelity (Invitrogen, Carlsbad, CA). Store in aliquots at −30°C (see Note 3).
3. Agarose-gel electrophoresis supplies and equipment.
4. Molecular biology-grade chemicals: phenol saturated with TE (pH 8.0) (see Note 4), chloroform, isopropanol, 70% ethanol.
5. Restriction enzymes: *Apa*I, *Bam*HI, *Bss*HII, *Eco*RI, *Hinc*II, *Kpn*I, *Nru*I, *Pst*I, *Sal*I, and *Xba*I (Takara Bio Inc., Japan). Store at −30°C.
6. BP Clonase® II Enzyme Mix and LR Clonase® II Plus Enzyme Mix (Invitrogen, Carlsbad, CA; see Note 5). Store in aliquots at −80°C (see Note 3).
7. MAX Efficiency® DH10B™ Competent Cells and One Shot® Mach1™ T1 Phage-Resistant Chemically Competent *E. coli* (Invitrogen, Carlsbad, CA). Store at −80°C.

8. Library Efficiency® DB3.1 Competent Cells (Invitrogen, Carlsbad, CA; see Note 6). Store at −80°C.
9. PureLink™ Quick Plasmid Miniprep kit and PureLink™ HiPure Plasmid Midiprep (Maxiprep) kit (Invitrogen, Carlsbad, CA; see Note 7).
10. Luria-Bertani (LB) agar plates containing the following: ampicillin (Amp; 50 μg/mL); kanamycin (Km; 50 μg/mL) plus chloramphenicol (Cm; 85 μg/mL); Km (50 μg/mL); Amp (50 μg/mL) plus Cm (85 μg/mL).
11. T4 DNA polymerase, T4 polynucleotide kinase, Calf intestine alkaline phosphatase (CIAP), and T4 DNA ligase (Takara Bio Inc., Japan). Store at −30°C.
12. BigDye® Terminator v3.1 Cycle Sequencing kit (Applied Biosystems Inc., Foster City, CA). Store at −30°C.
13. Sequencer and sequence analyzing software (see Note 8).
14. X-gal solution (20 mg/mL in dimethylformamide) and isopropylthio-ß-galactoside (IPTG) solution (100 mM in water). Store at −30°C. The tube containing X-gal solution should be stored in the dark or wrapped in aluminum foil.

3. Methods

3.1. Outline (See Fig. 1a, b)

1. Construct new vectors (custom pDONRs and pUC-DESTs) (Subheading 3.2).
2. PCR-amplify DNAs to be cloned with flanking *att*B sites and clone amplified DNAs into custom pDONR vectors to make pENTR clones (Subheading 3.3.1).
3. Assemble strings of DNA segments ($N = 3–4$) from pENTR clones into pUC-DEST vectors to make modular pEXP clones (Subheading 3.3.2).
4. Move strings from modular pEXP clones into custom pDONR vectors to make modular pENTR clones (Subheading 3.3.3).
5. Assemble string of strings of DNA segments ($N = 11$) from modular pENTR clones into standard pDEST vector (Subheading 3.4).

3.2. Construction of New Vectors

Ten custom pDONR and three pUC-DEST vectors are required for the procedures described below, with the final step resulting in 11 DNA segments being cloned into a standard Gateway destination vector. All the custom pDONR's are constructed with the same protocol and building blocks, as are all the pUC-DEST vectors. It is critical that every PCR product and vector be correct (both sequence and orientation) for the entire construction scheme to be successful.

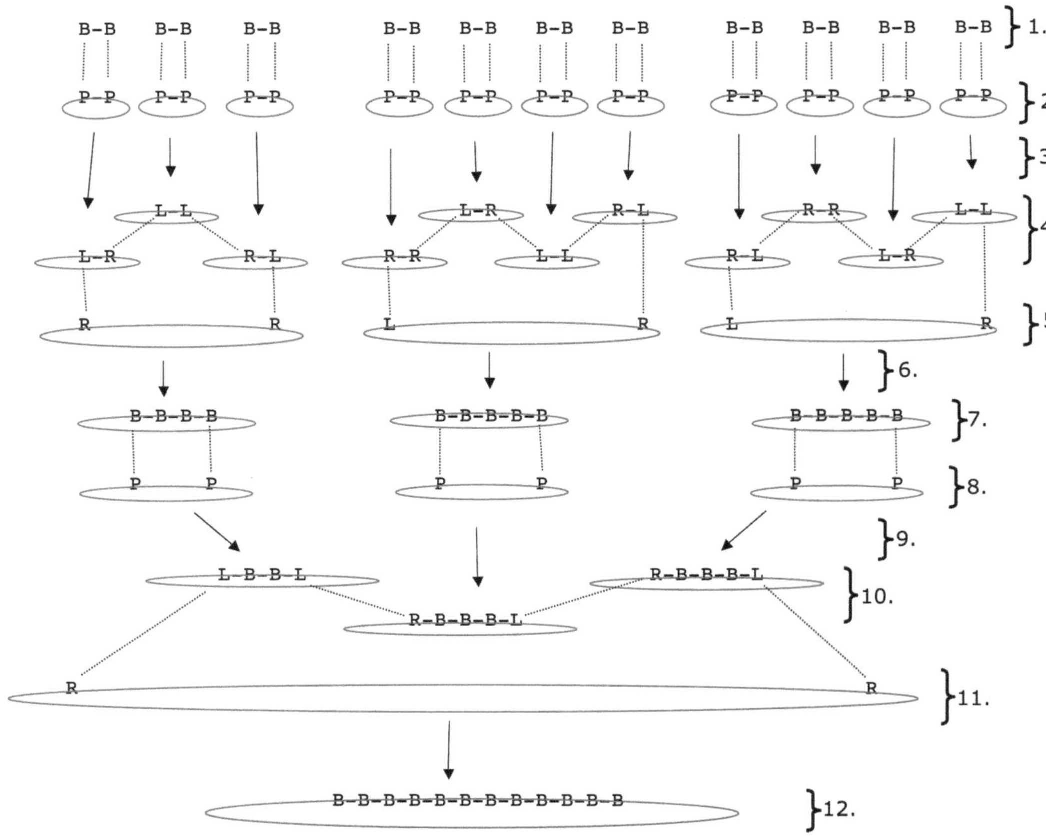

Fig. 1. Assembly of multiple DNA fragments by stepwise LR recombination reactions. (**a**) Simplified scheme showing how 11 PCR products are assembled into a single vector in an order and orientation dictated by the recombination sites in the PCR primers and in the vectors with which they react. 1: 11 attB PCR products. 2: pDONR vectors. 3: BP reactions. 4: Entry clones. 5: pUC-DEST vectors. 6: LR reactions. 7: Modular pEXPR clones. 8: pDONR vectors. 9: BP reactions. 10: Modular pENTR clones. 11: pDEST vector. 12: pEXPR clone containing 11 DNA segments. (**b**) Detailed construction of a multifragment pEXPR clone starting from PCR products.

3.2.1. Construction of Custom pDONR Vectors (See Fig. 2)

New pDONR vectors are constructed in two stages: new *att*P sites are made by overlap PCR and cloned into pUC plasmid (with restriction enzymes and ligase or with TOPO cloning) to make pUC-*att*P vectors, and the new *att*P sites are cut out of pUC and ligated to fragments from pDONR201 to form the new pDONR plasmids.

3.2.2. Overview of Construction of pUC-attP Vectors for Sources of attP Sites

*att*P sites are made up of an *att*L portion, an *att*R portion, and a core region (*see att*P sites in pDONR201, Fig. 2). Changes in the core region (the triangles in Fig. 2) give *att* sites their specificity. New *att*P sites are made by amplifying the *att*L and *att*R parts of an existing *att*P site using core primers that contain the appropriate base changes for each specificity, then combining the *att*L and *att*R PCR products and using them as templates (the *att*L and *att*R sequences) and primers (the core sequences) to make the desired *att*P products by overlap PCR.

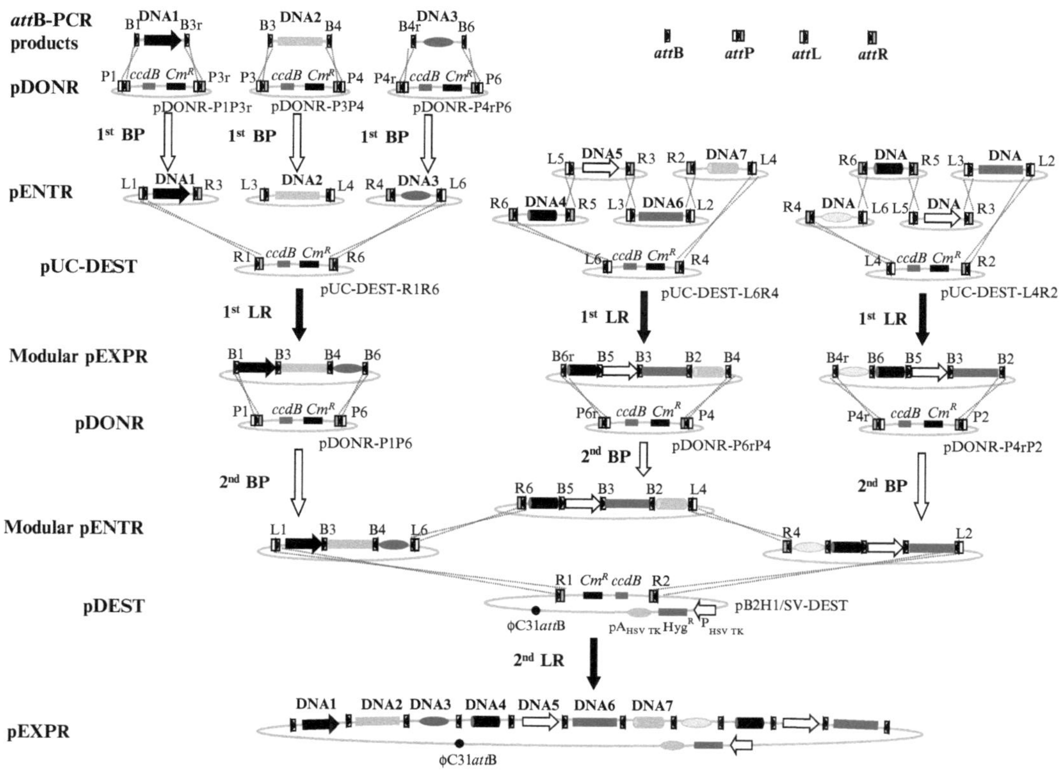

Fig. 1. (continued)

The names of primers used to construct the four new *att*P sites are shown in the top of Table 1A, and the sequences of those primers are shown in Table 1B. For example, to make the P3 *att* site, two PCR reactions are performed. Primers ApaI-L and B3-L are used in one reaction (with pDONR201 as the template) to amplify the *att*L segment, and primers B3-R and PstI-R are used in a second reaction (also with pDONR201 as template) to amplify the *att*R segment. The two PCR products (106 bp for *att*L, 527 pb for *att*R, Fig. 2) are then combined and cycled as described below. The resulting *att*P3 PCR product is then cloned into pUC to yield the desired pUC-*att*P vector.

Once the four new *att*P sites have been cloned into pUC and sequence verified, the other needed *att*P variants (reverse orientations (see Note 9) and different restriction sites for cloning at different positions in custom pDONRs) are amplified using forward and reverse primers and templates according to the bottom part of Table 1A.

Note: The protocol below uses blunt cloning into pUC with DNA ligase, but any other suitable cloning method (such as TOPO cloning) that preserves the *att*P and restriction sites, could be employed.

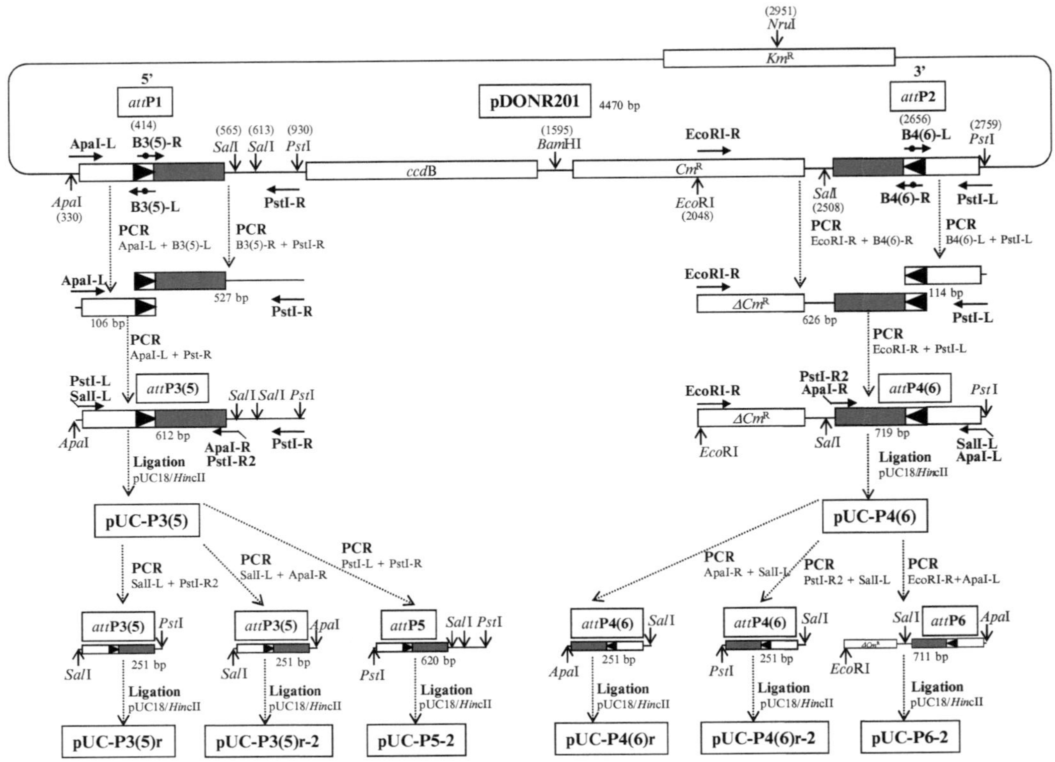

Fig. 2. Schematic drawing of pDONR201 and primers used to amplify *att*P signals. The map of pDONR201 with restriction sites used in Table 2 is drawn. The pDONR vector is used as a template to amplify *att*P3, *att*P4, *att*P5, and *att*6 signals and also the source of 5′ *att*P1 and 3′ *att*P2 signals. Although we cannot define which is 5′ in double-stranded circular DNA, we call the side of "*att*P1" in pDONR201 as "5′", because we conventionally clone transcribed genes in this orientation. The primers are indicated as *arrows*. Mutagenic primers are indicated as *arrows with a dot* at the middle, Primers with additional extension of restriction sites are indicated with *bended arrows*. For sequences of the primers, *see* Table 1B. For the combination of the primers and templates to amplify each *att*P signal, *see* Table 1A. The amplified *att*P signals are then cloned into pUC18 vector. The resulted pUC-*att*P vectors are then used as source plasmids for construction of customized pDONR vectors (*top half* of Table 2).

3.2.3. Construction of attP Vectors

1. Construct fragments containing *att*P3, *att*P4, *att*P5, and *att*P6 sites by doing PCR reactions using pDONR201 as template and primers shown in Table 1A, B. Perform two reactions containing 1 ng pDONR201 DNA as template and primers Fw1 + Rv2 and Fw2 + Rv1 with the following conditions: denature at 95°C for 2 min, followed by 25 cycles of 95°C for 30 s, 64°C for 1 min, and 68°C for 2 min, and final extension at 68°C for 5 min. Analyze 5 μL of each reaction by agarose gel electrophoresis. The sizes of the expected PCR products are shown on Fig. 2.
2. Five microliters each of the PCR products from Fw1-Rv2 and Fw2-Rv1 reactions are mixed with primers Fw1 and Rv1 (Table 1A, B) for the second round PCR in 100-μL scale under following condition: denature at 95°C for 2 min, followed by 5 cycles of 95°C for 30 s, 54°C for 1 min and 68°C for 2 min,

Table 1
List of primers. Names of the primers used in this chapter and their sequences

A.

*att*P signals			PCR primers				
***att*P**	**Location**	**Template**	**Fw1**	**Rv2**	**Fw2**	**Rv1**	**Resulted pUC-*att*P vector**
*att*P3	5′ end	pDONR201	ApaI-L	B3-L	B3-R	PstI-R	pUC-P3
*att*P4	3′ end	pDONR201	PstI-L	B4-L	B4-R	EcoRI-R	pUC-P4
*att*P5	5′ end	pDONR201	ApaI-L	B5-L	B5-R	PstI-R	pUC-P5
*att*P6	3′ end	pDONR201	PstI-L	B6-L	B6-R	EcoRI-R	pUC-P6
*att*P3r	3′ end	pUC-P3	SalI-L			PstI-R2	pUC-P3r
*att*P4r	5′ end	pUC-P4	SalI-L			ApaI-R	pUC-P4r
*att*P5r	3′ end	pUC-P5	SalI-L			PstI-R2	pUC-P5r
*att*P6r	5′ end	pUC-P6	SalI-L			ApaI-R	pUC-P6r
*att*P3r	5′ end	pUC-P3	SalI-L			ApaI-R	pUC-P3r-2
*att*P4r	3′ end	pUC-P4	SalI-L			PstI-R2	pUC-P4r-2
*att*P5r	5′ end	pUC-P5	SalI-L			ApaI-R	pUC-P5r-2
*att*P6r	3′ end	pUC-P6	SalI-L			PstI-R2	pUC-P6r-2
*att*P5	3′ end	pDONR-P5P2	PstI-L			PstI-R	pUC-P5-2
*att*P6	5′ end	pDONR-P1P6	ApaI-L			EcoRI-R	pUC-P6-2

B.

ApaI-L	CTC*GGGCCC*CAAATAATGATTTTATTTTG
PstI-L	GAG*CTGCAG*CTGGATGGCAAATAATGATT
SalI-L	CTCG*GTCGAC*AAATAATGATTTTATTTTG
PstI-R	AGT*CTGCAG*GTCGATACAGTAGAAATTAC
EcoRI-R	ACG*GAATTC*CGGATGAGCATTCATCAGGC
PstI-R2	TCG*CTGCAG*TACAGGTCACTAATACCATC
ApaI-R	TC*GGGCCC*CTACAGGTCACTAATACCATC
B3-L	**CAACTTTATTATACAAAGTTG**GCATTATAAAAAAG

(continued)

Table 1 (continued)

*att*P signals			PCR primers				
***att*P**	**Location**	**Template**	**Fw1**	**Rv2**	**Fw2**	**Rv1**	**Resulted pUC-*att*P vector**
B4-L	<u>**CAACTTTTCTATACAAAGTTG**</u>GCATTATAAGAAAG						
B5-L	<u>**CAACTTTTGTATACAAAGTTG**</u>GCATTATAAAAAAG						
B6-L	<u>**CAACTTTTTAATACAAAGTTG**</u>GCATTATAAGAAAG						
B3-R	<u>**CAACTTTGTATAATAAAGTTG**</u>AACGAGAAACGTAAAATG						
B4-R	<u>**CAACTTTGTATAGAAAAGTTG**</u>AACGAGAAACGTAAAATG						
B5-R	<u>**CAACTTTGTATACAAAAGTTG**</u>AACGAGAAACGTAAAATG						
B6-R	<u>**CAACTTTGTATTAAAAAGTTG**</u>AACGAGAAACGTAAAATG						
C.							
M13 Forward	CCCAGTCACGACGTTGTAAAACG						
M13 Reverse	AGCGGATAACAATTTCACACAGG						
SeqL-A	TCGCGTTAACGCTAGCATGGATCTC						
SeqL-B	GTAACATCAGAGATTTTGAGACAC						
M13 Reverse221	CCAGGAAACAGCTATGACCATG						

A. Combination of primers and templates for amplification of new *att*P signals. New *att*P signals are amplified by PCR using the indicated template DNA and PCR primers. Resulted PCR products are inserted into pUC18 vector to construct the pUC-*att*P vectors

B. PCR primers for amplification of *att* signals. These primers are used in site-directed mutagenesis and amplification of *att*P signals. The combinations of the primers used are summarized in Table 3. Restriction sites are indicated in ***italic and underlined font***

C. Sequencing primers. These primers are used in sequencing pUC-*att*P vectors and pENTR clones (see Note 7). The M13 Forward (−20) and M13 Reverse primers described in the manuals of MultiSite Gateway kits are also usable

then add primers Fw1 and Rv1 followed by 25 cycles of 95°C for 30 s, 64°C for 1 min and 68°C for 2 min, and carry out final extension at 68°C for 5 min.

3. Verify the length of the PCR fragments (Fig. 2) by agarose-gel electrophoresis, then purify the PCR fragments as follows (see Note 4). To gel-purify a DNA fragment, load the entire reaction mixture on a wide-well agarose-gel and separate DNA fragments by electrophoresis and excise the gel-slice containing the needed bands (see Note 10). Crush the gel-slice with pipette tip and mix with equal volume of phenol saturated with TE (pH 8.0) and then freeze it at −80°C. Extract DNA fragments first by phenol, then by chloroform, precipitate by isopropanol, wash by 70% ethanol, and then air-dry. Resuspend the DNA into 15 μL of distilled water or TE (pH 8.0). Estimate the concentration of the fragments by agarose-gel electrophoresis and ethidium bromide staining.
4. Blunt at least 250 ng of the purified PCR fragments by removal of 3′ extensions of deoxyadenine (dA) with T4 DNA polymerase and phosphorylate the 5′ ends with T4 polynucleotide kinase simultaneously. Add T4 DNA polymerase buffer to 1× final concentration, add mixed dNTPs to 0.25 mM, ATP to 1 mM. Then, add 5 units T4 polymerase and 10 units T4 kinase per μg PCR product and incubate at 37°C for 30 min. Inactivate the enzymes by phenol extraction followed by chloroform extraction. Ethanol-precipitate the linear and dephosphorylated DNA, dissolve in 10 μl of distilled water or TE (pH 8.0).
5. Cut 0.5 μg of pUC18 plasmid DNA in 50-μL scale to completion with HincII and verify complete digestion by agarose gel electrophoresis. Then, add 10 μL of 10× CIAP buffer and 39 μL of distilled water to the reaction, and incubate for 30 min at 55°C with ten units of CIAP enzyme. Inactivate the CIAP by phenol extraction followed by chloroform extraction. Ethanol-precipitate the linear and dephosphorylated DNA, dissolve in 10 μl of distilled water or TE (pH 8.0).
6. Mix 100 ng of blunt, kinased overlap PCR fragment with 50 ng of the linear and dephosphorylated pUC18 vector in 10-μl scale with 350 U of T4 DNA ligase and 1× buffer. Incubate at 16°C for 18 h using a heat–cool block.
7. Transform 5 μL of the ligation mixture into 25 μL of Max Efficiency DH10B Competent Cells. After 1-h incubation with 350 μL of S.O.C. Medium at 37°C, add 40 μL each of X-gal solution and IPTG solution, then spread it onto LB agar-plate containing Amp. Incubate at 37°C for 16 h (see Note 11).
8. Pick at least eight white colonies and purify plasmid DNAs of pUC-*att*Px with PureLink™ Quick Plasmid Miniprep kit (see Note 7). Verify the insertion of *att*P signal into pUC18 by

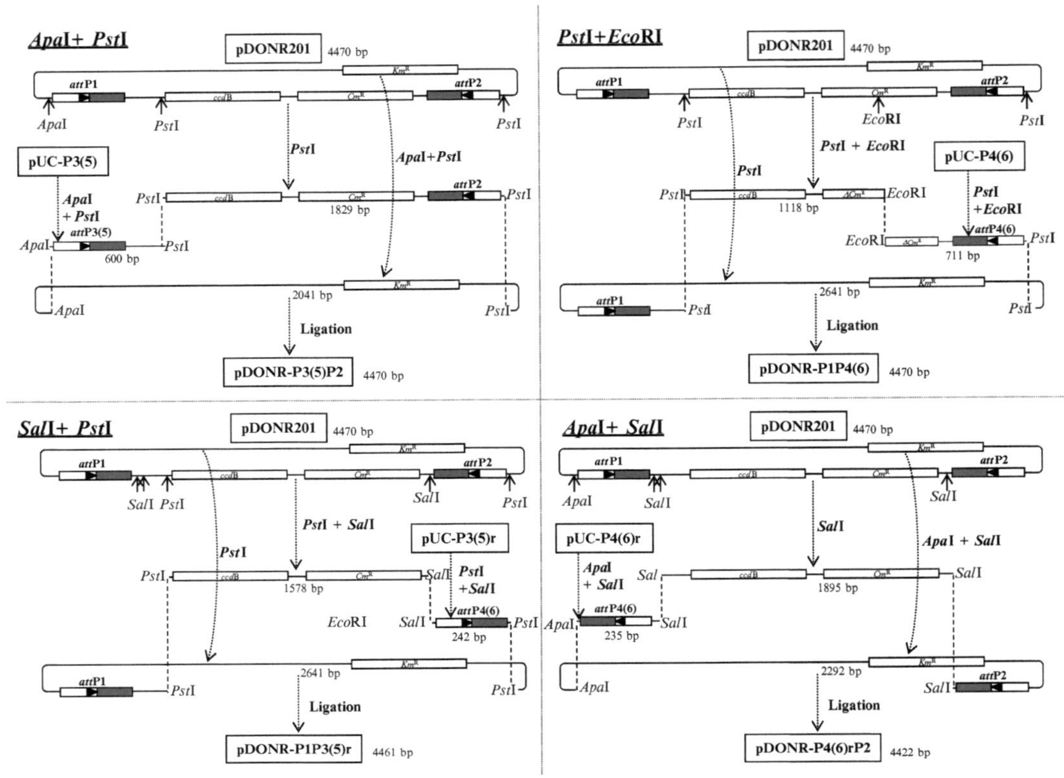

Fig. 3. Schematic drawing of construction of customized pDONR vectors from source plasmids. The four pattern of the restriction enzymes used are corresponding to that shown in top half of Table 2 (except for that of pDONR P4-P2). The source plasmids in *open box* are digested with the cocktail of the restriction enzymes described next to each *dotted arrows*, then gel-purified after electrophoresis. The three DNA fragments are mixed and ligated (see Note 12).

digesting with appropriate restriction enzymes, followed by agarose-gel electrophoresis.

9. Verify the new specificities of the *att*Px sites using sequencing primers M13 Forward and M13 Reverse (Table 1C) with BigDye® Terminator v3.1 Cycle Sequencing kit and analyze the result by a sequence analyzing software (see Note 8).
10. Using the pUC-*att*P clones (top of Table 1A) as templates, construct other pUC-*att*P clones (bottom of Table 1A), as in steps 1–9 above, except that the PCR amplification at sections in steps 1–2 is performed only once using primers Fw1 and Rv1 (Table 1A, B) and condition of first PCR (see also Fig. 2).

3.2.4. Construction of Custom pDONR Vectors

Custom pDONR vectors are constructed from pUC-*att*Px and pDONR201 (see Fig. 3).

1. To construct each customized pDONR vector, digest 500 ng of the source plasmids by indicated restriction enzymes in Table 2 and Fig. 3. For example, to construct pDONR-P1P3r, cut 500 ng of pDONR201 with PstI and purify the 2.6-kb

Table 2
Construction of customized pDONR from source plasmids

Customized pDONRs	Source plasmids: 5′ end *att*P	3′ end *att*P	Backbone	Restriction enzymes used	
pDONR-P3P2[b]	pUC-P3	pDONR201	pDONR201	*Apa*I	*Pst*I
pDONR-P5P2[b]	pUC-P5	pDONR201	pDONR201	*Apa*I	*Pst*I
pDONR-P1P4	pDONR201	pUC-P4	pDONR201	*Pst*I	*Eco*RI
pDONR-P1P6	pDONR201	pUC-P6	pDONR201	*Pst*I	*Eco*RI
pDONR-P3rP2	pUC-P3r-2	pDONR201	pDONR201	*Apa*I	*Sal*I
pDONR P4-P2[a]	pDONR P4-P1R	pDONR 201	pDONR P4-P1R/201	*Bam*HI	*Nru*I
pDONR-P4rP2	pUC-P4r	pDONR201	pDONR201	*Apa*I	*Sal*I
pDONR-P5rP2	pUC-P5r-2	pDONR201	pDONR201	*Apa*I	*Sal*I
pDONR-P6P2	pUC-P6-2	pDONR201	pDONR201	*Apa*I	*Sal*I
pDONR-P6rP2	pUC-P6r	pDONR201	pDONR201	*Apa*I	*Sal*I
pDONR P1-P3	pDONR201	pDONR P2R-P3	pDONR P2R-P3	*Apa*I	*Sal*I
pDONR-P1P3r[b]	pDONR201	pUC-P3r	pDONR201	*Sal*I	*Pst*I
pDONR-P1P4r	pDONR201	pUC-P4r-2	pDONR201	*Sal*I	*Pst*I
pDONR-P1P5	pDONR201	pUC-P5-2	pDONR201	*Sal*I	*Pst*I
pDONR-P1P5r[b]	pDONR201	pUC-P5r	pDONR201	*Sal*I	*Pst*I
pDONR-P1P6r	pDONR201	pUC-P6r-2	pDONR201	*Sal*I	*Pst*I
pDONR-P3P4r	pDONR-P3P2	pUC-P4r-2	pDONR-P3P2	*Sal*I	*Pst*I

(continued)

Table 2 (continued)

Customized pDONRs	Source plasmids: 5′ end *att*P	3′ end *att*P	Backbone	Restriction enzymes used		
pDONR-P3P6	pDONR-P3P2	pDONR-P1P6	pDONR-P1P6	*Apa*I	*Eco*RI	
pDONR-P5P4[b]	pDONR-P5P2	pDONR-P1P4	pDONR-P1P4	*Apa*I	*Eco*RI	
pDONR-P4rP3r[b]	pDONR-P4rP2	pDONR-P1P3r	pDONR-P1P3r	*Apa*I	*Eco*RI	
pDONR-P6rP5r	pDNR-P6rP2	pDONR-P1P5r	pDONR-P1P5r	*Apa*I	*Eco*RI	
pDONR-P6rP3r	pDNR-P6rP2	pDONR-P1P3r	pDONR-P1P3r	*Apa*I	*Eco*RI	
pDONR P3-P1r	pDONR-P3P2	pDONR P4-P1R	pDONR P4-P1R	*Apa*I	*Bam*HI	
pDONR P5-P1r	pDONR-P3P2	pDONR P4-P1R	pDONR P4-P1R	*Apa*I	*Bam*HI	
pDONR-P2rP4	pDONR P2R-P3	pDONR-P1P4	pDONR-P1P4	*Apa*I	*Bam*HI	
pDONR-P4rP6	pDONR-P4rP2	pDONR-P1P6	pDONR-P1P6	*Apa*I	*Bam*HI	
pDONR-P6rP4	pDONR-P6rP2	pDONR-P1P4	pDONR-P1P4	*Apa*I	*Bam*HI	
pDONR-P5P3r	pDONR-P5P2	pDONR-P1P3r	pDONR-P1P3r	*Apa*I	*Bam*HI	
pDONR-P3P4	pDONR-P3P2	pDONR-P1P4	pDONR-P1P4	*Apa*I	*Bam*HI	
pDONR P2r-P6	pDONR-P2rP4	pDONR-P1P6	pDONR P4-P1R	*Apa*I	*Bam*HI	*Pst*I
pDONR P3-P5r	pDONR-P3P2	pDONR-P1P5r	pDONR P4-P1R	*Apa*I	*Bam*HI	*Pst*I
pDONR P5-P6	pDONR-P5P2	pDONR-P1P6	pDONR P4-P1R	*Apa*I	*Bam*HI	*Pst*I

To construct customized pDONR vectors, the source plasmids are digested by indicated restriction enzymes and their fragments containing the indicated components (5′ end *att*P, 3′ end *att*P, Backbone) are ligated by T4 DNA ligase. For the location of the restriction sites in pDONR201, see Fig. 2.

[a]This pDONR vector has chimera backbone

[b]These pDONR vectors are already included in MultiSite Gateway Pro kits

*att*P1 – Km^R fragment, cut a second 500 ng of pDONR201 with PstI + SalI and purify the 1.5-kB *ccd*B – Cm^R fragment, and cut 500 ng of the pUC-P3r plasmid with SalI and PstI and purify the 242-bp fragment. Mix these fragments, ligate, and transform *E. coli* as below.

2. To gel-purify a DNA fragment from a restriction enzyme reaction, see step 3 in Subheading 3.2.3.
3. Ligate 100 ng of each gel purified DNA fragment in a 10-μL reaction using T4 DNA ligase (see Note 12). Incubate at 16°C for 6 h.
4. Transform 5 μL of the ligation reaction into 25 μL of Library Efficiency® DB3.1 Competent Cells (see Note 6). After 1-h incubation with 350 μL of S.O.C. Medium at 37°C, spread the entire 380 μL reaction onto an LB agar-plate containing Km and Cm. Incubate at 37°C for 16 h.
5. Pick at least four colonies to isolate plasmids with PureLink™ Quick Plasmid Miniprep kit (see Note 7). Verify the plasmids by digesting with appropriate restriction enzymes, followed by agarose-gel electrophoresis.
6. Validate the reactivity of each new Custom pDONR vector by reacting it with an appropriate *att*B PCR product (steps 1–3 in Subheading 3.3.1, below). For example, a new pDONR-P1P3r can be verified by reacting it in a BP reaction with an *att*B1-*att*B3r PCR product (Table 1A), transforming into *E. coli*, and selecting for Km-resistant (Km^R) colonies. As a control, cloning efficiency of a PCR product with different *att*B sites (such as *att*B1-*att*B2) using the same pDONR vector should be lower by at least 50-fold.

3.2.5. Construction of pUC-DEST Vectors

pUC-DEST plasmids are used to assemble strings of three or four DNA segments from corresponding pENTR plasmids (see Fig. 1a, line 5, and Fig. 1b). They are destination vectors because they contain a toxic gene that is replaced through LR reactions with inserts from multiple pENTR plasmids, but they are only assembly vectors and thus have no expression elements in their backbones. pUC-DEST plasmids are constructed in a two-step process (Fig. 4). For example, to make pUC-DEST-R1R6 (Fig. 1b), *att*Bx PCR products, such as *att*B1-stuffer-*att*B6, are produced using primers from Table 3, and are cloned with restriction enzymes and ligase into a pUC plasmid, such as pUC18. (The "stuffer" segment can be any arbitrary sequence because it will be lost in the next step.) Then the resulting plasmid is reacted in a BP reaction with a matching pDONR, such as pDONR-P1P6. The BP reaction is transformed into an *E. coli* strain that is resistant to the toxic gene *ccd*B (such as *E. coli* DB3.1 or *ccd*B Survival), and transformants are selected for pUC-DEST plasmids (the antibiotic resistance gene from pUC and the chloramphenicol-resistance gene from pDONR). The orientations

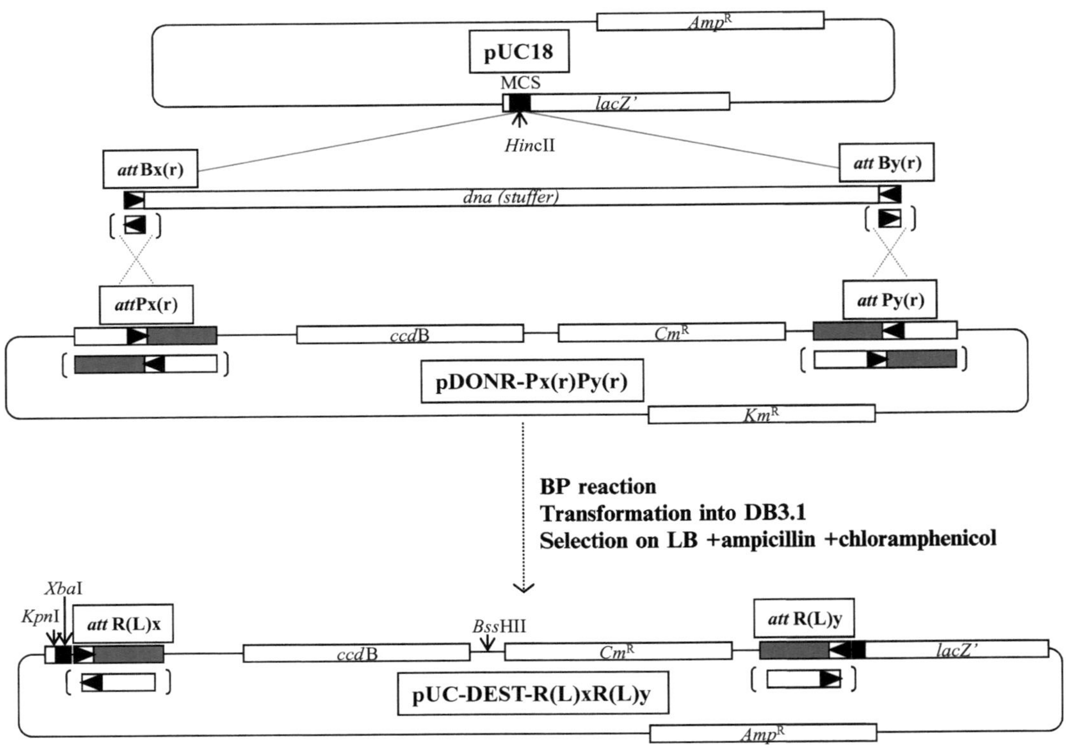

Fig. 4. Schematic drawing of construction of customized pUC-DEST vectors by BP reaction. *att* B-PCR fragment cloned into HincII digested pUC18 vector. The "*x*" or "*y*" represents an optional number of *att* signals. In the cases of reverse oriented *att* signals, they are shown in *parentheses*. In contrast to construct a pENTR clone by BP reaction, those *att* B or *att*P signals in forward orientation turn to *att*R signals and those *att*Br or *att*Pr signals in reverse orientation turn to *att*L signals in the resultant pUC-DEST vector (see Note 9).

Table 3
List of primers

A.		
B1-dna1-Fw	GGGG	ACA AGT TTG TAC AAA AAA GCA GGC ATG NNN NNN NNN NNN...
B3r-dna1-Rv	GGGG	CAA CTT TAT TAT ACA AAG TTG nnn nnn nnn nnn nnn...
B3-dna2-Fw	GGGGA	CAA CTT TGT ATA ATA AAG TTG ATG NNN NNN NNN NNN...
B4-dna2-Rv	GGGG	CAA CTT TGT ATA GAA AAG TTG nnn nnn nnn nnn nnn...
B4r-dna3-Fw	GGGG	CAA CTT TTC TAT ACA AAG TTG ATG NNN NNN NNN NNN...
B6-dna3-Rv	GGGG	CAA CTT TGT ATT AAA AAG TTG nnn nnn nnn nnn nnn...
B6r-dna4-Fw	GGGG	CAA CTT TTT AAT ACA AAG TTG ATG NNN NNN NNN NNN...

(continued)

Table 3 (continued)

B5r-dna4-Rv	GGGG	CAA CTT TTG TAT ACA AAG TTG nnn nnn nnn nnn nnn...
B5-dna5-Fw	GGGGA	CAA CTT TGT ATA CAA AAG TTG ATG NNN NNN NNN NNN...
B3r-dna5-Rv	GGGG	CAA CTT TAT TAT ACA AAG TTG nnn nnn nnn nnn nnn...
B3-dna6-Fw	GGGGA	CAA CTT TGT ATA ATA AAG TTG ATG NNN NNN NNN NNN...
B2-dna6-Rv	GGGGA	CCA CTT TGT ACA AGA AAG CTG nnn nnn nnn nnn nnn...
B2r-dna7-Fw	GGGG	ACA GCT TTC TTG TAC AAA GTG GCC ATG NNN NNN NNN NNN...
B4-dna7-Rv	GGGG	CAA CTT TGT ATA GAA AAG TTG nnn nnn nnn nnn nnn...
B4r-dna8-Fw	GGGG	CAA CTT TTC TAT ACA AAG TTG ATG NNN NNN NNN NNN...
B6-dna8-Rv	GGGG	CAA CTT TGT ATT AAA AAG TTG nnn nnn nnn nnn nnn...
B6r-dna9-Fw	GGGG	CAA CTT TTT AAT ACA AAG TTG ATG NNN NNN NNN NNN...
B5r-dna9-Rv	GGGG	CAA CTT TTG TAT ACA AAG TTG nnn nnn nnn nnn nnn...
B5-dna10-Fw	GGGGA	CAA CTT TGT ATA CAA AAG TTG ATG NNN NNN NNN NNN...
B3r-dna10-Rv	GGGG	CAA CTT TAT TAT ACA AAG TTG nnn nnn nnn nnn nnn...
B3-dna11-Fw	GGGGA	CAA CTT TGT ATA ATA AAG TTG ATG NNN NNN NNN NNN...
B2-dna11-Rv	GGGGA	CCA CTT TGT ACA AGA AAG CTG nnn nnn nnn nnn nnn...
B.		
B1-SDK-Fw	GGGG	ACA AGT TTG TAC AAA AAA GCA GGC TTC GAA GGA GAT AGA ACC
SDK-dna1-Fw		GGCTTC GAA GGA GAT AGA ACC ATG NNN NNN NNN NNN...

A. Primers for *att*B-flanked PCR products. These primers are used to amplify *att*B-flanked DNA fragments shown in Fig. 1b. To design *att*B-PCR primers for your DNA of interest, add four guanine (G) residues followed by 21 nucleotides core sequence of *att*Bx signals (***bolded and underlined***) to the 5′ end of the forward and reverse primers (see Note 10). If an optimal pair of these primers were designed to amlify a stuffer DNA, they can also be used for *att*B-flanked stuffer DNAs to construct pUC-DEST vectors

B. Primers for adding Shine-Dalgarno, Kozak, and *att*B1 motifs to a PCR product. SD and Kozak signals are indicated in *underlined font*

of the *att*B sites in the PCR products and the *att*P sites in the pDONR plasmids determine whether the *att* sites in the pUC-DEST plasmids are *att*L or *att*R sites (see Note 9).

1. Amplify a stuffer gene (such as eGFP) with *att*B primers listed in Table 3A (e.g., B1-stuffer-Fw and B6-stuffer-Rv). The

50 μL PCR reaction containing primers, 1 ng of stuffer template, and buffer, magnesium, and polymerase mix from the Platinum® Taq DNA polymerase High Fidelity kit is cycled as follows: denaturation at 98°C for 2 min, followed by 10 cycles of 98°C for 15 s, 55°C for 30 s and 68°C for 2 min, followed by 15 cycles of 98°C for 15 s and 68°C for 2 min, and final extension at 68°C for 5 min.

2. Purify, blunt, and phosphorylate the PCR product and perform ligation with pUC18 vector treated with HincII and CIAP according to steps 4–6 in Subheading 3.2.3, above.
3. Transform the ligation into competent *E. coli* cells such as DH5alpha or DH10B and plate for blue-white selection as in steps 7 and 8 in Subheading 3.2.3, above. Pick at least four white colonies and verify the plasmids by digesting with appropriate restriction enzymes, followed by agarose-gel electrophoresis (see Note 13).
4. Mix 20 fmoles of the *att*B-flanked DNA fragment in pUC18 vector (e.g., pUC-B1-dna-B6) and 20 fmoles of a pDONR vector with corresponding *att*P signals (e.g., pDONR-P1P6), then adjust the volume to 8 μL. Add 2 μL of BP Clonase® II Enzyme Mix and incubate at 25°C for 4 h (see Note 5). Add 1 μL of 2 μg/μL Proteinase K solution to each reaction. Incubate for 10 min at 37°C. Transform into Library Efficiency® DB3.1 Competent Cells (see Note 6) and spread on LB Amp and Cm plates.
5. Pick at least four colonies and verify the plasmids by digesting with appropriate restriction enzymes, followed by agarose-gel electrophoresis.
6. Verify the result of construction by using each candidate pUC-DEST vector (e.g., pUC-DEST R1R6) in an LR reaction with a pENTR clone with corresponding *att*L (or *att*R) signals (e.g., pENTR-L1-dna-L6). As a negative control, the number of colonies obtained with a mismatched pENTR (e.g., pENTR-L1-dna-L5) should be at least 50-fold lower. Confirm the sequence of the *att* sites, if mutation of an *att* site is doubted by obtaining too small number of colonies after LR reaction with a matched pENTR.
7. Once a series of pUC-DEST vectors are constructed as in Table 4 (upper part), pUC-DEST vector with other combination of *att*L or *att*R signals can be constructed by digestion of two pUC-DEST vectors with a pair of unique restriction enzymes (e.g., BssHII and XbaI), and ligation (Table 4, lower part).

3.3. Cloning Multiple Genes for Expression in Mammalian Cells

3.3.1. Construction of the Entry Clones

To design *att*B-PCR primers for DNA fragments of interest, such as genes from cDNA clones, first design template-specific forward and reverse primers (generally, 18–25 nucleotides) according to the standard PCR protocol adjusting their melting temperatures (Tm) to 55–65°C by Wallace's method (see Note 14). Then, add

Table 4
Construction of pUC-DEST vectors. The pUC-DEST vectors are constructed either by BP reaction between an *att*B-PCR fragment inserted in pUC18 vector and a pDONR vector, or by restrictive digestion and ligation of preexisting pUC-DEST vectors

pUC-DEST vectors	Source plasmids		Methods
pUC-DEST-R1L3	pUC-B1-EGFP-B3r	pDONR-P1P3r	BP reaction
pUC-DEST-R1R4	pUC-B1-EGFP-B4	pDONR-P1P4	BP reaction
pUC-DEST-R1L5	pUC-B1-EGFP-B5r	pDONR-P1P5r	BP reaction
pUC-DEST-R1R6	pUC-B1-EGFP-B6	pDONR-P1P6	BP reaction
pUC-DEST-R3R2	pUC-B3-EGFP-B2	pDONR-P3P2	BP reaction
pUC-DEST-L4R2	pUC-B4r-EGFP-B2	pDONR-P4rP2	BP reaction
pUC-DEST-R5R2	pUC-B5-EGFP-B2	pDONR-P5P2	BP reaction
pUC-DEST-L6R2	pUC-B6r-EGFP-B2	pDONR-P6rP2	BP reaction
pUC-DEST-R3L5	pUC-DEST-R3R2	pUC-DEST-R1L5	*Xba*I & *Bss*HIIcut & ligation
pUC-DEST-R3R4	pUC-DEST-R3R2	pUC-DEST-R1R4	*Xba*I & *Bss*HIIcut & ligation
pUC-DEST-R3R6	pUC-DEST-R3R2	pUC-DEST-R1R6	*Xba*I & *Bss*HIIcut & ligation
pUC-DEST-L4L3	pUC-DEST-L4R2	pUC-DEST-R1L3	*Xba*I & *Bss*HIIcut & ligation
pUC-DEST-L4R6	pUC-DEST-L4R2	pUC-DEST-R1R6	*Xba*I & *Bss*HIIcut & ligation
pUC-DEST-R5R4	pUC-DEST-R5R2	pUC-DEST-R1R4	*Kpn*I & *Bss*HIIcut & ligation
pUC-DEST-R5R6	pUC-DEST-R5R2	pUC-DEST-R1R6	*Kpn*I & *Bss*HIIcut & ligation
pUC-DEST-L6R4	pUC-DEST-L6R2	pUC-DEST-R1R4	*Kpn*I & *Bss*HIIcut & ligation

four guanine (G) residues followed by 21 nucleotides core sequence of *att*Bx signals (bolded and underlined in Table 3) to the 5′ end of the forward and reverse primers. For example, for a gene encoding a protein 5′ ATG NNN NNN NNN… on the top strand and 5′ nnn nnn nnn… on the bottom strand, to make a PCR product that contains *att*B1 at the 5′ end and *att*B3r at the 3′ end (as in Fig. 1a, upper left corner), order primers according to Table 1A: B1-dna1-Fw (GGGG ACA AGT TTG TAC AAA AAA GCA GGC ATG NNN NNN NNN NNN…; *att*B1 site underlined) and B3r-dna1-Rv (GGGG CAA CTT TAT TAT ACA AAG TTG nnn nnn nnn nnn nnn…; *att*B3r site underlined). In the case that the DNA fragment contains whole or a part of the gene that is translated into protein in vivo, care is necessary to ensure the correct reading frame (see Note 15). When two open reading frames (ORFs) are going to be connected by an *att*B site that is going to be translated in the final expression clone, the *att*B site connecting the two ORFs must be in the proper reading frame. Invitrogen has established the convention for the reading frames of *att*B sites that is shown in Table 3.

1. Amplify the *att*B-flanked DNA fragment of interest using primers with *att*B signals at their 5′ ends with Platinum® Taq DNA Polymerase High Fidelity in 25-μL scale under following condition: 1 μM each primer (For one-step adaptor PCR, the amount of inner primer should be one-tenth; see Note 15), 1 ng template DNA, denature at 98°C for 2 min, followed by 10 cycle of 98°C for 15 s, 55°C for 30 s and 68°C for 2 min, and then followed by 15 cycle of 98°C for 15 s and 68°C for 2 min, and carry out final extension at 68°C for 5 min.
2. Verify the length and amount of each PCR product by agarose-gel electrophoresis (see Note 16).
3. Mix 20 fmoles (see Note 17 for calculating moles of DNA) of each *att*B-flanked PCR product (e.g., Fig. 1a, top left, B1-dna1-B3r) and 20 fmoles of each pDONR vector with corresponding *att*P signals (e.g., Fig. 1a, top left, pDONR-P1P3r), then adjust the volume to 8 μL with distilled water or TE (pH 8.0). Add 2 μL of BP Clonase® II Enzyme Mix (see Note 5) and incubate at 25°C for 4 h using a heat–cool block.
4. Add 1 μL of 2 μg/μL Proteinase K solution to each reaction. Incubate for 10 min at 37°C.
5. Transform 5 μL of the reaction mixture into 25 μL of MAX Efficiency® DH10B™ Competent Cells, spread on LB Km plates.
6. Pick at least four colonies and verify the plasmids by digesting with appropriate restriction enzymes, followed by agarose-gel electrophoresis.
7. Verify the sequence of the pENTR clones using sequence primers listed in Table 1C (see Note 18) with BigDye® Terminator v3.1 Cycle Sequencing kit and analyze the result with sequence analysis software (see Note 8).

3.3.2. LR Reaction for Mixed Three or Four Genes

At this stage of assembly, the inserts from multiple pENTR clones are combined into pUC-DEST backbones to make Modular pEXPR clones (see Fig. 1a, b).

1. Quantify all the plasmids DNAs by A260 spectrometry and calculate their concentration in fmoles/μL (see Note 17).
2. To construct a modular pEXPR clone, mix 10 fmoles each of pENTR clones (e.g., Fig. 1a, upper left, pENTR-L1-dna1-R3, pENTR-L3-dna2-L4, and pENTR-R4-dna3-L6) and 20 fmoles of a pUC-DEST vector (e.g., Fig. 1a, upper left, pUC-DEST-R1R6) then adjust the volume to 8 μL. Add 2 μL of LR Clonase® II Plus Enzyme Mix and incubate at 25°C for 16 h (see Note 5).

3. Treat with Proteinase K for 10 min at 37°C. Transform 5 μL of the reaction mixture into 50 μL of One Shot® Mach1™ T1 Phage-Resistant Chemically Competent *E. coli* (see Note 19), spread on LB Amp plates.
4. Screen for Amp-resistant (Amp^R), Km-sensitive (Km^S) clones by using a sterile toothpick to transfer at least 16 colonies first to a Km LB agar plate, then to an Amp LB agar plate (see Note 20).
5. Pick at least eight colonies with phenotype Amp^R and Km^S into liquid medium containing Amp and verify the plasmids by digesting with appropriate restriction enzymes, followed by agarose-gel electrophoresis. Quantify the resulted modular pEXPR clone DNA by A260 spectrometry.

3.3.3. Construction of the Modular pENTR Clones

The Modular pEXPR clones constructed in the previous step contain three or four DNA segments separated by different *att*B sites. To combine these strings into larger assemblies the outer-most *att*B sites in each Modular EXPR clone must be converted back to *att*L or *att*R sites in accord with the overall cloning scheme. This is done by transferring each string into the appropriate pDONR molecule (refer to Fig. 1a, b).

1. Mix 20 fmoles of a modular pEXPR clone (e.g., Fig. 1a, left middle, pUC-B1-dna1-B3-dna2-B4-dna3-B6) and 20 fmoles of a pDONR vector with corresponding *att*P sites (e.g., Fig. 1a, left middle, pDONR-P1P6) then adjust the volume to 8 μL. Add 2 μL of BP Clonase® II Enzyme Mix and incubate at 25°C for 8 h (see Note 5).
2. After treatment with Proteinase K, transform 5 μL of the reaction mixture into One Shot® Mach1™ T1 Phage-Resistant Chemically Competent *E. coli* (see Note 19), spread on LB Km plates. Perform secondary antibiotic screening for Amp^R, Km^S colonies according to step 4 in Subheading 3.3.2, above (see Note 20).
3. Pick at least four colonies with phenotype Km^R and Amp^S, prepare plasmid DNA, verify the structures of the plasmids by digesting with appropriate restriction enzymes (using restriction sites in the genes in each string), followed by agarose-gel electrophoresis.
4. Purify each modular pENTR clone plasmid and quantify the DNA by A260 spectrometry.

3.4. Assembly of Multiple Genes into an Expression Vector

In the final step, strings of DNA segments from three modular pENTR clones are combined into a pEXPR vector.

1. To construct the final pEXPR clone, mix 10 femtomoles of each modular pENTR clone (e.g., Fig. 1a, line 10 and Fig. 1b lower part) and 20 fmoles of a pDEST vector (e.g., Fig. 1b, lower center, pB2H1/SV-DEST R1R2), then adjust the volume

to 8 μL. Add 2 μL of LR Clonase® II Plus Enzyme Mix and incubate at 25°C for 16 h (see Note 5).

2. After treatment with Proteinase K for 10 min at 37°C, transform 5 μL of the reaction mixture into 50 μL of One Shot® Mach1™ T1 Phage-Resistant Chemically Competent *E. coli* (see Note 19), spread on LB Amp plates. Perform secondary antibiotic screening (see Note 20).
3. Pick at least 12 colonies with phenotype Km^S and Amp^R, prepare plasmid DNA, and verify the plasmids by digesting with appropriate restriction enzymes, followed by agarose-gel electrophoresis. The confirmed pEXPR clone may now be incorporated into mammalian expression experiments (see Note 21).

4. Notes

1. Reading the manuals of MultiSite Gateway® Three-Fragment Vector Construction kit and MultiSite Gateway® Pro kits is quite helpful for your understanding. These manuals are available from Invitrogen's Web site. (http://tools.invitrogen.com/content/sfs/manuals/multisite_gateway_man.pdf, http://tools.invitrogen.com/content/sfs/manuals/multisite gateway pro_man.pdf).
2. Many types of pDEST vectors for mammalian expression systems are available from Invitrogen (Carlsbad, CA). An especially useful class of pDEST vectors is the pDEST vectors for constructing stable expression clones by site-specific genomic integration. There are pDEST vectors compatible for Flp-In™ system using Flp/FRT recombination system (9), Jump-In™ Fast sytem using φC31 reconbination system (10, 11) or Jump-In™ TI system using R4 recombination system (12).
3. Storing enzymes in aliquots decreases the chance of contamination and minimizes inactivation of the enzymes from repeated temperature changes.
4. Several kinds of DNA fragment purification kits (e.g., MagExtractor -PCR & Gel Clean up-; Toyobo Co. Ltd., Japan) may be used instead of the phenol–chloroform method described here.
5. These kits also contain 2 μg/μL Proteinase K solutions for termination of the reaction. Although recombination efficiency may decrease, the old versions of these enzymes (BP Clonase® Enzyme Mix and LR Clonase® Enzyme Mix; Invitrogen, Carlsbad, CA) are also usable. Reaction time of BP or LR reaction can be extended to 24 h in case of cloning large or many DNA fragments.

6. Library Efficiency® DB3.1 Competent Cells have been replaced recently with One Shot® *ccd*B Survival™ 2 T1R Competent Cells (Invitrogen, Carlsbad, CA).
7. Similar kits are available from many other companies. If you use the purified plasmid DNA for sequencing, final elution should be performed with distilled water instead of TE solution.
8. We generally use ABI PRISM® 3100 Genetic Analyzer (Applied Biosystems Inc., Foster City, CA) and Sequencher (Gene Codes, Ann Arbor, MI).
9. As the *att* signals are not palindromic sequences, they have orientations. Those *att*P or *att*B signals which turn to *att*L signals of a pENTR clone after BP reaction are defined as forward orientation and those *att*P or *att*B signals which turn to *att*R signals of a pENTR clone are defined as reverse orientation and designated with an "r." It is important to discriminate the *att*Bx (or *att*Px) and *att*Bxr (or *att*Pxr) signals correctly. For example, *att*B-PCR fragment flanked by *att*B3 and *att*B4r can only be cloned into pDONR-P3P4r, but not into pDONR-P3P4.
10. Exposure to UV light with wavelength shorter than 300 nm damages the DNA and cause decrease of ligation efficiency. Even if using UV light with longer wavelength, we recommend staining and exposing to UV only both edges of the gel and mark the band of interest, then excising the gel-slice containing the DNA fragment from the middle piece, which is not exposed to UV light.
11. Storing the plate at 4°C for 10 h following the 37°C incubation allows the blue color to develop more fully.
12. In the case of using nonunique restriction enzymes (SalI or PstI), you must ligate three DNA fragments to construct a pDONR vector (The 48-bp SalI fragment of SalI can be ignored.). Although the ligation efficiency of more than three fragments is not high, selection by both Km and Cm helps to obtain colonies with complete construct without treatment with CIAP. Orientations of the component fragments should be verified with the appropriate restriction enzymes.
13. Even if we might have mutations created by PCR reaction, the mutations contained in the area of new *att* sites should be eliminated by lower efficiency of BP-reaction. Thereby, it may not required to sequence at this stage of cloning the *att*B-PCR fragment into a pUC18. We confirm the sequence of the *att* sites at the stage of pUC-DEST vectors after BP-reaction, if mutation of an *att* site is doubted at step 6 of Subheading 3.2.5.
14. Tm for oligo-DNA less than 20 nucleotides is roughly calculated by Wallace's methods by the following expression: $Tm = 2 \times (T + A) + 4 \times (G + C)$.

15. If the DNA fragment is an ORF and needs to be expressed, addition of SD and Kozak signals (GAAGGAGATAGAACC) between the 5′ *att*B signal and ATG codon may help expression in both mammalian and bacterial systems. When using a set of overlapping forward primers (e.g., Table 3B, B1-SDK-Fw and SDK-EGFP-Fw) to amplify your ORF, one-step adapter PCR is useful (13). A key point of this method is to add normal amount of outer PCR primer and one-tenth amount of inner PCR primer. For detailed technique, see (13). If the DNA fragment was fused in frame with an N- or C-terminal tag, the primer may include additional one or two nucleotides to maintain the proper reading frame which Invitrogen recommends and no stop codon should appear in the frame. In Table 3, B1-SDK-Fw (in combination with SDK-dna1-Fw), B1-dna1-Fw, B2-dna6-Rv, B2r-dna7-Fw, and B2-dna11-Rv are the primers following the Invitrogen's instruction.
16. Purification of PCR fragment before BP reaction is usually unnecessary. But if the PCR fragment is larger than 300 bp in length, PEG precipitation by using 30% PEG solution included in BP Clonase® II Enzyme Mix helps to reduce primer dimers, which could be cloned into pDONR vectors instead of the *att*B-PCR product. In the case that you see many extra-bands, you may purify the proper band from the gel according to step 3 of Subheading 3.2.3.
17. According to Invitrogen's manuals for MultiSite Gateway kits, concentration in fmoles/μL (μM) of a plasmid DNA with N bp in length can be calculated by the following expression: fmoles/μL = (concentration in ng/μL) × 1,000,000/(N × 660).
18. Use SeqL-A and SeqL-B for pENTR clones with backbone of pDONR™201, and M13 Forward and M13 Reverse221 for the pENTR clones with backbone of pDONR™ 221 and its derivatives (pDONR™ P4-P1R and pDONR™ P2R-P3) (Table 1C). All the pDONR vectors included in the MultiSite Gateway kits are derivatives of pDONR™ 221. For other pDONRs constructed in this chapter, see Table 2.
19. Although MAX Efficiency® DH10B™ Competent Cells could be used, we strongly recommend using One Shot® Mach1™ T1 Phage-Resistant Chemically Competent *E. coli*. In case you could not obtain any colony by using DH10B after multiple LR recombination reaction, using Mach1 sometimes overcomes the problem.
20. Presumably because of transforming relatively large amount of complex mixture of plasmid DNAs (including substrate clones {pENTRs in LR reaction, modular pEXPRs in BP reaction}, substrate vectors {pDESTs in LR reaction, pDONRs in BP reaction}, complete recombination products {modular pEXPRs in LR reaction, modular pENTRs or modular pDESTs in BP

reaction}, intermediate recombination products {dimerized plasmids with single recombination}, and by-products), the colonies formed at the first screening often (about 50% of them) contains not only the desired clone but also the substrate clone or the intermediate recombination products of the recombination reaction. This secondary screening not only decreases the labor to pick such undesired colonies but also has more important meaning to remove the contamination of the substrate clones (even at the small amount), which will cause false positive colonies by harboring the same resistance gene for the antibiotics in the next steps.

21. For successful transfection into mammalian cells, a large amount of highly purified plasmid DNA of the expression clone is often required. For these purpose, midi- or maxiprep size of PureLink HiPure Plasmid kit (Invitrogen, Carlsbad, CA) is convenient.

Acknowledgments

The authors are grateful to Dr. Jonathan D. Chesnut (Lifetechnologies Corp.) for providing *att* signal sequences of MultiSite Gateway and information for construction.

References

1. Hartley, J. L., Temple, G. F., and Brasch, M. A. (2000) DNA cloning using in vitro site-specific recombination. Genome Res 10, 1788–95.
2. Sasaki, Y., Sone, T., Yoshida, S., Yahata, K., Hotta, J., Chesnut, J. D., Honda, T., and Imamoto, F. (2004) Evidence for high specificity and efficiency of multiple recombination signals in mixed DNA cloning by the Multisite Gateway system. J Biotechnol 107, 233–43.
3. Cheo, D. L., Titus, S. A., Byrd, D. R., Hartley, J. L., Temple, G. F., and Brasch, M. A. (2004) Concerted assembly and cloning of multiple DNA segments using in vitro site-specific recombination: functional analysis of multi-segment expression clones. Genome Res 14, 2111–20.
4. Sone, T., Yahata, K., Sasaki, Y., Hotta, J., Kishine, H., Chesnut, J. D., and Imamoto, F. (2008) Multi-gene gateway clone design for expression of multiple heterologous genes in living cells: modular construction of multiple cDNA expression elements using recombinant cloning. J Biotechnol 136, 113–21.
5. Sasaki, Y., Sone, T., Yahata, K., Kishine, H., Hotta, J., Chesnut, J. D., Honda, T., and Imamoto, F. (2008) Multi-gene gateway clone design for expression of multiple heterologous genes in living cells: eukaryotic clones containing two and three ORF multi-gene cassettes expressed from a single promoter. J Biotechnol 136, 103–12.
6. Yahata, K., Maeshima, K., Sone, T., Ando, T., Okabe, M., Imamoto, N., and Imamoto, F. (2007) cHS4 insulator-mediated alleviation of promoter interference during cell-based expression of tandemly associated transgenes. J Mol Biol 374, 580–90.
7. Sone, T., Nishiumi, F., Yahata, K., Sasaki, Y., Kishine, H., Andoh, T., Inoue, K., Thyagarajan, B., Chesnut, J. D., and Imamoto, F. (2009) 21. Cell engineering using integrase and recombinase systems. in Emerging Technology Platforms for Stem Cells (Lakshmipathy, U., Chesnut, J. D., and Thyagarajan, B., Eds.) pp 379–394, John Wiley & Sons, Hoboken.
8. Messing, J. (1991) Cloning in M13 phage or how to use biology at its best. Gene 100, 3–12.
9. O'Gorman, S., Fox, D. T., and Wahl, G. M. (1991) Recombinase-mediated gene activation and site-specific integration in mammalian cells. Science 251, 1351–1355.

10. Chalberg, T. W., Portlock, J. L., Olivares, E. C., Thyagarajan, B., Kirby, P. J., Hillman, R. T., Hoelters, J., and Calos, M. P. (2006) Integration specificity of phage phiC31 integrase in the human genome. J Mol Biol 357, 28–48.
11. Thyagarajan, B., Liu, Y., Shin, S., Lakshmipathy, U., Scheyhing, K., Xue, H., Ellerstrom, C., Strehl, R., Hyllner, J., Rao, M. S., and Chesnut, J. D. (2008) Creation of engineered human embryonic stem cell lines using phiC31 integrase. Stem Cells 26, 119–26.
12. Olivares, E. C., Hollis, R. P., and Calos, M. P. (2001) Phage R4 integrase mediates site-specific integration in human cells. Gene 278, 167–76.
13. Kagawa, N., Kemmochi, K., and Tanaka, S. (2004) One-step adapter PCR method for HTP Gateway technology cloning. QUEST **1**, 37–39.

Chapter 16

Optimizing Transient Recombinant Protein Expression in Mammalian Cells

Ralph F. Hopkins, Vanessa E. Wall, and Dominic Esposito

Abstract

Transient gene expression (TGE) in mammalian cells has become a routine process for expressing recombinant proteins in cell lines such as human embryonic kidney 293 and Chinese hamster ovary cells. The rapidly increasing need for recombinant proteins requires further improvements in TGE technology. While a great deal of focus has been directed toward optimizing the secretion of antibodies and other naturally secreted targets, much less work has been done on ways to improve cytoplasmic expression in mammalian cells. The benefits to protein production in mammalian cells, particularly for eukaryotic proteins, should be very significant – glycosylation and other posttranslational modifications will likely be native or near-native, solubility and protein folding would likely improve overexpression in heterologous hosts, and expression of proteins in their proper intracellular compartments is much more likely to occur. Improvements in this area have been slow, however, due to limited development of the cell culture processes needed for low-cost, higher-throughput expression in mammalian cells, and the relatively low diversity of DNA vectors for protein production in these systems. Here, we describe how the use of recombinational cloning, coupled with improvements in transfection protocols which increase speed and lower cost, can be combined to make mammalian cells much more amenable for routine recombinant protein expression.

Key words: Optimizing transient expression, Gateway cloning, Gateway Multisite, HEK293E, Polyethylenimine

1. Introduction

The development of nonviral gene transfer methods has had a positive influence on TGE, and today there are numerous transfection agents that can be used with various cell lines (1). This chapter focuses on optimizing transient recombinant protein (r-protein) expression in HEK-293 EBNA (293E) cells (2) using the cationic polymer polyethylenimine (PEI) as the nonliposomal DNA complexing agent (2–4). There are many transfection reagents that

James L. Hartley (ed.), *Protein Expression in Mammalian Cells: Methods and Protocols*, Methods in Molecular Biology, vol. 801, DOI 10.1007/978-1-61779-352-3_16, © Springer Science+Business Media, LLC 2012

Table 1
Reagent costs for transient transfection of 293E cells

Transfection reagent	Transfection efficiency in 293E cells (%)	Reagent cost per 1 L transfection
293 Fectin	21	$414
Superfect	36	$766
Escort V	37	$125
Gene juice	35	$750
PEI[a]	40	$0.10

[a]Polyethylenimine 25 kDa linear, Polysciences

work well for the 293E cell line; however, we have found that PEI gives comparable results at up to one thousand fold less the cost of other reagents (Table 1).

Many process variables can affect the level of TGE such as the health and density of the cells at the time of transfection, growth conditions, and culturing methods. In this chapter, we describe the transient transfection of 293E cells in various formats in an effort to optimize TGE. A 96-well plate format is described that is extremely easy perform and may be used to optimize TGE on a high-throughput basis. We also describe a 24-well deep block transient transfection procedure that yields higher levels of cells for small-scale purification scouting. Optimizing the large-scale process is also critical to achieving high levels of TGE, and we describe effective high-density culture method for large-scale expression of intracellular and secreted proteins at the multiple liter scale that uses low-cost, disposable containers.

In addition to cell culture parameters, optimization of the DNA constructs used for expression can play a major role in improving the levels of protein production. Variables such as promoters, enhancers, untranslated regions (UTRs), and episomal replication origins may all affect the amount of protein which can be expressed, and many of these variables may need to be tested for each individual protein to optimize expression constructs. In order to fully explore the space surrounding these variables, it is best to take advantage of a cloning platform that can quickly produce expression clones containing the different elements without other alterations in the vector background, which might confound the interpretation of data. The advent of recombinational cloning, and in particular the Gateway Multisite platform, permits high-throughput production of mammalian expression clones in which a series of elements can be combined quickly, easily, and at a low cost, and which allows screening of a large number of variables in a well controlled manner. The Gateway Multisite platform can be used in many configurations, and only one of those is described here. For additional information on the system, see the Invitrogen Web site at http://www.invitrogen.com or the cited references (5–7).

2. Materials

2.1. Gateway Cloning

1. 2× Phusion Master Mix HF (New England Biolabs, Beverley, MA).
2. DMSO: dimethyl sulfoxide (provided with Phusion Kit or available from Sigma, St. Louis, MO).
3. QIAQuick PCR Purification Kit (Qiagen,Valencia, CA).
4. Ready-Load 1-kb Plus DNA Ladder (Invitrogen, Carlsbad, CA).
5. ReadyAgarose Gels (BioRad, Hercules, CA).
6. BP Clonase II kit (Invitrogen, Carlsbad, CA; comes with BP Clonase® II enzyme mix, 2 μg/mL Proteinase K solution).
7. FastPlasmid DNA Kit (Eppendorf, Hamburg, Germany).
8. DH5α chemically competent cells (Invitrogen, Carlsbad, CA). Store at –80°C, do not reuse open vials.
9. SB medium: Superior Broth (AthenaES, Baltimore, MD), 40 g/L, autoclave for 20 min.
10. LB-agar plates (Teknova, Hollister, CA).
11. Falcon 2059 culture tubes (Fisher Scientific, Pittsburgh, PA).
12. 96-well V-bottom plates (PGC Scientific, Frederick, MD).
13. Air-pore Sealing Tape (Qiagen, Valencia, CA).
14. Supercoiled DNA Ladder (Invitrogen, Carlsbad, CA). Store at 4°C.
15. Turbo 96 Plasmid Prep Kit (Qiagen, Valencia, CA).
16. Gateway® LR Clonase™ Plus Enzyme Mix (contains Gateway® LR Clonase™ Plus Enzyme Mix, 5× LR Clonase™ Plus Reaction Buffer and 2 μg/μL Proteinase K Solution), (Invitrogen, Carlsbad, CA).
17. QIAprep® Spin Miniprep Kit (Qiagen Inc., Valencia, CA).
18. *BsrGI* Restriction Enzyme (New England Biolabs® Inc., Ipswich, MA).

2.2. Cell Culture

1. 293 C18 Human Embryonic Kidney cells (HEK 293EBNA), (CRL-10852, American Type Culture Collection, Manasas, VA) (see Note 1).

2. FreeStyle 293 Expression Medium (Gibco/Invitrogen, Carlsbad, CA) (see Note 2).
3. SFM4 HEK293 medium (HyClone, Logen, UT).
4. Trypan Blue 0.4% solution (Gibco/Invitrogen, Carlsbad, CA).
5. Geneticin 50 mg/mL (Gibco/Invitrogen, Carlsbad, CA).

2.3. Cell Culture Equipment and Instruments

1. Standard cell culture plasticware, including 490 cm^2 roller bottle and 3-L Erlenmeyer flasks with vent (Corning, Corning, NY), CO_2 incubator, centrifuge.
2. Innova shaker model 2100 (New Brunswick Scientific, Edison, NJ).
3. Infors Multitron II Orbital Shaker incubator with CO_2 control (Infors, Bottmingen, Switzerland).
4. TallBoys microplate shaker (Henry Troemner LLC, Thorofare, NJ).
5. Cedex Cell Analysis System for counting cells (Innovatis, Bielefeld, Germany).

2.4. Transient Transfection

1. Polyethylenimine (PEI), linear 25 kDa (Polysciences, Inc. Warrington, PA) (see Note 3).
2. Normal saline (9 g NaCl/L) (Quality Biologicals, Gaithersburg, MD).
3. Polypropylene 24-well deep block plates, round bottom (Thomson Instrument Co., Oceanside, CA).
4. Polystyrene 96-well flat bottom, round bottom, and black flat-bottom sterile plates (Costar, Corning, NY).
5. Culture ware described in Subheading 2.2 above could also be used for large-scale transfections.
6. TE buffer consisting of 10 mM Tris–HCl and 0.1 mM EDTA.

2.5. Evaluation of Protein Expression

1. M-PER® Mammalian Protein Extraction Reagent (PIERCE, Rockford, IL).
2. NuPAGE® LDS Sample Buffer (4×) (Invitrogen, Carlsbad, CA).
3. Bond-Breaker® TCEP Solution (PIERCE, Rockford, IL).
4. Criterion™ Precast Gel 4–12 and 10–20% Tris–HCl, 1.0 mm, 26-well comb, 15 μL, 1.0 mm (Bio-Rad Laboratories, Inc., Hercules, CA).
5. 10× TGS (Tris/Glycine/SDS Buffer), diluted to 1× working concentration (Bio-Rad Laboratories, Inc., Hercules, CA).
6. Benchmark™ Prestained Protein Ladder (Invitrogen, Carlsbad, CA).

7. iBlot™ Gel Transfer Stack Nitrocellulose, Regular (Invitrogen, Carlsbad, CA).
8. 20× TBS-Tween®, pH 7.4, diluted to 1× working concentration (USB®, Cleveland, OH).
9. Blotting Grade Blocker Nonfat Dry Milk (Bio-Rad laboratories, Inc., Hercules, CA).
10. SuperSignal West Pico Chemiluminescent Substrate (PIERCE, Rockford, IL).

3. Methods

To use the Gateway Multisite technology one constructs a series of Entry clones which contain sequence verified elements, including promoters, enhancers, UTRs, reporters, or target genes. Each element is flanked by recombination sites, called *att* sites, and these *att* sites are arranged to recombine with each other or with *att* sites in an expression vector. Critically, all recombination reactions are conservative (no nucleotides are added or lost), directional (orientations of segments in new clones are fixed according to the orientations of the *att* sites), and concerted (all strand exchanges occur in protein-DNA complexes, no free ends are available for side reactions). Standard Gateway uses the *att*1 and *att*2 sites exclusively. Multisite Gateway adds an additional number of different specificities to allow recombination to occur in various different formats. In this way, multiple genes can be linked together in a particular order (see Fig. 1). In the configuration described here, Entry clones are generated with promoter sequences flanked by *att*4 and *att*1 sites, while the Entry clones containing a GFP reporter are flanked by *att*1 and *att*2 sites. When these Entry clones are recombined with a Destination Vector containing *att*R4 and *att*R2 sites, the final clone is produced with the promoter linked to the reporter. The order of the recombination reactions is specific and directional, and illegitimate recombination cannot occur. In this way, libraries of promoters and reporters can be easily combined, all in the same exact vector background. Expansion of this system to include three, four, and even as many as eight different fragments has been reported (8, 9).

3.1. Gateway Entry Clone Construction

1. To clone most elements, 18–21 bp of gene/promoter specific 5′ and 3′ sequences are used for primer annealing. For simple Entry clones, which do not contain large amounts of additional features, the recombination signal sequences given in Table 2 can be added directly to the gene-specific primers. (see Note 4).

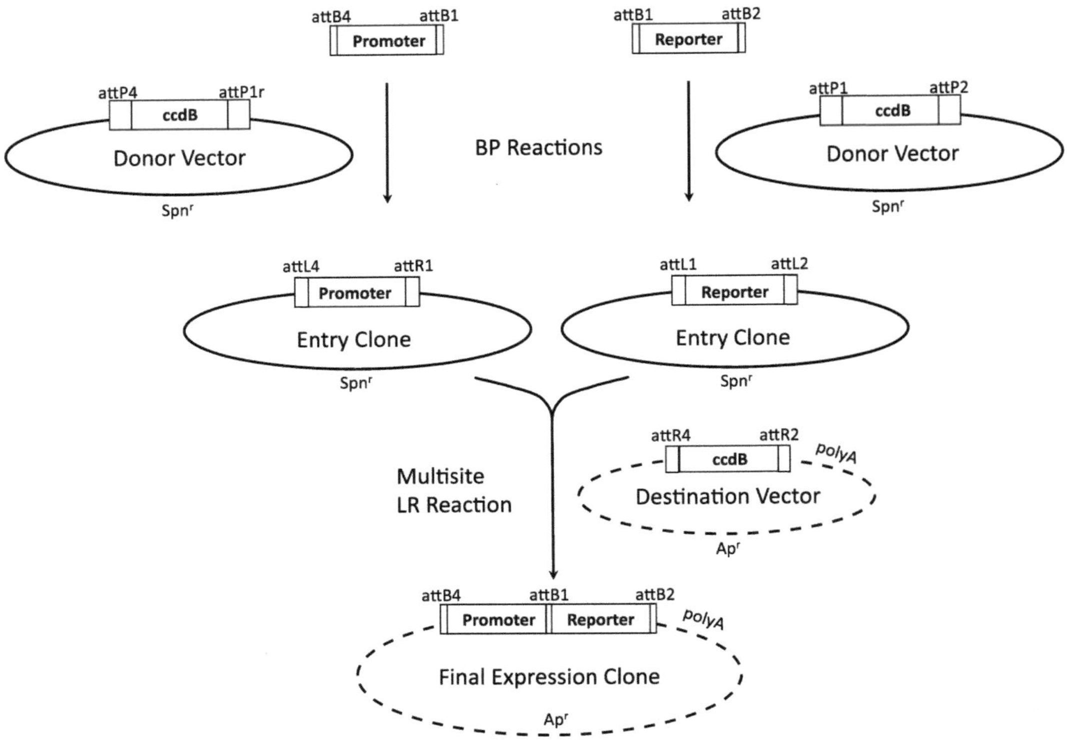

Fig. 1. General Multisite Gateway cloning scheme. This diagram presents a schematic representation of the Multisite Gateway reactions, identifying the different plasmid types found in the BP and LR reactions. The scheme shown involves a promoter containing Entry clone and a reporter containing Entry clone, as described in Subheading 3.

Table 2
Oligonucleotide sequences for Gateway sites

Standard Gateway primers

attB4 (promoter forward)	5′-GGGGACAACTTTGTATAGAAAAGTTGGC – gsp
*attB1*rev (promoter reverse)	5′-GGGGCCAACTTTTTTGTACAAAGTTG – gsp
attB1 (reporter forward)	5′-GGGGACAACTTTGTACAAAAAAGTTGGCACC *ATG* – gsp
attB2 (reporter reverse)	5′-GGGGACAACTTTGTACAAGAAAGTTGG *CTA* – gsp (reverse comp)

gsp, gene-specific primer (should contain 18–21 bp of 5′ or 3′ of gene, in reverse primers this sequence must be the reverse complement of the sense strand of the gene). *Underlined* sequences in the *att* sites identify the actual recombination sites which provide specificity. The *italicized* sequences represent the start (ATG) and stop (TAG) sites for the reporter clones

2. To a 200 μL thin-walled PCR tube, add 1 μL of each 10 μM oligonucleotide primer, 1.5 μL DMSO (see Note 5), 100–200 ng template DNA (see Note 6), and water to 25 μL final volume.
3. Add 25 μL 2× Phusion Master Mix HF, mix well, and carry out the PCR amplification using the following parameters: initial denaturation at 98°C for 30 s, 20 cycles of (98°C for 10 s, 55°C for 30 s, and 72°C for 30 s per kb of the expected product), followed by a 10 min elongation at 72°C, and cooling to 4°C (see Note 7).
4. If only a small amount of template DNA is available, increase the number of overall cycles from 20 to 30. This increases the likelihood of PCR errors, but should improve PCR product yield.
5. After cycling, load 5 μL of the PCR product on a 1% agarose gel to verify size by comparison to a linear DNA standard such as the ReadyLoad 1-kb DNA ladder.
6. Purify the PCR product using the QIAQuick PCR Purification Kit following the manufacturer's protocol and elute the DNA in 50 μL (see Note 8).
7. Add the following reagents to a microcentrifuge tube in the following order (the total reaction volume should be 10 μL): 1–6 μL H_2O, *attB* flanked PCR fragment (15–150 ng, see Note 9), 150 ng Donor vector (see Note 10), and 2 μL of BP Clonase II. A master mix can be used for all reagents except for BP Clonase II, which must be added last. Mix briefly by gentle vortexing.
8. Incubate the reaction mixture for at least 1 h at 30°C.
9. Add 1 μL of 2 mg/mL Proteinase K to inactivate the BP Clonase and incubate for 15 min at 37°C.
10. Add 1 μL of the BP reaction to a microcentrifuge tube containing 20 μL of chemically competent *E. coli* DH5α and incubate on ice for 5–10 min (see Note 11).
11. Heat-shock the cells in 42°C water bath for 45 s and immediately add 80 μL of SOC medium. Shake the reaction for 1 h at 37°C.
12. Spread 100 μL of the transformation mix on LB agar plates containing the proper antibiotic (see Note 12) and incubate overnight at 37°C. A good BP cloning result with a standard length (1 kb) insert should yield greater than 200 colonies per transformation.
13. Pick two to four separate Entry clone colonies into Falcon 2059 culture tubes containing 2 mL of SB medium with antibiotic and grow overnight at 37°C with 200 rpm shaking.
14. Spin 1 mL of the culture in a microcentrifuge to pellet the cells, and isolate plasmid using the FastPlasmid kit, eluting the DNA in 75 μL of elution buffer. (see Note 13).

15. Verify the size of the plasmid using agarose gel electrophoresis. Load 1 μL of the purified Entry clone DNA on a 1% agarose gel and compare sizes to the Supercoiled DNA ladder.
16. Properly sized Entry clones should be sequence-verified to ensure that no oligonucleotide or PCR generated errors have been introduced.
17. Glycerol stocks of the *E. coli* strains containing Entry clones should be made by adding 100 μL sterile filtered 60% glycerol to 400 μL of culture. After mixing and incubation at room temperature for 5 min, these stocks can be frozen at −80°C and used to start new cultures if more Entry clone DNA is required in the future.

3.2. Multisite Gateway Expression Clone Construction

1. Add the following reagents to a microcentrifuge tube in the following order (the total reaction volume should be 10 μL): 1–5 μL H_2O, first Entry clone DNA (50 ng, see Note 14), second Entry clone DNA (50 ng), Destination vector DNA (100 ng, see Note 15), and 2 μL LR Clonase II Plus.
2. Incubate the reaction mixture overnight at 25°C (see Note 16).
3. Add 1 μL of 2 mg/mL Proteinase K to inactivate the LR Clonase II Plus and incubate for 15 min at 37°C (see Note 17).
4. Add 1 μL of the LR II Plus reaction to a microcentrifuge tube containing 20 μL of chemically competent *E.coli* DH5α and incubate on ice for 20 min (see Note 11).
5. Heat-shock the cells in a 42°C water bath for 45 s and immediately add 80 μL of SOC medium. Shake the reaction for 1 h at 37°C.
6. Spread 100 μL of the transformation mix onto an LB agar plate containing the correct antibiotic (often ampicillin, but check the Destination vector information) and incubate overnight at 37°C.
7. Pick two separate colonies into Falcon 2059 culture tubes containing 2 mL of SB medium containing the correct antibiotics and grow overnight at 37°C with 200 rpm shaking.
8. Spin 1 mL of the culture in a microcentrifuge tube to pellet the cells and isolate plasmid using the Qiagen miniprep kit, eluting the DNA in 50 μL of elution buffer (see Note 18).
9. Verify the size of the plasmid using agarose gel electrophoresis. Load 1 μL of the purified expression clone DNA on a 1% agarose gel and compare sizes to the Supercoiled DNA ladder.
10. Additional confirmation of the expression clone should be carried out by restriction enzyme analysis (see Note 19). 1 μL of expression clone DNA can be digested using *BsrGI* restriction endonuclease for 1 h at 37°C.

11. Glycerol stocks of the *E. coli* strains containing Multisite clones should be made by adding 100 μL sterile filtered 60% glycerol to 400 μL of culture. After mixing and incubating at room temperature for 5 min, these stocks can be frozen at −80°C and used to start new cultures if more expression clone DNA is required in the future.
12. In some cases, miniprep DNA is of sufficient volume and quality for small scale mammalian transfection. Where large quantities of DNA are required, or where higher quality endotoxin-free DNA is required (see Note 20), 50 mL cultures of expression clones can be grown using SB media and the correct selective antibiotics. Plasmid preparation from these cultures can be done using commercially available Maxiprep DNA kits. We commonly use Sigma GenElute HP kits, which are rapid, are inexpensive, and produce high-quality DNA.
13. DNA for transfection should be concentrated to at least 300–500 μg/mL to minimize the volume of DNA required in the transfection. Concentration can be accomplished by vacuum centrifugation in a SpeedVac apparatus.

3.3. Cell Cultivation

1. The 293E cells are grown at 37°C in a humidified 5% CO_2 incubator in a growth medium consisting of a 50/50 mix of the Gibco FreeStyle medium and the HyClone SFX4HEK medium containing 100 μg/mL of G418. This mixed media formulation resulted in faster growth rates and ultimately much higher cell densities than cells grown in either of the individual formulations (Fig. 2). The higher cell density resulted in increased TGE due to a fivefold increase in cell number (Table 3).
2. Maintenance cultures are maintained at 200 mL in Corning 490 cm^2 roller bottles on an Innova 2100 orbital shaker at 135 orbits per minute (see Note 21). The cultures are transferred twice a week by first performing viable cell counts using the Cedex cell counter followed by centrifuging the appropriate number of cells to seed the new 200 mL culture at $4–5 \times 10^5$ cells/mL.

3.4. PEI Preparation

1. Dissolve the PEI in distilled water at a concentration of 1 mg/mL. Mild heat helps the PEI dissolve. The PEI may not fully dissolve until the pH is adjusted.
2. Adjust the pH of the solution to 7.0 using 1 M HCl.
3. Filter-sterilize through a 0.22 μm filter, aliquot and store at −80°C (see Note 22).

3.5. 96-Well Transient Transfection

1. Add 200 ng of plasmid DNA in 5 μL of TE buffer to each well of a Costar round bottom plate.
2. Using a multiwell pipettor and fresh tips for each addition, add 15 μL of PEI stock (1 mg/mL) diluted 1:15 in saline to each well followed by gentle mixing.

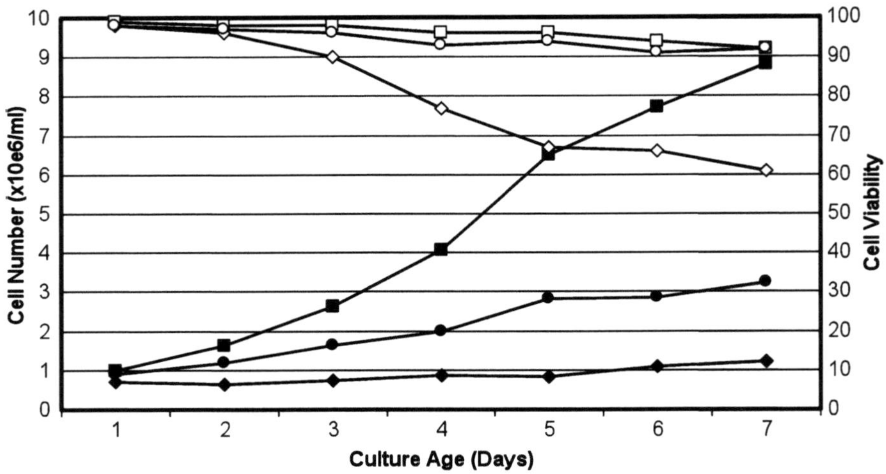

Fig. 2. The influence of various media formulations on the growth of 293E cells. One liter cultures of 293E were set at an initial cell density of 5×10^5/mL in three different media formulations and cell counts (*solid symbols*) and viabilities (*empty symbols*) were determined over a period of 7 days. The Hyclone medium is represented by a *filled circle*, the Gibco/Invitrogen medium by a *filled diamond*, and the 50/50 mixed formulation by a *filled square*. The data show that the mixed formulation of the two media is superior for high-density growth of 293E cells.

Table 3
Increasing expression yield using high-density 293E culture

Initial cell count in 500 mL	Cell count after media addition to 1 L	Cell count at harvest	Percentage of GFP-positive cells	Total number of GFP-positive cells
1×10^6/mL	5×10^5/mL	1.26×10^6/mL	49	6.17×10^8/mL
6×10^6/mL	3×10^6/mL	6.52×10^6/mL	48	3.12×10^9/mL fivefold increase in GFP(+) cells

3. Incubate the plate for 15–20 min at room temperature.
4. During the incubation wash 293E cells in the FreeStyle medium prepare a cell suspension (10 mL per plate) in FreeStyle medium at a concentration of 5.0×10^5 cells/mL (see Notes 2 and 23).
5. Following the incubation period add 100 μL of the 293E suspension (5×10^4 cells) to each well using fresh pipet tips for each addition. Mix the DNA–PEI and cells gently and transfer 100 μL to either Costar black flat-bottom or clear plates depending on what analysis is to be performed. Use clear plates

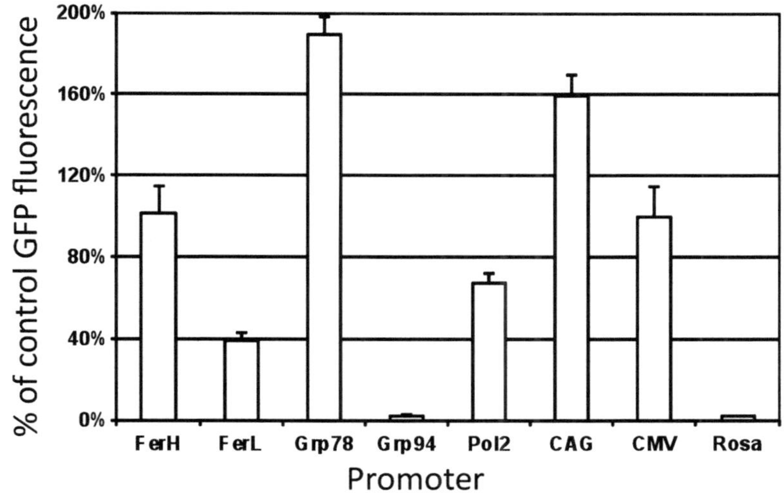

Fig. 3. Expression of GFP reporter using various promoter expression constructs. Data shown represent averages and standard deviations of quadruplicate measurements of clones containing eGFP produced from a variety of different promoters in the pDest-302 backbone. Values are presented as a percentage of the fluorescence observed with the CMV promoter. Promoters used were: FerH, human ferritin heavy chain promoter; FerL, human ferritin light chain promoter; Grp78, hamster glucose response protein 78 promoter; Grp94, hamster glucose response protein 94 promoter; Pol2, murine RNA polymerase 2 promoter; CAG, chicken beta-actin promoter/CMV enhancer; CMV, human cytomegalovirus promoter; Rosa, murin Rosa26 promoter.

when lysates are being prepared and black plates for reading fluorescent proteins on a plate reader.

6. Incubate the plates for 4 h at 37°C in a humidified incubator, and then add 100 μL of HyClone SFX medium.

3.6. Evaluation of Protein Expression by Fluorimetry

For fluorescent reporters or fusions to fluorescent proteins, analysis of expression by fluorimetry can be carried out.

1. Forty-eight hours posttransfection, the fluorescence intensity of the 96-well plate is measured for GFP (excitation 485 nm, emission 520 nm) and/or RFP (excitation 584 nm, emission 620 nm) expression on a FLUOstar Omega plate reader (BMG Technologies, Germany).
2. Normalized fluorescence intensity data can be used to generate graphs indicative of relative protein expression. Owing to variability in transfection efficiency and fluorescence readings, it is best to perform such experiments at least in triplicate (three separate wells of DNA per clone) and average the results. Generally, we find standard deviations of these readings to be within 5–10%.
3. Figure 3 shows results of data obtained from a 96-well plate assay of GFP reporter expression driven by a series of different promoters.

3.7. Evaluation of Protein Expression by Immunoblotting

For proteins with epitope tags, or where antibody to the protein is available, expression can be monitored by immunoblotting.

1. Remove media from transiently transfected adherent cells by pipetting for single plates or by gently decanting for 96-well plates and then blotting on paper towels.
2. Add 50 μL per well of a 96-well plate of M-PER® Mammalian Protein Extraction Reagent (see Note 24). Shake the reaction at 600 rpm for 5 min at 37°C. Replicate protein samples are now pooled into single tubes.
3. Prepare 4× SDS-PAGE sample buffer: 100 μL H_2O, 150 μL TCEP, and 750 μL Nu-PAGE. Add 5 μL of the mixture to 10 μL of each protein sample.
4. Heat the protein samples to 90°C for 5 min prior to SDS-PAGE analysis.
5. Set up a Criterion SDS-PAGE gel of a proper percentage range for the proteins of interest (see Note 25).
6. Fill the upper and lower chambers of the gel apparatus with 1× TGS buffer.
7. Load 15 μL of sample in each well, along with 2 μL of Benchmark™ Prestained Protein Ladder as a molecular size standard.
8. The protein samples are then subjected to electrophoresis at 200 V for 55 min.
9. Upon completion of the run, transfer the SDS-PAGE gel to a nitrocellulose membrane using a iBlot™ Dry Transfer System and iBlot™ Gel transfer stacks (see Note 26).
10. Once transferred, rinse the nitrocellulose membrane briefly with ultrapure water in a plastic tray before blocking by incubation in TBS-T blocking buffer for 1 h at room temperature (see Note 27).
11. Wash the nitrocellulose membrane with TBS-T (see Note 28) and then incubate overnight at 4°C in the presence of a primary antibody diluted in TBS-T blocking buffer (see Note 29).
12. Wash the membrane with TBS-T and then incubate for 1 h in the presence of an appropriate secondary antibody diluted in TBS-T blocking buffer, at room temperature.
13. For chemiluminescent detection, incubate the membrane in SuperSignal West Pico Chemiluminescent Substrate and detect luminescence using a Fujifilm Intelligent Dark Box LAS-4000 with ImageReader LAS-4000 software.

3.8. 24-Well Deep Block Plate Transient Transfection

1. Prepare the 293E cells as above (Subheading 3.5), except that 4×10^6 cells are required for each of the 24 wells.
2. Add 100 μL of saline to each well followed by 4 μg of plasmid DNA. Mix, then add 20 μL of stock PEI (1 mg/mL) and mix

by pipetting to form the DNA–PEI complexes. Incubate the plate for 15–20 min at room temperature.

3. Add 2 mL of 293E cell suspension at 2×10^6 cells/mL in FreeStyle medium to each well and incubate the plate at 37°C on a microplate shaker at 525 orbits per minute.
4. Following 4 h of incubation, feed each well with 2 mL of HyClone SFX medium. Return the plate to the shaker for an additional 48 h and collect either supernatant or cells (or both if inefficient secretion is suspected) for sample analysis.

3.9. Large-Scale Transient Transfection of 293E Cells

1. Perform liter-scale transfections in Corning 3 L Erlenmeyer flasks at an initial volume of 500 mL of FreeStyle medium containing 3×10^9 total viable cells (see Notes 2 and 23). The Innova shaker speed is set at 100–110 orbits per minutes.
2. Add 1 mg of DNA to 25 mL of saline in a 50 mL centrifuge tube followed by 5 mL of the PEI stock solution. Vortex this mixture briefly and incubate for 15–20 min at room temperature. Add the DNA–PEI mixture directly to the 3 L Erlenmeyer flask which is then returned to the shaker at 37°C for 4 h.
3. Following the 4 h incubation, increase the volume to 1 L by adding 500 mL of HyClone SFX medium. Return the flask to the shaker.
4. Collect cells or supernatant at 48–72 h posttransfection for analysis or processing.

4. Notes

1. For simplicity purposes the designation for the 293 C18/HEK 293 EBNA cell line is abbreviated to 293E. The parental strain of 293E, HEK 293, was developed in 1977 by transformation of normal human embryonic kidney cells with sheared adenovirus 5 DNA (10). Since then, 293-derived cell lines have been developed such as 293E expressing the Epstein-Barr virus (EBV) EBNA-1 protein and the 293T expressing the SV-40 virus large T-antigen. The 293E derivative was created by transfecting the parental HEK 293 cell line with the EBV EBNA-1 gene and the gene for neomycin resistance. Stable transfectants are maintained by growing the cells in the presence of G-418 at 100 μg/mL. One interesting note is that there is a controversy on the origin of the HEK 293 cell line. In 2002, Shaw et al. (11) published a report with evidence that the cell line has neuronal traits.
2. Many serum-free media formulations used for growing 293E cells totally inhibit PEI-mediated transfection. This is also the case for many of the lipid-based reagents. The formulations of these media are proprietary, and therefore, the chemical(s)

responsible for this inhibition have not been clearly identified. Many suspect it may be an additive which reduces cell clumping in suspension cultures, possibly heparin sulfate or dextran sulfate. The FreeStyle medium is, however, one medium that does not inhibit PEI-mediated transfection of 293E cells, and thus, it is the primary medium used during the initial transfection event. Cells for the transfection are centrifuged at $300 \times g$ and then washed once in the FreeStyle medium before preparing cells for transfection.

3. It is our experience that PEI–DNA complexes formed with the linear form of the molecule result in higher transfection efficiencies than those complexes formed with branched PEI.
4. Longer PCR primers are required for more complicated tagging in Entry clones. Introduction of protease cleavage sites or epitope tags often leads to oligonucleotide lengths in excess of 60 bp. In our experience, such long oligonucleotides often are of reduced quality containing higher numbers of mutations or deletions. To avoid this problem, a technique known as adapter PCR can be utilized. Adapter PCR involves the use of multiple nested primers to add long 5′ or 3′ sequences to the gene of interest. First, a primer that contains the gene-specific portion and part or all of a tag sequence (such as a TEV protease site) is added to the PCR reaction. After a few rounds of amplification, a second primer is added, which contains the attB recombination signal and 12–18 bp of overlapping sequence with the first primer. A mixture of PCR products is produced, but only the full-length product has the attB recombination sites necessary for recombination to occur.
5. The addition of 3% DMSO to the PCR reaction can help to minimize the effects of very GC rich primers or template. Often, it is not required, but we have seen no detrimental effect from including it in most PCR reactions. If templates are very AT rich, omit the DMSO.
6. The use of high amounts of template DNA helps to dramatically reduce PCR errors by forcing the use of original template molecules for subsequent PCR cycles rather than PCR products which may contain errors. If limited template is available, this amount can be reduced by 10–20-fold, but PCR errors increase.
7. Phusion polymerase has become the standard PCR reagent in our lab due to its robust activity and extremely high fidelity. Other polymerases (KOD, Pfu) can also be used, but we recommend using only high fidelity enzymes to limit PCR mutations, particularly with long genes or promoters. The suggested conditions are optimized for use on BioRad or Applied Biosystems PCR machines; some optimization may be required if PCR machines with slower ramp times are used.

8. Column purification of PCR products is only successful for products >150 bp in length. For smaller products, a PEG precipitation can be carried out as detailed in the Gateway product manuals. Failure to purify the PCR products generally leads to a large amount of primer-dimers (small fragments caused by primer misannealing), which clone very efficiently in the BP reaction. In extreme cases, gel purification may be necessary to eliminate these products completely.
9. Generally, the more PCR product used in the reaction, the higher the efficiency is. Particularly with long PCR products (>5 kb), the higher end of the concentration range should be used. Be aware that with adapter PCRs, the effective concentration of PCR product with both *attB* sites may be lower than the total concentration.
10. There are numerous Donor vectors available – these are the *att*P containing vectors which become the backbone of the Entry clone after the BP reaction. pDonr-221 is a common vector for use with *att*1-*att*2 clones. Other vectors are required for different combinations, and can be obtained from Invitrogen or the authors. Note that Donor vectors must be propagated in E. coli *ccd*B Survival or another strain which is resistant to the *ccd*B toxin.
11. DH5α is a recommended *E. coli* strain for Gateway reactions. However, it can be substituted with any other *recA*⁻ *endA*⁻ strain (such as TOP10 or DH10B) if necessary. Strains used must not carry the F′ episome, as it interferes with the Gateway negative selection. For good results, be sure that the competent cells have a transformation efficiency of at least 1×10^8 cfu/μg. Electrocompetent cells can also be used instead of chemically competent cells; however, the only advantage would be in the case of a very low efficiency reaction (such as with a very long gene) – usually the number of colonies obtained with standard chemically competent cells are more than sufficient.
12. Most donor vectors contain either kanamycin or spectinomycin resistance markers. Both antibiotics should be used at 50 μg/mL final concentration.
13. Many commercial kits are available for generating plasmid DNA from *E. coli*. We prefer the FastPlasmid kit for routine plasmid preps due to its high speed and consistent results. FastPlasmid can only be used for DNA generated in *endA*⁻ hosts, as it does not remove nucleases, which could affect downstream processes. A standard alkaline lysis plasmid prep also works in most cases.
14. In Multisite recombination, the ratio of Entry clones to Destination vector is crucial. Addition of extra Entry clone may improve efficiency, but this likely also increases the chance of cotransformation of Expression clone DNA and Entry clone

DNA into the same cell. We suggest a maximum of 50 ng per Entry clone and 100 ng of destination vector to avoid cotransformation problems. However, increases in efficiency, if needed, can be achieved by increasing the amount of Destination Vector.

15. Destination vectors can be added as either supercoiled plasmids, or as plasmids that have been linearized within the Gateway cassette. Contrary to the manufacturer's assertion, using linearized Destination vector actually improves LR II Plus efficiency two to fivefold. However, due to the high efficiency of the LR II Plus reaction, we do not generally find the extra effort required to prepare and purify linearized Destination vector to be worthwhile.
16. Owing to the complexity of the LR II Plus reaction, an overnight incubation at 25°C is required to increase recombination efficiency and produces more than sufficient number of colonies. Shorter incubation times do not produce enough colonies on a consistent basis.
17. Failure to Proteinase-K-treat the LR II Plus reactions results in dramatically reduced colony counts due to the inability of the DNA to transform while coated with Clonase proteins.
18. Many commercial kits are available for generating plasmid DNA from *E. coli*. The Qiagen miniprep kit is used here as it provides both DNA of a reasonable quality and yield, and also produces DNA with a relatively low level of endotoxin.
19. Gateway reactions are usually so efficient and accurate that further confirmation of Expression clones is not necessary. However, the slight possibility of unusual recombination byproducts can occur with Multisite reactions and so restriction enzyme analysis is often worthwhile. Some of the Gateway *att*B sites can be cleaved with the restriction enzyme *BsrGI*, which generally yields diagnostic bands to allow verification of insert sizes. Alternately, other restriction sites can be employed within the genes or flanking regions.
20. For transfection into common lab cell lines such as HEK293, it is not necessary to have extremely low endotoxin levels in DNA preparations. In these cases, standard high-speed plasmid preps such as the Sigma GenElute or Qiagen Spin preps are suffice. For some cell lines, much lower levels of endotoxin may be required, and in these cases one of the commercially available low endotoxin plasmid prep kits is strongly suggested.
21. The orbit diameters of shakers can vary considerably, and therefore, cell growth and maintenance of a uniform cell suspension can vary. For example, the orbit diameter of the Innova 2100 is 1 in. and the Infors Multitron is 2 in. The Tallboys microplate shaker has an orbit of 3 mm. When working with a new shaker, optimum orbit speeds should be determined that maintain a uniform suspension without undue stress on the cells.

22. Even though the PEI is prepared the same way each time, newly prepared lots should be titrated by setting up a transient transfection with control plasmid DNA expressing an easily detectable reporter protein such as GFP. Transfection efficiencies can vary between DNA–PEI ratios of 1:4–1:5. A good test would be to set four small transient transfections with DNA–PEI ratios of 1:3, 1:4, 1:5, and 1:6 μg/μg. Transfection efficiencies can be determined after 48 h by calculating the percentage of GFP-positive cells using a hemocytometer on a fluorescence microscope.
23. Twenty-four hours prior to transfecting cells, they should be split 1:2 or 1:3 to keep the cells in exponential growth. On the day of transfection, the cells should not be above a cell density of 6×10^6/mL.
24. M-PER® reagent is a nondenaturing detergent used to extract cellular proteins, leaving them in their native conformations, from the adherent cells without having to scrape them from the bottom of the tissue culture dish. This enables protein extraction from 24- and 96-well plates.
25. Criterion gels offer a wide range of polyacrylamide concentrations to optimize separation of proteins of various molecular weights. We prefer the wide-range 4–12 and 10–20% gels for proteins in the 40–150 and 10–80 kDa ranges, respectively. Criterion offers 26-well gels that have high resolution and fast run times; however, other SDS gel systems can be substituted, and voltages and run times may vary by manufacturer.
26. The iBlot™ Dry Transfer System quickly and efficiently transfers proteins from SDS-PAGE gels in 7 min. This approach has proven to be fast and of high quality, and yields reproducible results.
27. Nonspecific binding is reduced by blocking unoccupied membrane binding sites with an inert protein. We achieve this by using 5% nonfat milk diluted in 1× TBS-T (Tris Buffered Saline-Tween), which has a greater affinity for the membrane than the primary antibody.
28. Washing the membrane removes any unbound blocking agent and/or antibody which would otherwise cause high background and poor detection. We wash the membrane in a series of three washes: 15, 5, and 5 min.
29. Antibodies are diluted in blocking buffer to prevent nonspecific binding to the membrane. We recommend an overnight incubation with primary antibody although less time can be used. If using an overnight incubation, then the membrane must be kept at 4°C to prevent contamination and subsequent destruction of your protein of interest. Agitation is also recommended, as this enables homogeneous rather than uneven binding of the antibody.

Acknowledgments

The authors thank Cammi Bittner, Veronica Roberts, Leslie Garvey, and Kelly Esposito for assistance in protocol development. This work has been funded in whole or in part with federal funds from the National Cancer Institute, National Institutes of Health, under contract HHSN261200800001E. The content of this publication does not necessarily reflect the views or policies of the Department of Health and Human Services, nor does mention of trade names, commercial products, or organizations imply endorsement by the US Government.

References

1. Peel, A., ed. Transfection of mammalian cells, in *Methods*, vol. 33, No 1, 2004.
2. Durocher, Y., Perret, S., and Kamen, A. (2002) High-level and high-throughput recombinant protein production by transient transfection of suspension-growing human 293-EBNA1 cells. *Nucleic Acids Res.* **30**, 2–9.
3. Boussif, O., Lezoualc'h, F., Zanta, M. A., Mergny, M. D., Scherman, D., Demeneix, B., and Behr, J.-P. (1995) A versatile vector for gene and oligonucleotide transfer into cells in culture and *in vivo*: Polyethylenimine. *Proc. Natl. Acad. Sci.* **92**, 7297–7301.
4. Schlaeger, E.-J. and Christensen, K. (1999) Transient gene expression in mammalian cells grown in serum-free suspension culture. *Cytotechnology* **30**, 71–83.
5. Hartley, J.L., Temple, G.F., and Brasch, M.A. (2000) DNA cloning using in vitro site-specific recombination. *Genome Res* 10, 1788–95.
6. Esposito, D., Garvey, L.A., and Chakiath C.S. (2009) Gateway cloning for protein expression. *Methods Mol Biol* 498, 31–54.
7. Cheo, D.L., Titus, S.A., Byrd, D.R., Hartley, J.L., Temple G.F., and Brasch, M.A. (2004) Concerted assembly and cloning of multiple DNA segments using in vitro site-specific recombination: functional analysis of multi-segment expression clones. *Genome Res* 14, 2111–20.
8. Sasaki Y, Sone T, Yoshida S, Yahata K, Hotta J, Chesnut JD, Honda T, Imamoto F. (2004) Evidence for high specificity and efficiency of multiple recombination signals in mixed DNA cloning by the Multisite Gateway system. *J. Biotechnol.* 107, 233–43.
9. Sasaki, Y., Sone, T., Yahata K., Kishine H., Hotta, J., Chesnut, J.D., Honda, T., and Imamoto, F. (2008) Multi-gene gateway clone design for expression of multiple heterologous genes in living cells: eukaryotic clones containing two and three ORF multi-gene cassettes expressed from a single promoter. *J. Biotechnol.* 136, 103–12.
10. Graham, F.L., Smiley, J., Russell, W.C., and Nairn, R. (1977). Characteristics of a human cell line transformed by DNA from human adenovirus type 5. *J. Gen. Virol.* **36**, 59–74.
11. Shaw, G., Morse, S., Ararat, M., and Graham, F.L. (2002) Preferential transformation of human neuronal cells by human adenoviruses and the origin of HEK 293 cells. *FASEB J.* **16**, 869–871.

Index

James L. Hartley (ed.), *Protein Expression in Mammalian Cells: Methods and Protocols*, Methods in Molecular Biology, vol. 801, DOI 10.1007/978-1-61779-352-3, © Springer Science+Business Media, LLC 2012

D

T

U

V

W

MIX
Papier aus verantwortungsvollen Quellen
Paper from responsible sources
FSC® C105338

If you have any concerns about our products,
you can contact us on
ProductSafety@springernature.com

In case Publisher is established outside the EU,
the EU authorized representative is:
Springer Nature Customer Service Center GmbH
Europaplatz 3, 69115 Heidelberg, Germany

Printed by Libri Plureos GmbH
in Hamburg, Germany